Smart Plant Factory

Toyoki Kozai
Editor

Smart Plant Factory

The Next Generation Indoor Vertical Farms

Editor
Toyoki Kozai
Japan Plant Factory Association (NPO)
Kashiwa, Chiba, Japan

ISBN 978-981-13-1064-5   ISBN 978-981-13-1065-2 (eBook)
https://doi.org/10.1007/978-981-13-1065-2

Library of Congress Control Number: 2018956169

© Springer Nature Singapore Pte Ltd. 2018
This work is subject to copyright. All rights are reserved by the Publisher, whether the whole or part of the material is concerned, specifically the rights of translation, reprinting, reuse of illustrations, recitation, broadcasting, reproduction on microfilms or in any other physical way, and transmission or information storage and retrieval, electronic adaptation, computer software, or by similar or dissimilar methodology now known or hereafter developed.
The use of general descriptive names, registered names, trademarks, service marks, etc. in this publication does not imply, even in the absence of a specific statement, that such names are exempt from the relevant protective laws and regulations and therefore free for general use.
The publisher, the authors and the editors are safe to assume that the advice and information in this book are believed to be true and accurate at the date of publication. Neither the publisher nor the authors or the editors give a warranty, express or implied, with respect to the material contained herein or for any errors or omissions that may have been made. The publisher remains neutral with regard to jurisdictional claims in published maps and institutional affiliations.

This Springer imprint is published by the registered company Springer Nature Singapore Pte Ltd.
The registered company address is: 152 Beach Road, #21-01/04 Gateway East, Singapore 189721, Singapore

Acknowledgements

The editor would like to thank Ms. Tokuko Takano for her editorial assistance and dedication. Thanks are also extended to Professors T. Maruo, M. Takagaki, T. Yamaguchi, Y. Shinohara, and S. Tsukagoshi of Chiba University for their academic guidance, and to Ms. Mariko Ando and other administrative staff members of Japan Plant Factory Association (nonprofit organization). We also appreciate the guidance and assistance provided by Dr. Mei Hann Lee, Ms. Momoko Asawa, and Mr. Vinoth Kuppan of Springer and Ms. Sudantra Devi of SPi.

Contents

Part I Characteristics of Current Plant Factories and Next Generation Plant Factory

1 **Current Status of Plant Factories with Artificial Lighting (PFALs) and Smart PFALs**.................................... 3
   Toyoki Kozai

2 **Plant Factories with Artificial Lighting (PFALs): Benefits, Problems, and Challenges**................................ 15
   Toyoki Kozai

3 **Protocols, Issues and Potential Improvements of Current Cultivation Systems**.................................. 31
   Na Lu and Shigeharu Shimamura

4 **Design and Control of Smart Plant Factory**.................. 51
   Yoshihiro Nakabo

5 **Designing a Cultivation System Module (CSM) Considering the Cost Performance: A Step Toward Smart PFALs**........... 57
   Toyoki Kozai

Part II Recent Outcomes in Development and Business

6 **Business Planning on Efficiency, Productivity, and Profitability**... 83
   Kaz Uraisami

7 **Renewable Energy Makes Plant Factory "Smart"**.............. 119
   Kaz Uraisami

8 **Total Indoor Farming Concepts for Large-Scale Production**...... 125
   Marc Kreuger, Lianne Meeuws, and Gertjan Meeuws

| 9 | **SAIBAIX: Production Process Management System** .... | 137 |

Shunsuke Sakaguchi

| 10 | **Air Distribution and Its Uniformity** .... | 153 |

Ying Zhang and Murat Kacira

Part III Re-considerations of Photosynthesis, LEDs, Units and Terminology

| 11 | **Reconsidering the Fundamental Characteristics of Photosynthesis and LEDs** .... | 169 |

Toyoki Kozai and Masayuki Nozue

| 12 | **Reconsidering the Terminology and Units for Light and Nutrient Solution** .... | 183 |

Toyoki Kozai, Satoru Tsukagoshi, and Shunsuke Sakaguchi

Part IV Advances in Research on LED Lighting

| 13 | **Usefulness of Broad-Spectrum White LEDs to Envision Future Plant Factory** .... | 197 |

Hatsumi Nozue and Masao Gomi

| 14 | **LED Lighting Technique to Control Plant Growth and Morphology** .... | 211 |

Tomohiro Jishi

| 15 | **The Quality and Quality Shifting of the Night Interruption Light Affect the Morphogenesis and Flowering in Floricultural Plants** .... | 223 |

Yoo Gyeong Park and Byoung Ryong Jeong

Part V Advanced Technologies to Be Implemented in the Smart Plant Factory

| 16 | **Mechanization of Agriculture Considering Its Business model** .... | 241 |

Tamio Tanikawa

| 17 | **Quantifying the Environmental and Energy Benefits of Food Growth in the Urban Environment** .... | 245 |

Rebecca Ward, Melanie Jans-Singh, and Ruchi Choudhary

| 18 | **Applications for Breeding and High-Wire Tomato Production in Plant Factory** .... | 289 |

Marc Kreuger, Lianne Meeuws, and Gertjan Meeuws

| 19 | **Molecular Breeding for Plant Factory: Strategies and Technology** .... | 301 |

Richalynn Leong and Daisuke Urano

Contents

20 Production of Value-Added Plants 325
Shoko Hikosaka

21 Chemical Inquiry into Herbal Medicines and Food Additives 353
Natsuko Kagawa

**22 Detection and Utilization of Biological Rhythms
in Plant Factories** 367
Hirokazu Fukuda, Yusuke Tanigaki, and Shogo Moriyuki

**23 Automated Characterization of Plant Growth and Flowering
Dynamics Using RGB Images** 385
Wei Guo

**24 Toward Nutrient Solution Composition Control
in Hydroponic System** 395
Toyoki Kozai, Satoru Tsukagoshi, and Shunsuke Sakaguchi

**25 Phenotyping- and AI-Based Environmental Control
and Breeding for PFAL** 405
Eri Hayashi and Toyoki Kozai

26 Plant Cohort Research and Its Application 413
Toyoki Kozai, Na Lu, Rikuo Hasegawa, Osamu Nunomura,
Tomomi Nozaki, Yumiko Amagai, and Eri Hayashi

27 Concluding Remarks 433
Toyoki Kozai

Index 443

Part I
Characteristics of Current Plant Factories and Next Generation Plant Factory

Chapter 1
Current Status of Plant Factories with Artificial Lighting (PFALs) and Smart PFALs

Toyoki Kozai

Abstract The necessity of plant factories with artificial lighting (PFALs) is discussed in relation to the food, resource, and environment trilemma. The definition and main components of the PFAL are described. The number of PFALs in the world, the production cost and wholesale price of the produce, and the plants suited to PFALs are briefly introduced. The main objectives of this book are described, and the image and expected ultimate functions of the smart PFAL are discussed.

Keywords Global technology · Local technology · PFAL (plant factory with artificial lighting) · Smart PFAL · Trilemma

1.1 Introduction: Food, Resource, and Environment Trilemma

1.1.1 Contribution of Plant Factories with Artificial Lighting (PFALs) to Solving the Trilemma

We are facing a trilemma in which there are three almost equally undesirable alternatives: (1) shortage and/or unstable supply of food, (2) shortage of resources, and (3) degradation of the environment (Fig. 1.1). This trilemma is occurring at the global as well as local and national level amid an increasing urban population and a decreasing and/or aging agricultural population.

The resource shortage includes arable land area, water for irrigation, fertilizers, fossil fuel, and farmers/growers. The environmental degradation includes pollution of soil, atmosphere, and water, accumulation of salt on the soil surface, desertification, and recent climate change causing droughts, high/low air temperatures, high/low solar radiation, heavy rain, flooding, strong wind, and spread of pest insects, all of which result in unstable market prices for vegetables.

T. Kozai (✉)
Japan Plant Factory Association (NPO), Kashiwa, Chiba, Japan
e-mail: kozai@faculty.chiba-u.jp

© Springer Nature Singapore Pte Ltd. 2018
T. Kozai (ed.), *Smart Plant Factory*, https://doi.org/10.1007/978-981-13-1065-2_1

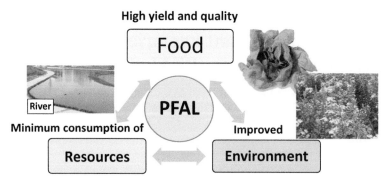

Fig. 1.1 Plant factories with artificial lighting (PFALs) are considered to play a significant role in solving the food-resource-environment trilemma

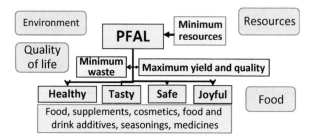

Fig. 1.2 Mission and goals of plant factories with artificial lighting (PFALs)

To help solve this trilemma, transdisciplinary methodologies based on new concepts need to be developed by which the yield and quality of food are substantially improved with less resource consumption and environmental degradation compared to current plant production systems. Plant factories with artificial lighting (PFALs) are one such system expected to achieve this mission (Fig. 1.2). The benefits of the PFAL include high resource use efficiency, high annual productivity per unit land area, and production of high-quality plants without using pesticides (Kozai et al. 2015). Current major problems to be solved are the high initial investment and the electricity and labor costs.

The next-generation "smart" plant factories with artificial lighting (smart PFALs) are expected to help solve food, resource, and environment issues concurrently, by significantly reducing the initial and operation costs.

People living in urban areas consume large amounts of food daily. At the same time, they produce large amounts of waste including CO_2, biomass garbage, wastewater, and heat energy. As a result, significant amounts of resources are consumed to process the waste/garbage to keep the urban environment clean (Despommier 2010). This is another aspect of the food, resource, and environment trilemma to be solved.

1.1.2 Reducing the Loss of Fresh Vegetables in Urban Areas Using PFALs

Roughly one-third of the food produced in the world for human consumption is wasted every year – approximately 1.3 billion tons. Vegetables have one of the highest wastage rates of any food. In developing countries, 40% of this loss occurs at the postharvest and processing levels, while in industrialized countries, more than 40% of the loss occurs at the retail and consumer levels. Food loss and waste also amount to a major squandering of resources, including water, land, energy, labor, and capital, and needlessly produce greenhouse gas emissions, contributing to global warming and climate change (http://www.fao.org/save-food/resources/keyfindings/en/). Thus, local production for local consumption is important for fresh vegetables with water content of around 90% to reduce vegetable loss and conserve resources. Fresh vegetables are heavy and easily damaged during transportation.

1.1.3 Waste Produced in Urban Areas Can Be Used as Essential Resources for Plant Production

It is well known that plants with green leaves grow by photosynthesis, and the essential resources for photosynthetic growth at moderate temperatures are water, CO_2, light energy, and inorganic fertilizer consisting of 13 nutrient elements (including N, P, K, Ca, and Mg).

It should be noted that a considerable portion of the waste produced in urban areas can be converted, in theory, to essential resources for plant photosynthesis and growth. The emitted CO_2 is used as a carbon source for photosynthesis. The wastewater can be used as irrigation water after proper treatment to make it cleaner than river, lake, and groundwater. The garbage (raw materials of organic fertilizer) can be converted to inorganic fertilizer through decomposition using specific microorganisms. The heat energy at temperatures of 30–60 °C released at restaurants, offices, and various types of industrial factories can be used as a heat source for greenhouse heating in winter, food and material drying, etc. Thus, the amount of resource consumption and waste in urban areas can be significantly reduced by growing plants for food and other uses.

The only insufficient and expensive resource for PFAL plant production is light energy, which needs to be generated using lamps and electric energy. On the other hand, surplus electric energy is available at nighttime in urban areas and can be used for lighting and air conditioning in the PFAL, often at a lower cost than in the daytime.

The electricity consumption, and thus its cost, in the PFAL has been decreasing by 30–40% or more with the use of LED (light-emitting diode) compared to using fluorescent lamps. The cost will be further reduced with the development of the "smart" LED lighting system. Moreover, the cost for electric energy generation using natural sources such as solar, wind, and biomass energy has been decreasing year by year. Then, the PFAL, especially the smart PFAL, can be a resource-saving and environmentally friendly high-yield plant production system in urban areas and is expected to help solve the trilemma.

1.2 Characteristics and Main Components of the PFAL

In most Asian countries, plant factories include both PFALs and plant factories with solar lighting (PFSLs). However, since the PFSL is not discussed in this book except in Sect. 1.2.2 of this chapter, the term "**plant factory**" is used to mean "**plant factory with artificial lighting (PFAL)**" only and does not include the PFSL. Plant production systems like the PFAL are also called vertical farms (Despommier 2010), indoor farms, and closed plant production systems.

1.2.1 Characteristics of the PFAL

The PFAL, not the smart PFAL, is defined in this book as one type of closed plant production system with artificial lighting (CPPS; Kozai et al. 2015). The cultivation room (or culture room) of the PFAL is thermally well insulated and airtight and kept clean. Since the interior of the PFAL is not influenced by the weather due to its airtightness and thermal insulation, the environment can be controlled at the desired set points if the control units are well designed and properly operated.

1.2.2 Difference Between PFAL and PFSL

Since the walls of the PFAL cultivation room are optically opaque, thermally insulated and airtight, environmental control units such as heaters, shading screens, thermal screens, insect screens, and roof/fan ventilators or an evaporative cooling system are not required. On the other hand, these control units are required in PFSLs and controlled-environment greenhouses. In the PFAL, the heat generated from the lamps needs to be removed by air conditioners to keep the air temperature at a set point even on winter nights. This means that a heating system is not required in the PFAL. Thus, the difference between the electricity cost for lighting and cooling in the PFAL and the heating/cooling/insect screen costs in the PFSL is one of the factors to be considered with respect to environmental control. Other factors are described in Chap. 2.

1.2.3 Main Components of the PFAL

The main components (or units) of the PFAL cultivation room are listed in Table 1.1 and are schematically shown in Fig. 1.3 (see Kozai et al. (2015) for more details).

Currently, most large-scale PFALs with a floor area of 5000 m^2 or more are equipped with machines or facilities for automating the seeding, transplanting, transporting, and packing operations. PFALs with a floor area of 1000 m^2 or less are not equipped with machines for automation, and those with a floor area of

Table 1.1 Main components (or units) of a typical PFAL for commercial production. The number in the left-hand column corresponds to the number in Fig. 1.3

No.	Main components for plant cultivation
1	Cultivation room, which is thermally insulated, airtight, and clean
2	Racks with horizontal multi-tiers in the cultivation room. Hydroponic cultivation beds are placed on the surface of each tier (vertical distance between tiers is 40–100 cm) (PFALs with vertical, moving cultivation panels/pipes have recently been developed)
3	Cultivation space sandwiched between the cultivation beds and the ceiling of each tier
4	Lighting unit (with reflectors) installed on the ceiling
5	Nutrient solution delivery/circulation unit for the cultivation beds, consisting of water pumps, nutrient solution sterilizer with filters, nutrient solution tank, stock nutrient solution tanks, and piping
6	Air conditioning/dehumidifying unit with air mixing and filtering fans
7	CO_2 supply unit consisting of pure CO_2 tanks, gas valves, and piping
8	Data acquisition and control unit for all control variables
Main components for plant and equipment handling and worker welfare	
Machines and/or spaces for seeding, transplanting, transporting, harvesting, trimming, weighing, packing in bags, packaging in boxes, storing, shipping, etc.	
Rooms for entrance, locker/clothing change, air shower, handwashing, various operations such as trimming, storage, meeting, pre-cooling, healthcare, etc.	

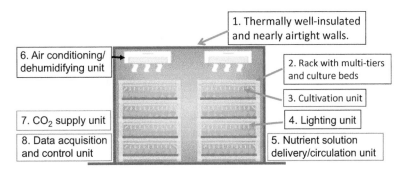

Fig. 1.3 Configuration of the culture room in a plant factory with artificial lighting (PFAL) consisting of eight main components (units). Machines for automation are not shown in the figure. Most electric energy is consumed by lighting and air conditioning units. The numbers in the diagram correspond to those in Table 1.1. This is a revised version of Fig. 4.1 in Chapter 4 of Kozai et al. (2015)

1000–5000 m^2 are partially automated. Thus, automation is not a necessary condition of the PFAL. In any PFAL, the detection of physiological disorders on leaves and removal/trimming of these leaves are still conducted manually by workers. A tool or machine is used for cleaning the cultivation panels and beds and the cultivation room floor, but these operations are not fully automated.

1.3 Current Status of PFALs

1.3.1 Estimated Number of PFALs

As of September, 2018, the number of PFALs for commercial production is roughly estimated at over 200 in Japan, about 100 in Taiwan, and over 500 in the world. The numbers of PFALs in China, the USA, and Korea have been increasing significantly since 2015. During 2013–2017, one to several PFALs were built in Singapore, Panama, Mongolia, Russia, France, Vietnam, the Netherlands, India, the UK, Malaysia, and a few Middle East countries including Dubai. Thailand and several other Southeast Asian countries are planning to build one or more PFALs by around 2020. The PFAL business has been emerging and will continue to grow in the forthcoming years. However, it is not yet big enough to be called a PFAL industry.

1.3.2 Production Capacity, Production Cost, and Wholesale Price

As of 2018, the largest PFAL in the USA and in China is said to have a planned daily production capacity of around 5000 kg fresh weight of leafy greens. The daily production capacity of the largest PFAL company in Japan is around 4000 kg of lettuce plants (grown at two separate PFALs).

The percentage, by component, of the production cost in Japan is roughly 30% for depreciation of initial investment, 20% for electricity (lighting and air conditioning), 20% for labor, 20% for supplies (seeds, water, supports (substrates), fertilizer, CO_2, plastic bags, delivery boxes, etc.), and 10% for maintenance and security management. The percentage, by component, of the production cost and the wholesale price largely depend on the local economic, social, cultural, and technological situation.

The wholesale price of leafy greens in Japan ranges roughly between 10 and 20 USD per kg. It is said that about 90% of the produce at full production capacity needs to be sold to recover the initial investment within 5–6 years.

1.3.3 Plants Suited to PFALs

Most PFALs for commercial production produce functional or specialty plants. Currently, leafy vegetables such as lettuce, rocket salad (or rucola; *Eruca sativa*), and kale and herbs such as basil are commonly produced in the PFAL. Microgreens (mixture of several young leafy greens harvested about 20 days after sowing) are widely produced in the USA. Commercial production of medicinal plants, edible flowers, fruit-vegetables such as cherry tomatoes and strawberry plants, and dwarf root crops such as mini carrots is still produced on a small scale or just as a trial. On the other hand, the demand for PFAL-grown herbs has been increasing for use as medicine, seasoning, as well as additives and ingredients for food, drinks, cosmetics, health supplements, etc.

In Japan, different kinds of transplants including grafted tomatoes, cucumbers, watermelons, and eggplants are widely produced in specially designed PFALs or CPPSs (Kozai et al. 2015). One type of specially designed PFAL was recently used to produce pharmaceuticals (e.g., for periodontal disease) from genetically modified plants.

PFALs are not used for commercial production of staple crops such as wheat, rice, and maize. This is because the market price per kg (dry matter) of functional plants is generally 10–100 times higher than that of staple crops, while the cost for producing 1 kg (dry matter) of functional plants is more or less the same as that of staple crops. Potato and sweet potato plants can be economically produced if most parts of the leaves, petioles, stems, and roots are usable and salable. On the other hand, PFALs are going to be used for breeding both functional and staple crops to shorten the required time period.

A simple index to evaluate the economic suitability of plants to PFALs is the electric energy productivity, P_E in units of kg/kWh, which is the yield of marketable produce (kg) per unit electric energy consumption (kWh = 1000 × 3600 s/h = 3.6 MJ) for lighting and air conditioning. The electric energy productivity on a monetary basis, P_M, can be roughly estimated by (P_E/U_E) where U_E denotes unit electricity cost ($/kWh). The electric energy consumption is roughly proportional to the days of cultivation, photoperiod (hours per day), and PPFD (photosynthetic photon flux density).

1.4 Main Objectives and Outline of This Book

The first objective of this book is to provide the ideas, concepts, methodologies, technologies, and the recent and upcoming research and development (R&D) to be implemented in the smart PFAL and to discuss the potential and actualized benefits of the PFAL, as well as the problems to be solved and the challenges.

The second objective is to provide a concrete image of the next-generation "smart" PFAL. The book also explains the reasons why the smart PFAL is expected

Fig. 1.4 Simplified roadmap toward smart PFALs, the details of which are described in this book

to achieve a much higher yield with a higher quality of plants compared to the current PFAL, with less resource consumption and environmental pollution, resulting in lower production costs. A discussion is also made on the possibility of the smart PFAL as a tool to accelerate the breeding and seed propagation processes of any kind of plant and as an educational and self-learning tool at school, home, and the community center.

The third objective is to introduce recent trends in research, development, management, and marketing of the PFAL and to present a roadmap toward the smart PFAL starting from the current PFAL (Fig. 1.4).

1.5 Image of the Smart PFAL

The term "smart PFAL" implies an intelligent or cognitive computing PFAL with the ability to make almost all decisions and solve almost all problems without human intervention. The smart PFAL must (1) adapt its behavior based on experience (learning), (2) not be totally dependent on instructions from people (learn on its own), and (3) be able to respond to unanticipated events (https://www.gartner.com/it-glossary/smart-machines).

In that regard, there are no smart PFALs in operation in the world as of 2017. In this book, the smart PFAL is loosely defined as an intelligent PFAL that uses artificial intelligence (AI) with a big database in clouds, the Internet of Things (IoT), light-emitting diodes (LEDs), a phenotyping unit with cameras for noninvasive measurement of plant traits, robots, etc. (Fig. 1.5).

It is important to remember that the history of PFAL R&D is only half a century, and the third wave of PFAL with LEDs started around 2010, to be followed by the fourth wave around 2020 (Fig. 1.6). The technologies to be introduced in the fourth-generation PFALs will be different in many aspects from those of the current third-generation PFAL, although LEDs will continue to be the main light source. In any case, the history of PFALs is only a fraction of that of the greenhouse horticulture industry, which started more than a century ago, and that of agriculture, which started around 10,000 years ago.

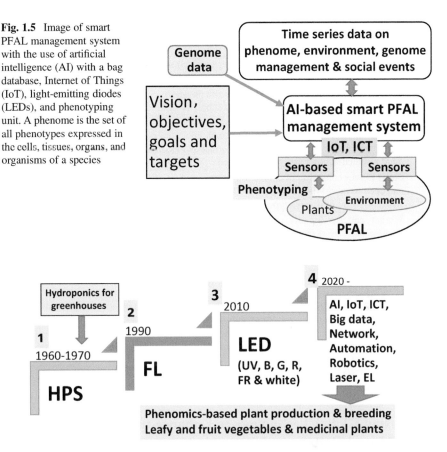

Fig. 1.5 Image of smart PFAL management system with the use of artificial intelligence (AI) with a bag database, Internet of Things (IoT), light-emitting diodes (LEDs), and phenotyping unit. A phenome is the set of all phenotypes expressed in the cells, tissues, organs, and organisms of a species

Fig. 1.6 The fourth wave of PFAL R&D and business will start around 2020
HPS high-pressure sodium lamp, *FL* fluorescent lamp, *LED* light-emitting diode lamp, *UV B, G, R, and FR* ultraviolet, blue, green, red, and far-red, respectively. *AI* artificial intelligence, *IoT* Internet of Things, *ICT* information and communication technology. Phenomics: methodology of plant trait measurement and its applications

1.6 Expected Ultimate Functions of the Smart PFAL

Expected ultimate functions of the smart PFAL are listed in Table 1.2. Such functions will be realized step by step, with a major part of them being realized within 5–10 years, if its R&D is conducted with a clear vision, mission, goals, and numerical targets. Some concrete goals and targets are proposed in Chap. 5.

Table 1.2 Expected ultimate functions of smart PFALs

No.	Expected ultimate functions of smart PFALs
1	Contribution to solving the food, resource, and environment trilemma at the personal, local, regional, national, and global levels
2	Contribution to improving our quality of life physically, mentally, and spiritually, in addition to food security including stable supply of nutritious and safe foods for our health
3	Energy-autonomous, ecologically sustainable, and economically viable PFALs that achieve the highest yield and quality of produce with minimum resource consumption and maximum use of solar, biomass, fluid, and thermal and mechanical energy, resulting in minimum production costs and waste emission
4	Smart PFALs consisting of cultivation system modules (CSMs) networked with each other and open to most PFAL users via the Internet. Hardware/software subunits of CSM standardized to fit diverse types of CSMs
5	Worker- and user-friendly, enabling plant production for and by anyone anywhere for any plant species. Users can choose the mode of production process control considering the purpose of use, interest, skills level, etc.
6	Easy integration with other biological systems such as mushroom cultivation and aquaculture systems and with electric energy generation systems using natural energy easily obtained locally
7	Assisting users in creating new ideas for service, products, and markets. The products and produce do not compete with those being produced and sold in the current markets
8	Assisting users in understanding the principles and mechanisms of plant growth, plant production process control, and energy/substance balance in closed ecosystems, for educational and self-learning purposes
9	Nurturing and empowering people to adopt the basic skills for culturing plants, culturing minds, and caring for plants, food, mind, and body as one

1.7 Using Global Technology to Enhance the Quality of Local Culture and Technology

Local production of food for local consumption through the use of PFALs is slowly gaining in popularity for improving food security in urban areas. Local production for local consumption is important not only for food security but also for local agriculture, which is strongly related to the local culture of the mind, lifestyle, history, and landscape there (Fig. 1.7). Furthermore, local agriculture and technologies are all influenced by the local climate, soil, landscape, traditions, and history. PFALs can play a role in producing food and other plant-based products to meet the local culture and personal needs, even though the PFAL itself is being developed using global technologies.

The local culture and technology can be enhanced, improved, and evolved by using recent advanced but inexpensive global technologies such as the smartphone with camera, global positioning system (GPS), Internet, artificial intelligence (AI), etc. (Fig. 1.8). Although the basic functions of smartphones have been developed by using global technologies, users can customize the phone by downloading their favorite application software and inputting their own data to fit their personal and local needs. We can download numerous types of free application software to personalize the contents of our smartphones. Similarly, PFAL users can customize their PFALs for their particular personal and local needs using the global technologies.

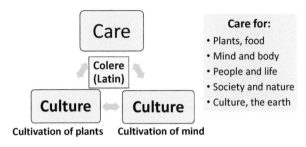

Fig. 1.7 Philosophical background of smart PFALs: *colere* (Latin) as the origin of care, culture as the cultivation of plants, and culture as the cultivation of the mind

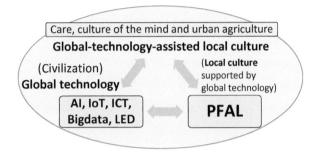

Fig. 1.8 Scheme showing the revival of local culture supported by global technology

These improvements to local/personal life and local technologies using global technologies can be achieved with a small personal investment. To realize the expected ultimate functions shown in Table 1.2, recent advanced but inexpensive technologies such as those implemented in smartphones need to be properly introduced.

1.8 Conclusions

Key factors for developing the next-generation "smart" PFALs include (1) creative and innovative minds; (2) deep understanding of plant physiology, dynamic balance of energy and materials, and management; (3) skills for big data collection, analyses, and systematic integration of the analyzed results; and (4) implementation of recent advanced technologies with a clear vision, mission, goals, and numerical targets. It is a big but worthwhile challenge to develop the smart PFAL that will play a key role in solving the food, resource, and environment trilemma in our diverse societies in the forthcoming decades.

References

Despommier D (2010) The vertical farm. St. Martin's Press, New York, 305 pp
Kozai T, Niu G, Takagaki M (eds) (2015) Plant factory: an indoor vertical farming system for efficient quality food production. Academic, Amsterdam, 405 pages

Chapter 2
Plant Factories with Artificial Lighting (PFALs): Benefits, Problems, and Challenges

Toyoki Kozai

Abstract The benefits, unsolved problems, and challenges for plant factories with artificial lighting (PFALs) are discussed. The remarkable benefits are high resource use efficiency, high annual productivity per unit land area, and production of high-quality plants without using pesticides. Major unsolved problems are high initial investment, electricity cost, and labor cost. A major challenge for the next-generation smart PFAL is the introduction of advanced technologies such as artificial intelligence with the use of big data, genomics, and phenomics (or methodologies and protocols for noninvasive measurement of plant-specific traits related to plant structure and function).

Keywords Artificial intelligence · Annual productivity · Cultivation system module (CSM) · Phenotyping · Resource use efficiency (RUE) · Smart LED lighting system · Standardization

2.1 Introduction

The Dutch glass greenhouse technology is currently the most advanced in the world. The average yield of greenhouse tomatoes in the Netherlands was 60 kg m^{-2} in 2008, is estimated at around 70 kg m^{-2} for 2017, and will approach 100 kg m^{-2} in the near future by making full use of advanced technologies such as light-emitting diodes (LEDs) for supplemental lighting. Looking back on the past, the greenhouse tomato yield in the Netherlands was 9.5 kg m^{-2} in 1960, 20 kg m^{-2} in 1970, 29 kg m^{-2} in 1980, 44 kg m^{-2} in 1990, and 55 kg m^{-2} in 2000 (Heuvelink 2006). The yield in 2017 is about sevenfold than in 1960. The development of the Dutch greenhouse technology has progressed over the past 50 years through the active collaboration of Dutch industries, public institutions and governments, and a very limited number of

T. Kozai (✉)
Japan Plant Factory Association (NPO), Kashiwa, Chiba, Japan
e-mail: kozai@faculty.chiba-u.jp

private companies in neighboring countries such as Denmark, Sweden, Belgium, and Germany.

The current technology for plant factories with artificial lighting (PFALs) can be compared to the Dutch glass greenhouse technology of around 1980. Thus, the yield of the PFAL can probably be doubled, tripled, or even more by actualizing the potential benefits, solving the current problems, and taking on the challenges of the next-generation smart PFALs.

The PFAL technology with the use of LEDs has been extensively developed since around 2010 mainly in Southeast Asian regions (Japan, Taiwan, China, and Korea), the USA, and the Netherlands. It is also noted that many private companies in the fields of information technology (IT), electronics, mechatronics, housing, food, environmental control engineering, chemical engineering, and venture capital have recently become involved in PFAL research and development (R&D) and the PFAL business. Thus, the progression of PFAL technology in the next 10 years or so may be as fast as that of the Dutch greenhouse technology in the past 40 or 50 years, if the major technologies are well documented, standardized, and opened and shared with many people.

2.2 Potential and Actualized Benefits of the PFAL

The PFALs that have actualized most of the potential benefits described in this section are making a profit and expanding their production capacity; however, the number of such profitable PFALs is currently limited. To actualize the potential benefits of the PFAL, the concepts behind the benefits and the methodology for their actualization must be thoroughly understood before designing and operating the PFAL. At the same time, the vision, mission, and goals of the PFAL design, operation, and business model must be clearly established and shared with the team members.

(1) *High resource use efficiency (RUE)*, which reduces resource consumption and waste and thus lowers production costs. RUE is defined as the amount of resource fixed or utilized in the plants (F) divided by the amount of resource supplied to the PFAL (S). Namely, RUE = F/S or (S − R)/S where R is the amount of resource that is released and wasted by escaping outside the PFAL (Kozai et al. 2015). Essential resources to be supplied regularly to grow plants in the PFAL are light energy, CO_2, water, fertilizer (nutrients), seeds/transplants, and labor only. The light energy, CO_2, water, and fertilizer (nutrients) are essential resources for seeds/transplants to grow by photosynthesis.

Water use efficiency (WUE) is the amount of water fixed or held in the plants divided by the net amount of water supplied to the culture beds and absorbed by the plant roots. Since about 5% of the water absorbed by plants is fixed in the plants and the remaining about 95% is either transpired from the leaves or drained somewhere without being absorbed by the roots, the WUE for irrigation to the greenhouse is 0.05 or lower.

On the other hand, in the airtight PFAL, almost all the transpired water is condensed and collected at the cooling panels of the air conditioners and returned to the nutrient solution tank. Then, the net consumption of water is the difference in amount between the irrigated water and the water returned to the nutrient solution tank. WUE for the PFAL is around 0.95 (= the fresh weight of plants divided by the difference in weight between the irrigated water and the water returned to the nutrient solution tank) provided that there is no leakage of water from the cultivation beds or nutrient solution tank (Kozai et al. 2015). This high WUE (95% water saving) of the PFAL is a big advantage in arid regions and other water-scarce areas.

CO_2 and fertilizer efficiencies of the PFAL are also relatively high (0.80–0.90) compared with those of CO_2-enriched and soil cultivation greenhouses with ventilators closed (0.5–0.6 for both) (Kozai et al. 2015). The CO_2 emitted by plant respiration during the dark period accumulates in the air of the cultivation room and is absorbed by plants during the photoperiod. The nutrient solution is not discharged to the drain, except in emergency cases such as the accumulation of ions (Cl^-, Na^+, etc.), which are not well absorbed by the plants.

Light energy use efficiency of the PFAL, however, is still very low (0.032–0.043), although it is higher than that of the greenhouse (0.017) (Kozai et al. 2015). The use efficiencies of electric energy, light energy, space, and labor in the current PFALs need to be considerably improved in the next-generation PFALs through the application of LEDs, intelligent lighting systems, better environmental control, and the introduction of new cultivars that grow well under low photosynthetic photon flux densities (PPFD).

A recent simulation of the RUEs for lettuce production in a PFAL and a greenhouse in the Netherlands and two other climate regions provided very useful academic and practical results (Graamans et al. 2018). This type of work will become increasingly important for efficient use of resources for plant production and for a better choice from various types of plant factories and greenhouses in a given locality.

(2) *High annual productivity* per unit land area. Over 100-fold annual productivity per unit land area can be achieved without the use of pesticides, compared with the annual productivity per unit land area in open fields, mainly due to the use of multilayers (ten tiers on average), shortened cultivation period (often by half) by optimal environmental control, high land area use efficiency (no vacant cultivation space throughout the year), high planting density, and virtually no damage by weather and pest insects. The annual productivity of the PFAL with 15 tiers is roughly estimated to be 200 kg m^{-2} (fresh weight of marketable produce). It would be interesting to estimate the maximum annual productivity of the PFAL under the optimal environment using a simulation model.

Land area use efficiency is defined as (A_u × n × N) divided by (365 × A_t) where A_u is the area of a unit cultivation space, n is the number of units of cultivation space in the PFAL, N is the average number of days per year during which the unit cultivation space is occupied by plants under cultivation, and A_t is the land area occupied by the floor of the PFAL. The unit cultivation space can be a cultivation panel, a tier, or a rack consisting of more than one tier. The unit of "space" is m^2 in the case of horizontally placed cultivation panels but can be m^3 in the case of vertically placed cultivation panels/tubes.

Major components of the production cost in the PFAL are electricity, labor, and depreciation of initial investment (the sum of these three cost components account for 75–80% of the total production cost). Thus, electric energy productivity (kg of produce per kWh of electricity consumption), labor productivity (kg of produce per labor hour), and space productivity (kg of produce per floor area or cultivation area) are important indices for analyzing and improving the productivity of the PFAL.

(3) *High weight percentage of marketable parts* over the whole-plant biomass. In other words, a low percentage of trimmed/damaged plant parts as waste over the whole-plant biomass. The percentage of marketable part can be increased by the proper environmental control method, cultivation system, and cultivar selection. Currently, the fresh weight percentages of marketable leafy parts and trimmed leafy or root parts of leaf lettuce plants are, respectively, estimated to be 77–80% and 20–23% in most PFALs in Japan.

Leafy parts of root crops such as carrot, turnip, and radish need to be edible, tasty, nutritious, and presentable to improve the weight percentage of the marketable parts of root crops. For example, the leafy parts and the root parts of such crops are often tasty and nutritious when harvested 15–20 days earlier than the conventional harvesting date.

(4) *High-quality plants* can be produced as scheduled by proper environmental control, cultivar selection, and cultivation system (Kozai et al. 2016). Shape/appearance, taste, and mouth sensation, as well as composition/contents of functional components such as vitamins, polyphenols, and minerals, can be controlled. However, most of such factors are still controlled by trial-and-error based on past experiences. On the other hand, researchers are systematically conducting a series of experiments to produce consistent functional components.

(5) *High controllability of plant environment.* Controlled aerial environmental factors include PPFD, VPD (water vapor pressure deficit), air temperature, CO_2 concentration, light quality (spectral distribution), lighting cycle (photoperiod/dark period), and air current speed. Controlled hydroponic culture factors include strength, composition, temperature, pH (potential of hydrogen), and dissolved O_2 concentration and flow rate of the nutrient solution.

(6) *High reproducibility and predictability* of yield and quality. Because of the high controllability of the environment all year round, scheduled and/or on-demand plant production is possible regardless of the weather. The quality of produce and the yield can be controlled by controlling the environment. For example, mouth sensation, taste, color, and flavor of lettuce can be finely

controlled to suit its use in salads, sandwiches, or hamburgers by environmental control.

(7) *High traceability* throughout the supply chain of PFAL industry, which enables a high level of risk management.

(8) *High adaptability for the location* (near or in food/meal delivery shops, etc.). The PFAL can be built without any problems in shaded areas, on contaminated or infertile soil, and in vacant rooms/buildings/land in urban areas. The PFAL can also be built in very cold, arid, or hot areas. For example, there are no heating costs for the PFAL even when the outside air temperature is below -40 °C because the walls and floor are thermally well insulated and heat is generated by the lamps in the culture room.

The PFAL is most suited to urban areas where the production site is close to the consumption site (local production for local consumption). This saves fuel, time, and labor for transportation of fresh produce and creates job opportunities for handicapped, elderly, and young people in or near their residential area.

(9) *High controllability of sanitary conditions*. Because of this high controllability, pesticide-free and other contaminant-free plants are produced. Global GAP (Good Agricultural Practice) and/or HACCP (Hazard Analysis and Critical Control Point) can be introduced relatively easily, and so a high level of risk management can be achieved.

(10) *Long shelf life* due to low CFU (colony-forming units of microorganisms) per gram, which decreases the amount of vegetable garbage or loss at home and in shops. The shelf life is estimated to be around two times longer for lettuce plants grown in the PFAL than for those grown in the field. Due to this advantage, the market price of PFAL-grown vegetables is often 20–30% higher than that of field-grown and greenhouse-grown vegetables.

(11) *No need to wash* or cook before serving, if packed in a sealed package after harvest in the culture room. This reduces the consumption of water for washing, electricity/city gas/fuel for boiling and stir-frying, and labor for washing and cooking. On the other hand, when eating fresh vegetables, the CFU per gram of vegetable needs to be lower than around 300.

(12) *Easy measurement* of hourly and/or daily rates of resource supply, production, and waste. The RUEs can be estimated online. Based on the estimation, the production costs can be predicted and subsequently reduced using the data on RUEs.

(13) *Stepwise improvement* of the RUEs, productivity and economic value of the plants is possible by visualizing the flow of energy, substances and workers in the production process and related costs/sales. To do so, models of plant growth, energy/substance balance, and production process scheduling are necessary.

(14) *Light and safe work* under comfortable air temperature and moderate air movement. There are still some problems to be solved to further improve the working environment both for large-scale PFALs with automation and small-scale PFALs in order to increase job opportunities.

(15) *Design and environment control are simpler in the PFAL than in the greenhouse* due to its airtightness, high thermal insulation of walls and floor and no solar light transmission to the cultivation space. Global standardization of the PFAL design (except for building design) is easier than that of the greenhouse design. To use solar light, which is free of charge, the greenhouse needs heating, shading and venting/cooling systems, insect screens, thermal screens for saving on heating costs, and transparent covering materials such as glass and plastic film. Those systems are not required in the PFAL. The environment in the PFAL is not influenced by the weather.

(16) *A small PFAL* with a floor area of $0.1–10\ m^2$ is a wonderful way to learn the principles of life science, engineering, and technology at home, school, or community center, especially when the PFAL is connected via the Internet to other small PFALs and a PFAL database for the exchange of information and opinions (Harper and Siller 2015). Through an interdisciplinary approach via hands-on experience, users acquire an understanding of the functions and mechanisms of the ecosystem, energy, and material conversion and circulation and learn the basic skills of growing plants and using advanced technologies.

2.3 Current Unsolved Problems of PFALs

2.3.1 Actions Required for Solving the Problems

(1) *Drastic reduction in initial investment and operation costs.* The operation cost per kg of fresh produce needs to be reduced by 30–50%, and the current initial cost per annual production capacity needs to be reduced by about 30% by the year 2020–2022, compared with the costs in 2017.

(2) *Sustainable production.* Improvement of the cultivation system and its operation to reduce, recycle and reuse resources, and use natural energy is essential. Energy-autonomous PFALs need to be designed, operated, and commercialized. The area of solar panels necessary for supplying all the electricity to the PFAL is currently estimated in Japan to be about eightfold that of the flat roof area of a PFAL with ten tiers, when a battery is installed at the PFAL. The necessary area of solar panels would be significantly lower in arid regions than in Japan. In addition, the solar panel area will decrease steadily year by year due to the technological advancements of solar panels and LEDs.

(3) *Advanced technologies* including artificial intelligence (AI), big data, the Internet of Things (IoT), bioinformatics, genomics, and phenomics need to be introduced to improve the resource use efficiency and cost performance of the PFAL (Fig. 2.1). Phenomics is an emerging research field along with the methodologies and protocols for the noninvasive measurement of cellular- to canopy-level plant-specific traits related to plant structure and function.

(4) *Robotics and flexible automation* need to be introduced to reduce the amount of heavy, dangerous, simple, and/or troublesome manual work.

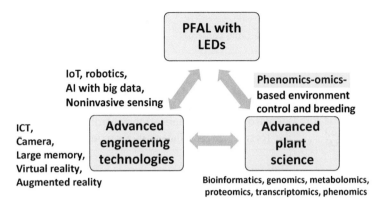

Fig. 2.1 Recent advanced technologies to be introduced in the next-generation smart PFAL. *AI* artificial intelligence, *ICT* information and communication technology, and *IoT* Internet of Things

(5) *Medicinal plants* for high-quality health care and cosmetics products need to be produced at low cost. Genetically engineered plants for production of *pharmaceuticals* such as vaccines for influenza and other viruses need to be produced in a specially designed PFAL.
(6) *Worldwide active organizations* of plant factory industries and academics need to be established for better global communication and information sharing.
(7) *Organic hydroponic systems* for PFALs that are easy to handle and economically viable need to be developed. Symbiosis of plants with microorganisms will benefit plant growth in the PFAL. Organic fertilizer can be produced from fish waste, vegetable garbage, mushroom waste, and other types of biomass.

2.3.2 Some Specific Technical Problems

(1) *Efficient use of white LEDs*, which emit a significant amount of *green light* (20–40% of total light energy). The optimal spectral distribution of white LEDs to meet a specific requirement is still unknown.
(2) *Green light effect* on photosynthesis, growth, development, secondary metabolite production, disease resistance, and human health in the PFAL has become an emerging research topic in relation to white LEDs.
(3) *Net photosynthesis, transpiration, and dark respiration* of plants in the PFAL can be continuously measured. Efficient methods for utilizing this data need to be developed.
(4) *Energy and mass (substance) balance* in the PFAL can be continuously measured. Efficient methods for utilizing this data need to be developed.
(5) *Resource use efficiency (RUE) and cost performance* of the PFAL can be measured, visualized, and controlled (Fig. 2.2). Efficient methods for utilizing this data need to be developed.

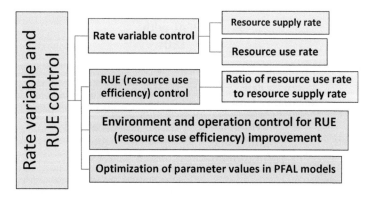

Fig. 2.2 Online measurement and control of rate variables and resource use efficiencies (RUEs)

(6) *Algae growth inhibition* in hydroponics. Also, the occurrence of *intumescence* (or edema) and *tip burn* symptoms on the leaves of leafy vegetables needs to be inhibited by proper environmental control and cultivar selection.
(7) *Microbiological ecosystems* in the culture beds are currently unknown and uncontrolled. Organic acids produced by plant roots, dead roots, dead and living algae, and many kinds of microorganisms including pathogens should be present in the culture beds. Beneficial and stable microbiological ecosystems need to be established.

2.4 Actions Required for Enhancing PFAL R&D and Business

(1) *Rational, powerful, and clear messages on the vision, mission, and goals of the PFAL.* People are increasingly interested in the potential benefits of the PFAL and are expecting further progress in R&D on smart PFALs.
(2) *Open database and open-source business planning and management system.*
(3) *Human resource development for PFAL managers and workers.* Human resource development programs for capacity building of PFAL managers are crucially needed. Well-edited books, manuals, and guidelines need to be published. Software/hardware systems for managing the complicated cause-and-effect relationships in the PFAL need to be developed.
(4) *Tools, facilities, and guidelines* for worker safety, labor saving, and quality operations. Compact and safe systems for efficient seeding, transplanting, harvesting, transporting, and packaging.
(5) *Software with database for minimizing the electricity costs* for lighting and air conditioning under a given lighting schedule.
(6) *Software with database for "smart" environmental control* for production of targeted functional components of a plant species under economic, botanical,

and engineering constraints. A computer-assisted support system for sensing, data analysis, control, visualization, and decision-making needs to be developed.
(7) *Well-designed floor plan and equipment layout* to maximize labor and space productivity.
(8) *Marketing* for creating new markets for health care. New products that do not compete with currently used products are necessary.
(9) *Understanding and anticipation* by local residents regarding the potential of the PFAL. Actual working/virtual models showing people the future are necessary.
(10) *Breeding the plant cultivars for the PFAL.* The characteristics of plants suited to cultivation in the PFAL are (1) fast growth under relatively low PPFD, high CO_2 concentration, and high planting density; (2) fast growth under low stresses of water, temperature, and pest insects/pathogens; (3) fast growth without physiological disorders; (4) secondary metabolite production sensitive to environmental conditions or stresses; (5) dwarf fruit venetagles and medicinal plants, and (6) high economic value per kg of produce due to qualitative plant traits. Molecular breeding can be a powerful tool.

At present, the plant cultivars grown in the PFAL are bred to suit open-field and greenhouse conditions where the environment varies greatly with time. Basically, the genetic characteristics of these cultivars are not suited to the environment in the PFAL. New cultivars bred for the PFAL environment are expected to significantly change the cost performance of the PFAL business.

(11) *Guidelines and manuals* for sanitary control, food and worker safety, and LED lighting.
(12) *Standardization of terminology and units* for the basic properties of light, lamps, and nutrient solution.
(13) *Standardization of PFAL components.* The design and method of operation of current PFALs are diverse, and the specifications of each component are not standardized. This diversity is a result of the creative and tireless efforts by many researchers and developers in the past several decades. The diversity of hardware and software, on the other hand, may delay domestic/international standardization of safety, salinity, and hardware and software parts. The diversity may also delay collaborative research and development with public institutions. It is also causing the high cost of each component, lack of a standard cultivation system, and lack of information and opinion exchange in the industry. On the other hand, the standardization should not restrict creative challenges.

2.5 Challenges for the Smart PFAL

Challenges for developing the software/hardware units or systems to be implemented in the smart PFAL as the next-generation PFAL include:

(1) The PFAL as *an essential unit to be integrated with other biological systems to improve the sustainability* of a building or a city.
(2) The PFAL for *large-scale production and breeding* of high-wire tomatoes and other fruit vegetables and berries such as strawberry and blueberry.
(3) *Cultivation system module* (or unit) as a minimum component of the plant cultivation space in the PFAL, which is easily connected to other basic module units to make a larger PFAL.
(4) *Hydroponic system without the use of substrate (supports) and a nutrient solution circulation unit without the drainage of nutrient solution* from the culture beds. Then, the total volume of nutrient solution in the culture beds, piping, and nutrient solution unit is greatly reduced. Also, the structure of the hydroponic system is simplified, and the physical weight of the system containing the nutrient solution is reduced. However, such a hydroponic system for widespread commercial use in the PFAL does not exist.
(5) *Phenotyping unit* for continuous and nondestructive (or noninvasive) measurement of plant traits such as fresh weight, leaf area, number of leaves, leaf angle, three-dimensional plant community architecture, leaf surface temperature, optical properties and chemical components of the plants, and physiological disorders such as tip burn and intumescence (or edema) of the leaves. The measured data on plant traits is used as input data for the phenome-genome-environment model to determine the set points of environmental factors and/or the selection of elite plants for breeding (Fig. 2.3). A small and inexpensive

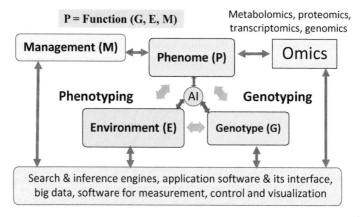

Fig. 2.3 Scheme showing phenotyping- and AI (artificial intelligence)-based environmental control and breeding for PFALs

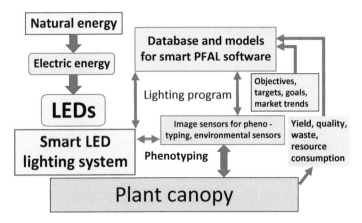

Fig. 2.4 Scheme showing smart LED lighting system and its peripherals

phenotyping unit that can be placed adjacent (1–50 cm) to the plants is preferable.

(6) *Periodic movement of plants* due to circadian rhythms (biological clock), water stress, air current patterns, etc. and their effects on hormonal balance, photosynthesis, transpiration, and growth of plants.

(7) *Smart LED lighting unit* for maximizing the cost performance (product of the unit economic value and the yield of produce divided by the operating cost) by time-dependent control of light environment factors such as light quality, photosynthetic photon flux density (PPFD), lighting cycle (photo-/dark period), and lighting direction (Fig. 2.4).

(8) *Ion concentration control unit* for the hydroponic system. The concentration of each major ion type (NO_3^-, K^+, Mg^{2+}, Ca^{2+}, Na^+, NH_4^+, Mg^{2+}, Cl^-, PO_4^{3-}, and SO_4^{2-}) in the nutrient solution is separately measured or estimated and controlled.

(9) Software unit for *discriminating the effects of spatial variations of the environment* on the spatial variations of individual plant growth from the effects of genetic variations among the plants on the spatial variations of individual plant growth. The spatial variation of plant growth is obtained from the data measured by the phenotyping unit.

(10) *Hardware/software unit for minimizing the spatial variations* of the air temperature, vapor pressure deficit (VPD), and air current speed by controlling the spatial distributions of PPFD and air current speed under the given LED lighting system and three-dimensional plant canopy architecture. The spatial variations of the environment are obtained from the distributed environmental sensors.

(11) *Software unit for automatically determining the set points of environmental factors* to meet the objectives of PFAL operation under the given constraints, using the phenome data and other data.

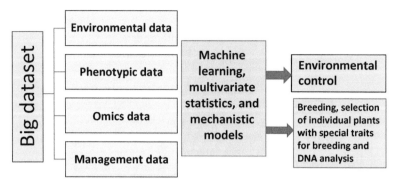

Fig. 2.5 Big dataset with low noise obtained in the PFAL is useful for machine learning, multivariate statistics and mechanistic models

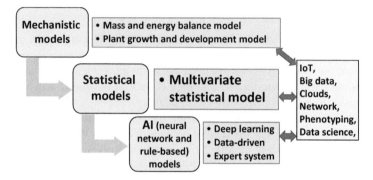

Fig. 2.6 Three types of models for PFAL environment control

(12) *Deep learning unit for searching for a function (G, E, M)* or relationship among the phenome, genome, and environment datasets. P = function (G, E, M) where P is phenome data, G is genome data, E is environment data, and M is management data. Using the big datasets of P, G, E, and M, the function (G, E, M) is found by deep learning. In the PFAL with a controlled environment, the datasets of P, E, and M can be collected relatively accurately and easily. Thus, the genome dataset is known, and the function (G, E, M) can be found relatively easily by deep learning. Figure 2.5 is a scheme showing the phenotyping-, AI-, and big dataset-based environmental control and breeding.

(13) Software/hardware unit for searching for *DNA expressions/markers* driven by environmental changes, using the genome, phenome, and environment datasets, and for determining the set points of environmental factors using the above dataset. Deep learning unit using the phenome, genome, and environment datasets would become a powerful breeding tool.

(14) *Integration of the deep learning model with mechanistic, multivariate statistical, and behavior (or surrogate) models*. Figure 2.6 is a scheme showing three types of models to be used for the PFAL environment control. Figure 2.7 is the

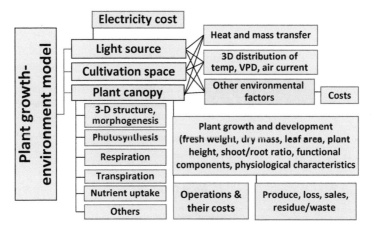

Fig. 2.7 General scheme of plant growth-environment model for maximizing the cost performance (CP)

general scheme of a plant-environment model for maximizing the cost performance of the PFAL.

(15) *Speed breeding.* Watson et al. (2018) proposed a method of "speed breeding" which greatly shortens generation time and accelerates breeding and research programs. They envisage great potential for integration speed breeding with other crop breeding technologies, including high-throughput genotyping, genome editing, and genomic selection, for accelerating the rate of crop improvement.

(16) *Dual (virtual/actual) PFAL,* a pair of virtual and actual PFALs. *The virtual PFAL* placed in the cloud is used to simulate the actual PFAL output using the data input to the actual PFAL. The parameter values in the virtual PFAL are adjusted automatically by using the input and output data of the actual PFAL (Figs. 2.8 and 2.9). The virtual PFAL can be used for training, self-learning, education, fun, and research and development of the PFAL.

2.6 Conclusion

To actualize the potential benefits of the PFAL, a considerable amount of systematic research, development, and marketing with the appropriate vision, mission, strategy, and methodologies are necessary. On the other hand, actualization of the potential benefits is relatively easy, because the energy and material balance and the plant-environment relationship in the PFAL are much simpler than in the greenhouse. Thus, the methods for actualizing the benefits are relatively straightforward. The issues described in this chapter are discussed in more detail in the following chapters of this book.

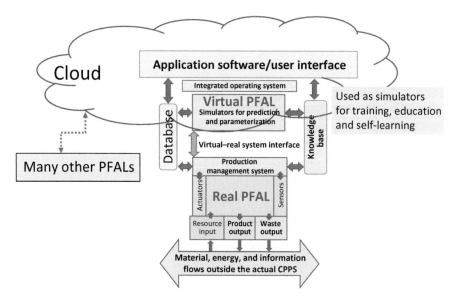

Fig. 2.8 Scheme showing dual (real and virtual) PFAL and its network. (Revised from Kozai et al. 2016)

Fig. 2.9 Software configuration of the dual PFAL

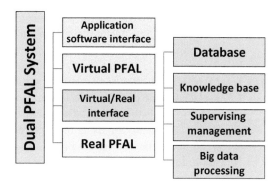

References

Graamans L, Baeza E, van den Dobbelsteen A, Tsafaras I, Stanghellini C (2018) Plant factories versus greenhouses: comparison of resource use efficiency. Agric Syst 160:31–43

Harper C, Siller M (2015) Open Ag: a globally distributed network of food computing. Pervasive Comput 14(4):24–27

Heuvelink E (ed) (2006) Tomatoes (crop production science in horticulture series). CAB International, Wallingford, 339 pages

Kozai T, Niu G, Takagaki M (eds) (2015) Plant factory: an indoor vertical farming system for efficient quality food production. Academic, Amsterdam, 405 pages

Kozai T, Fujiwara K, Runkle E (eds) (2016) LED lighting for urban agriculture. Springer, Singapore, p 454

Watson A, Ghosh S, Williams MJ, Cuddy WS, Simmonds J, Rey HMAM, Hinchliffe A, Steed A, Reynolds D, Adamski NM, Breakspear A, Korolev A, Rayner T, Dixon LE, Riaz A, Martin W, Ryan M, Edwards D, Batley J, Raman H, Carter J, Rogers C, Domoney C, Moore G, Harwood W, Nicholson P, Dieters MJ, DeLacy IH, Zhou J, Uauy C, Boden SA, Park RF, Wulff BBH, Hickey LT (2018) Speed breeding is a powerful tool to accelerate crop research and breeding. Nature Plants 4:23–29

Chapter 3
Protocols, Issues and Potential Improvements of Current Cultivation Systems

Na Lu and Shigeharu Shimamura

Abstract This chapter outlines the existing cultivation systems, protocols, management approaches as well as issues encountered in plant factory operations. Comparisons, suggestions and potential solutions based on academic data and practical experiences are introduced. Those who are going to start a business or research in a plant factory and those who are already operating a plant factory have a general idea about the basic requirements of plants in various environmental conditions and the acceptable set points of such conditions. These experiences may not always maximize production but will reduce the risks of failure and loss in operation. More advanced technologies for the next generation of plant factories are expected on the basis of the current situation.

Keywords Hydroponic systems · Issue control · Light aspects · Overall management · Set points

3.1 Introduction

Plant factory (with artificial lighting) plays a key role in food security and is being considered as an emerging business opportunity in urban areas of many countries. Many commercialized plant factories and successful business case studies have emerged due to the continuous efforts towards the development of plant cultivation knowledge, environmental control technology as well as marketing management. An increasing number of investors, companies and researchers have become involved in plant factory projects. The differences in companies, countries, crops

---

N. Lu (✉)
Center for Environment, Health and Field Sciences, Chiba University, Kashiwa, Chiba, Japan
e-mail: na.lu@chiba-u.jp

S. Shimamura
HANMO CO., Ltd, Tokyo, Japan
e-mail: s.shimamura@hanmo.jp

© Springer Nature Singapore Pte Ltd. 2018
T. Kozai (ed.), *Smart Plant Factory*, https://doi.org/10.1007/978-981-13-1065-2_3

and market needs have resulted in diversified cultivation systems and environmental control protocols that are uniquely rewarding. There is no strict standard for the management of plant factories. However, when considering plants as the key focus, there are some common principles that are to be followed in order to meet plant growth requirements. The knowledge based on current situation and operation experiences are introduced in this chapter. It is expected that this information will guide plant factory owners as well as the individuals, who are planning to start a plant factory in the future.

3.2 Hydroponic System

Hydroponics is a method of growing plants in a water-based, nutrient-rich solution. In plant factories, a hydroponic system commonly uses small sponge cubes to fix the plant in the cultivation panel, which directly exposes the plant shoots to the air and the plant roots to the nutrient solution. Although sponges are used in most cases, sometimes, according to different application purposes (e.g. a unique crop, different design requirements, easier management and whole-plant-sell product), a medium, such as rock wool, clay pellets, perlite, peat moss or vermiculite, can also be used for supporting the plant root system.

3.2.1 Irrigation Methods

The following are the main irrigation systems currently used in plant factories to fertilize the plants.

3.2.1.1 The Nutrient Film Technique (NFT) System

In this system, a shallow film (2–3 mm) of nutrient solution flows over the plant roots, ensuring that they are irrigated but not completely soaked. The upper part of the roots is exposed to the air and has access to oxygen. The recommended slope for a NFT system is commonly a 1:70 to 1:100 ratio. This means that a 1 cm drop (slope) is recommended for every 70 to 100 cm of horizontal length. However, this ratio can slightly be adjusted according to different cultivation conditions. The NFT system requires a pump to circulate the nutrient solution from the nutrient tank or reservoir to the cultivation beds or channels at an appropriate flow rate. The basic concept of an NFT system needs a cultivation bed, nutrient solution tank, water pump, filter and pipes (Fig. 3.1).

Fig. 3.1 Sketch of conventional NFT hydroponic system

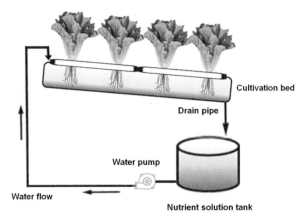

Fig. 3.2 Sketch of conventional DFT hydroponic system

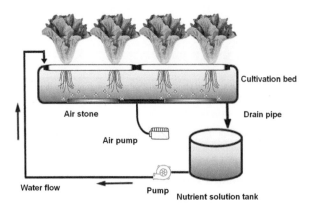

3.2.1.2 The Deep Flow Technique (DFT) System

In the DFT system, the plant roots are soaked in a deep flow of nutrient solution rather than a film. This system also requires constant recirculation of the nutrient solution to introduce oxygen along the entire length of the growing bed or channel, thus ensuring that oxygen content in the root zone is continually high enough for root growth. Some systems even use an additional air pump for oxygen supply. Usually, the panel for supporting the plants floats on water. The nutrient solution is pumped by a water pump from one side of the planting bed and flows back to the tank from the other side. One of the advantages of the DFT system is that the nutrient solution remains available for plant growth even in the event of a power outage. A demerit of this system is that the cultivation bed is relatively heavy. Moreover, the floatable materials, such as foam boards used as cultivating panels, are prone to dust and algal growth, and its reflection rate quickly declines after using it for a few times. The basic concept of a DFT system requires a cultivation bed, nutrient solution tank, water pump, filter, pipes and at times an air pump (Fig. 3.2).

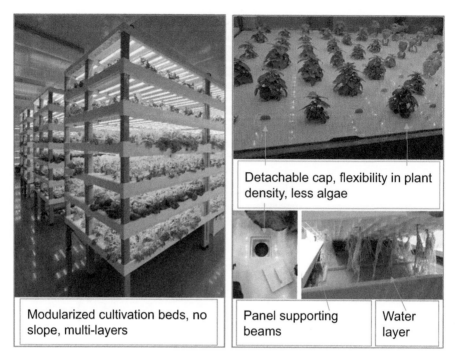

Fig. 3.3 Images of the modified hybrid hydroponic system in a plant factory (Photos provided by Fujian Sanan Sino-Science Photobiotech Co., Ltd)

3.2.1.3 Modified Hybrid System

A modified hybrid system, as the name suggests, is a combination of the NFT and DFT systems. This system does not use pipes, which makes its installation faster and convenient. It includes beams at two different heights that are integrated into the cultivation bed and several sluice doors to adjust the water level. It allows an easy switch between the NFT and DFT systems according to the cultivation objectives. It is designed without a slope and is easier to be built as multilayers to save space. The panels are supported by premoulded beams (not floating on water) that provide a better option when selecting the material of the cultivating panel (i.e. more choices than floating foams). A white-coloured (with high reflection rate), anti-contaminative and anticorrosive material is recommended. The panel can be designed with multiple holes and detachable caps ensuring an efficient use of space and light, thus allowing flexibility in plant density for different crops and their growth stages (Fig. 3.3).

Fig. 3.4 Sketch of spray hydroponic system

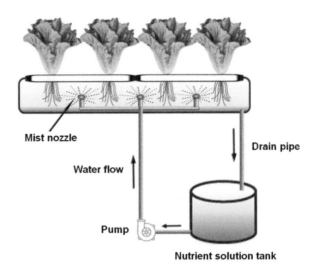

3.2.1.4 Spray System

Spray system, also called as aeroponics system, atomizes the nutrient solution and sprays it on the plant roots that are exposed to the air. In order to prevent the roots from drying out, a continuous spray is required. This method keeps the plant roots wet and while being well aerated, allowing high respiratory activity of the roots. Moreover, the cultivation bed is light in weight because the amount of nutrient solution in the cultivation bed is relatively smaller.

On the other hand, there are also some risks or disadvantages of this system. If the pump stops for some reason, such as a blackout or a malfunction of the pump, the roots would be subjected to drought stress in a short period. Another disadvantage is clogging of the spray nozzle that can occur due to plant residues, crystallization of fertilizers and carcasses of microorganisms. It is necessary to introduce a cleaning system to regularly deal with such issues. Additionally, deeper cultivating beds are required for spraying when compared with other systems. It is relatively inefficient in terms of optimal use of space (Fig. 3.4).

3.2.1.5 Ebb and Flow System

This system is similar to the NFT and DFT systems (Figs. 3.1 and 3.2). Instead of continuous recirculation of the nutrient solution, this system pumps the nutrient solution in the cultivation bed at certain intervals per day. The plants are usually grown in pots with substrates or in rock wool cubes that can hold water for a few hours to support plant growth. The basic concept of an ebb and flow system requires substrates, cultivation bed, nutrient solution tank, water pump, filter, pipes and timer.

Fig. 3.5 Sketch of drip hydroponic system

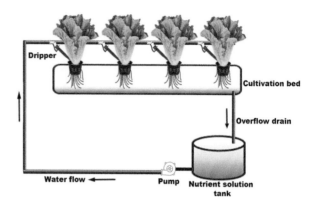

3.2.1.6 Drip Irrigation System

This system works in the presence of a growth medium or substrates and is ideal mainly for potted plants, flowers or strawberries. It avoids the plant roots being soaked in water. Therefore, this system can be lighter than a DFT system, allowing easier plant movement together with the pot (Fig. 3.5).

3.2.1.7 Wicking System

This system does not require electricity because the water is absorbed by wick ropes or a fabric sheet at the base; it is suitable for young potted plants, seedlings and leafy herbs. The basic concept of a wicking system needs a nutrient solution tank, growth substrate and wick rope or fabric sheet at the base. The nutrient solution is added to the reservoir tank (planter), as required, when the water level is below the minimum water-level mark, and the unwoven fabric sheet at the base will absorb it while the water is consumed by the plants (Fig. 3.6).

This system allows the roots to absorb water as per their needs, and it also ensures a high oxygen concentration around the root zone. It is economic and eco-friendly because there is no wastage of water or nutrient solution from the system.

The NFT, DFT, hybrid, spray, shallow water, ebb and flow and drip systems are active systems, meaning that they rely on pumps to work. Wicking systems is a passive system that does not require pumps. Additionally, growth medium is required when growing some medicinal plants, such as *Huperzia serrata*, *Anoectochilus formosanus*, *Dendrobium kingianum* and some other orchids that are sensitive to water content.

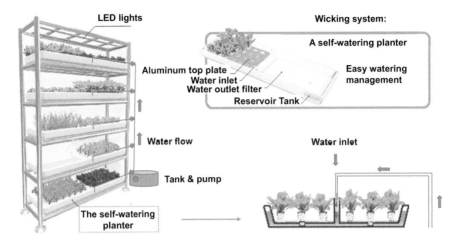

Fig. 3.6 Sketch of a wicking system (self-watering system) (Data provided by Planet Co., Ltd)

3.2.2 Nutrient Solutions

Nutrients are one of the basic requirements of any hydroponic system. For a fertilizer to be incorporated into a hydroponic system, it must be soluble in water. The type and concentration of fertilizers can easily be controlled using hydroponics. Hydroponic nutrients can be a complex issue or as simple as mixing and pouring. Anyone not familiar with hydroponic nutrients should use only a proven formula from a reputable manufacturer.

3.2.2.1 Nutrient Composition

There are over 20 elements that are required for optimal plant growth. Inadequate or excessive amount of any nutrient can result in poor crop performance. There are 17 essential elements: carbon (C), hydrogen (H), oxygen (O), phosphorus (P), potassium (K), nitrogen (N), calcium (Ca), magnesium (Mg), sulphur (S), iron (Fe), boron (B), manganese (Mn), copper (Cu), zinc (Zn), molybdenum (Mo), chlorine (Cl) and nickel (Ni). C, H and O are absorbed from the air and water during photosynthesis. The other macro- and micronutrients are supplied using preformulated commercial hydroponic nutrients. Some proven nutrient formulae are listed in Table 3.1.

Table 3.1 Some proven nutrient solution formulae for leafy vegetables

Formula	Elemental concentration of macronutrients (in ppm: mg kg^{-1} or mg L^{-1})						
	NO_3^-	NH_4^+	PO_4^{3+}	K^+	Ca^{2+}	Mg^{2+}	SO_4^{2-}
Enshi	992.0	23.4	126.7	312.8	160.4	48.6	192.2
Yamazaki (lettuce)	372.0	9.0	47.5	156.4	40.1	12.2	48.1
Chiba-U Saladana	744.0	23.4	126.7	312.8	80.2	24.3	96.1

	Elemental concentration of micronutrients (in ppm: mg kg^{-1} or mg L^{-1})						
	Fe	B	Mn	Cu	Zn	Mo	
Enshi	3.0	0.5	0.5	0.02	0.05	0.01	

3.2.2.2 Electrical Conductivity (EC)

The unit of EC is S m^{-1} (siemens per metre) and the designated unit is dS m^{-1}. For instance, 1.2 mS cm^{-1} = 0.12 S m^{-1} = 1.2 dS m^{-1}. A high EC indicates high solution concentration. The EC value indicates total ionic concentration while not representing a specific ion's concentration. The nutrient solution formulae in Table 3.1 are basically designed for controlling the EC or ion concentration. Ion concentration management is commonly used in hydroponic systems, and a setting point of 0.8–1.5 dS m^{-1} for seedlings and 1.5–2.5 dS m^{-1} for the growth stage of leafy plants is acceptable. Another method, known as quantitative management, is also proposed (Tsukagoshi and Shinohara 2015). It is assumed that the demand for different mineral ions varies with plant growth stages, and in leafy vegetables, the total quantity of each mineral ion is more important than the total ionic concentration (Maneejantra et al. 2016). This method is expected to avoid excessive absorption of nutrients and can increase the efficiency of fertilizer use.

3.2.2.3 Management of pH

In general, a pH (potential of hydrogen) value of 5.0–7.0 is an optimum range for plant growth. NaOH or KOH can be used to increase the pH, while H_2SO_4, H_3PO_4 or HNO_3 can be used to decrease it.

3.2.2.4 Nutrient Solution Temperature

It is important to keep the root zone at an appropriate temperature; generally, a temperature ranging from 18 to 22 °C is suitable for most of the crops that are grown in plant factories. Plants with nutrient temperatures too high or too low can limit application of other environmental factors, thus limiting plant growth. Higher temperatures increase root respiration rate, Ca deficit, transplanting injury, risk of disease and decrease water uptake. Lower temperatures increase anthocyanin synthesis but decrease plant growth and uptake of Mg and P. There are some applications of organic fertilizers in a hydroponic system, but they have not been discussed in this chapter.

3.3 Lighting System and Light Aspects

A good lighting system is essential to ensure sufficient light and efficient energy use in a plant factory. The core light aspects, which are important for a lighting system, include a well-designed light source (such as high electro-light conversion rate, long life span, wide beam angle and waterproofness), appropriate light spectrum, light intensity, light uniformity, light/dark period, lighting direction, light distribution and a methodology combining the above factors to obtain a desired objective (a so-called a light recipe), to maximize the production using the least energy input.

3.3.1 Photosynthetic Photon Flux Density (PPFD) and Light Period

An average PPFD of 100–300 µmol m^{-2} s^{-1} with a 10–18 hours photoperiod is applicable for most of the leafy vegetables grown in a plant factory. A PPFD <100 µmol m^{-2} s^{-1} would cause the plants to stretch excessively or cause an improper shape with light colour. But for microgreens and sprouts, a PPFD <100 µmol m^{-2} s^{-1} may be acceptable. A PPFD >300 µmol m^{-2} s^{-1} acts positively on the growth of most of the plants but will increase the initial cost on lighting equipment and may not economically be feasible for commercial plant factories. A minimum DLI (daily light integral) is recommended for various crops. For instance, a DLI of 12–17 mol m^{-2}d^{-1} has been recommended for lettuce and leafy crops (Kubota 2015). A target DLI of 12 mol m^{-2}d^{-1} can be achieved by a PPFD of 300 µmol m^{-2} s^{-1} for an 11-h photoperiod or a PPFD of 210 µmol m^{-2} s^{-1} for a 16-h photoperiod. Sometimes a 24-h photoperiod for a specific crop variety or intermittent lighting methods for avoiding physiological disorders can also be designed in practical cases. The lighting time is scheduled mainly during the night to take advantage of the off-peak electricity rates.

3.3.2 Light Spectrum

A light spectrum that may effectively influence plant growth can be any single colour of light or combination of lights among the photosynthetically active radiation

(400–700 nm), UV (ultraviolet) (280–400 nm) and far-red light (700–780 nm). Several studies have highlighted the effects of light spectra on plant growth, morphology and accumulation of secondary metabolites. One of the most important roles of light is in photosynthesis; however, not all wavelengths of artificial light are efficient in photosynthesis. For beginners who are trying to grow plants in a plant factory, a typical blue and red light (5–30% of blue light and 70–95% of red light) or a whitish light with even-distributed peaks is acceptable for almost all vegetables without encountering a major problem. This does not mean the plants do not need other colours of light. Generally, UV and blue light are more efficient in accumulating secondary metabolites (that affect plant characteristics such as colour turning; deeper colouring; compact, hard and firm leaves; thickness and strong taste). Red light is more efficient in photosynthesis, and the far-red/red light ratio is more effective in controlling flowering, stretch of stem and leaves and other morphological characteristics (such as producing looser, soft, light-colour leaves). Green light can be effective in producing a dense canopy of plants with similar colour of leaves, such as iceberg lettuce or cabbage, since green light has a higher penetration rate than blue and red light.

3.4 Set Points of Air Temperature, VPD, CO_2 Concentration and Air Current Speed

An air temperature of 18–25 °C is acceptable for most of the crops in a plant factory. The ideal range for VPD (vapour pressure deficit) is 0.8–0.95 kPa, with an optimal setting of around 0.85 kPa (Niu et al., 2015). Concentration of CO_2 can be set to 500–2000 ppm when there is an effective air circulation system continuously allowing CO_2 evenly distributed inside the plant canopy; otherwise a higher CO_2 setting is recommended. The air current speed is usually controlled between 0.5 and 1.0 ms^{-1} using air circulation fans to promote gas exchange.

3.5 Air Conditioning System and Air Circulation

3.5.1 Air Conditioning System

Air conditioning (AC) system design is an extremely important part of plant factories. Controlling air temperature inside the plant factory almost completely depends on AC system. It is impossible to keep a suitable temperature for plant growth if the cooling capacity of AC is insufficient, resulting in a huge influence on plant cultivation.

The load on the AC comes mainly from the heat generated by lighting system and the exchanged heat from outside air temperature. If the facility is adequately insulated, about 80% of the AC load will be because of the heat from illumination. In such a case, even if the outside air temperature is below freezing (-0 °C to ~ -40 °C), the AC inside a plant factory would still be operational in cooling mode when the lighting is on. Therefore, it would be better to use light sources (e.g. LED lamps) that generate less heat to reduce the load on AC system. Scheduling the lighting cycle can be another method to reduce the heat load. Instead of lighting up all the cultivation areas at the same time, the heat load for the AC system can be reduced by shifting the lighting time for each cultivation area and ensure that a certain amount of lighting is constantly in off mode. Generally, the average lifetime of an AC (in the case of office use) is 10–13 years. However, there are several instances where it is shortened to about 7–8 years when used in plant factories because of the harsh conditions and when they are used continually for 24 h a day all year round.

3.5.2 Air Circulation

Two types of AC are usually seen in the market, one is the floor stand type and the other is the suspension type. In most cases, suspension-type ACs are used and hung near the ceiling area since the warm air (mainly heated by lighting) goes towards the ceiling. Meanwhile, fans are used in combination with ACs to aid air circulation. There are two main reasons for using circulation fans in plant factories: (1) to stabilize the room temperature and (2) to allow better air flow in order to improve plant growth. The selection and application of the fans vary with their purpose. In order to stabilize the room temperature, fans with high power/capacity/size are required to aid faster air movement. When ACs are installed on the floor, an appropriate air flow needs to be sought out by running some tests. On the other hand, when aiming the air flow into the plant canopy, a very strong flow would become a stress for plant growth, so fans with relatively smaller capacity should be selected.

3.6 Salinity Control and Sterilization of Nutrient Solution

3.6.1 Salinity Control

The ground water in the coastal areas and remote islands usually contains a high content of sodium (Na^+) ions. Since Na^+ ions can adversely affect plant growth, it is necessary to remove them from the groundwater before making the nutrient solution. There are two methods that can be utilized to remove salts from ground water: using reverse osmosis membranes and using ion exchange resins. In areas with adequate rainfall, it is also possible to use rainwater because it does not contain Na^+ ions, and it is a low-cost method that can avoid salt damage. The use of rainwater can be considered as a valid method when the average rainfall is predicted over 100 mm a month. Rainwater can be collected and stored in a big tank for preparing nutrient solution when needed. At the same time, installation of a sterilization system will ensure water quality and easy maintenance. Since water is a valuable resource, it also should be used as efficiently as possible. Another way is to reuse the AC drain water for preparation of the nutrient solution. Almost all of the drained water collected by the AC is the water transpired by plants, so it is nearly pure and without the presence of salts.

3.6.2 Sterilization of Nutrient Solution

There are various methods to sterilize the nutrient solution. Methods such as UV, ozone and heat sterilization, silver/titanium oxide method and sand filtration have been shown to be promising. However, each method has its own advantages and disadvantages, and there is no perfect or ideal alternative at the moment.

3.6.2.1 The UV Sterilization Method

This method uses light from UV lamps to kill microorganisms in the nutrient solution. It is possible to sterilize a large amount of nutrient solution at once, as long as the UV light can reach it. On the other hand, as a disadvantage, it causes precipitation of Mn and Fe. After a certain period, the nutrient solution becomes turbid and weakens the UV light's permeation into the nutrient solution, resulting in a decrease in its sterilizing capacity.

3.6.2.2 The Ozone (O_3) Sterilization Method

In this method, ozone is dissolved into the nutrient solution, and the microorganisms are killed by oxidizing capacity. O_3 has high sterilizing power, and it is relatively effective. On the other hand, as a disadvantage, like UV, it also causes precipitation of Mn and Fe. Moreover, as the organic matter in the solution increases, O_3 decomposes organic compounds, which decreases its sterilizing ability. A high concentration of O_3 increases the sterilizing ability but can also damage the plants.

3.6.2.3 The Heat Sterilization Method

This is the most common method that includes heating of the nutrient solution in a short time by passing the nutrient solution through heated pipes. Unlike UV or O_3 sterilization method, heat sterilization method does not precipitate Mn or Fe. As a disadvantage, the sterilizing potential is relatively lower, and only small quantities of the solution can be handled at once, in addition to extra energy cost required for overheating.

3.6.2.4 Silver/Titanium Oxide Utilization Method

In this method, sterilization is carried out by directly contacting the microorganisms with silver or titanium oxide. In general, we sterilize the nutrient solution by kneading the silver or titanium oxide in the filter and passing the nutrient solution through the filter. The disadvantage is that nutrient solution cannot be sterilized without direct contact with silver or titanium oxide. Therefore, it is difficult to kill the microorganisms stored in the cultivation bed. Also, when silver dissolves into the nutrient solution, it may be absorbed by the plant and accumulate inside it.

3.6.2.5 Sand Filtration Method

In this method, sand is packed in a cylindrical tube, and the nutrient solution is poured from the upper side of the tube resulting in the removal of microorganisms, and a sterile nutrient solution flows from the other end of the tube. A biofilter is installed on the surface of the sand filtration device that removes the microorganisms. The advantage of this method is that material cost is low, and it is possible to filter a large amount of nutrient solution in a short period. It also saves energy because it uses the gravitational force to filter the nutrient solution. On the other hand, this method requires cleaning of the sand and its periodic replacement. Secondly, the Mn in the solution can be absorbed by the sand, which can affect the balance of the nutrient composition.

3.6.2.6 Other Sterilization Methods

Besides the above sterilization methods, another potential method is the oxygen (O_2) bubbling method. This is a method of confining air or O_2 gas into fine bubbles (called microbubbles or nano-bubbles) and mixing them into the circulation of the nutrient solution. Oxygen suppresses the growth of anaerobic microorganisms in the nutrient solution. Since O_2 has weaker oxidizing power than O_3, it does not damage the plant when its roots are exposed to O_2. Therefore, it is possible to apply it to the plant cultivation beds. Oxygen also improves the respiratory activity of the roots, thus strengthening the plant growth. However, since O_2 has a weaker sterilizing ability than other sterilization methods, it is not used for the purpose of sterilizing microorganisms in the culture solution, but of suppressing the growth of microorganisms.

3.7 Floor Layout

A plant factory is basically a closed building with several rooms for preparation, cultivation and packaging and shipping. Plant cultivation can roughly be divided into three stages, i.e. the germination stage (seed sowing to budding), the nursery stage (seedlings to planting) and the growth stage (planting to harvesting).

The germination period lasts for several days (from 2 days up to 1 week depending on the crop); the nursery stage usually takes several weeks, while the growth stage lasts for around 2 weeks until harvest. The time taken from the germination stage to the nursery stage is about 50–70% of the total cultivation period, but the space used for these two stages is only about 20–25% of the total space. Therefore, the floor layout needs to be well designed to ensure that necessary space is allotted for each stage. Production demand determines the number of seeds that will be sowed and seedlings for the nursery. Thus, the space required for germination and nursery can be calculated if the space for growing stage is known.

3.8 Plant Species and Cultivars

In plant factories, the plant species (such as leafy lettuce, herbs, microgreens and some medicinal plants) having short growth cycles, high added values and less wastes in produce are preferred. The demand for exclusive plant species and cultivars is significantly increasing year by year. While growing plants in traditional open fields, pest control and disease control are the top priorities for breeding. However, in plant factories, plants are cultivated in a closed environment where the risks of insects and diseases are significantly reduced; hence, the issues of developing resistance to pests and diseases no longer hold importance for breeding.

On the other hand, since the environment can be controlled to be suitable for growing plants, the plant growth is accelerated that can easily cause physiological disorders, such as tip burn. The objective of breeding in a plant factory should be to overcome these physiological disorders. In addition, high-temperature-resistant cultivars that can grow without much cooling can be cultivated in order to reduce the cost of AC system. For these reasons, the current cultivars may not be suitable for plant factory production, and new breeding methodologies should be developed. Cultivars with much higher plant production and lower loss are expected to be developed for plant factory. A comparison experiment on different lettuce cultivars was conducted, and the results showed large differences in production, morphology and tip burn occurrence among cultivars (Fig. 3.7). For example, the plant fresh weight (FW) was about 90–100 g for cultivar V_2 and V_7, 60–70 g for V_1 and V_5 and 50–60 g for V_3, V_4 and V_6. The production differed by a factor of two between the highest and the lowest cultivars even though they were grown under the exact same conditions. Moreover, the new variety, which could achieve similar high production

3 Protocols, Issues and Potential Improvements of Current Cultivation Systems

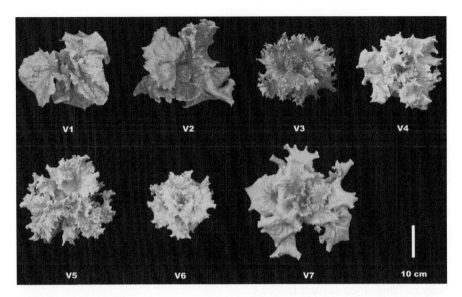

Fig. 3.7 Cultivar differences in plant production, morphology and tip burn occurrence of seven types of lettuce plants grown in a plant factory (Data were taken on the 35th days after sowing)

with conventional variety, had no tip burn (cultivars V_1, V_2 and V_3 showed high tip burn occurrences, and no tip burn was observed in others).

3.9 Issues to Be Solved

The above sections introduced the necessary factors required by a plant factory and knowledge of their setting points/ranges ensuring a workable start-up. However, it does not guarantee a smooth functioning of the plant factory. There can be several factors that can hinder its smooth operation.

3.9.1 Light

Some cultivators have experienced that the plant production goes smoothly and is stable as scheduled and predicted at the beginning. However, after a certain period, the production declines without visible changes in the cultivation system. One of the possible reasons can be that the light intensity on the plant canopy dropped (e.g. 30% drop in production probably caused by 30% reduction in light intensity). This may not only be because of the quality of the lamp but may be due to stained surface of the lamp and reflectors due to plant leaves or nutrient solution during daily operation. Thus, a weekly or biweekly check of the light intensity is recommended. It helps the cultivators to make a regular plan to maintain and clean lighting equipment and reflectors.

3.9.2 Temperature

The changes in the air and nutrient solution temperature are not easily detected, and this can influence plant production and plant growth. Especially in the large-scale plant factories, minor changes such as malfunctioning of one to two ACs or setting errors are not noticeable. But this can have impacts on the plant growth, quality and even the whole production schedule. The most dangerous situation is during a blackout at night or in weekends that may lead to plant damage due to a sudden increase of water temperature. Therefore, daily checks on the system operation and water temperature are important for commercial-size plant factories. Additionally, the temperature settings of the air conditioners should also be changed according to seasonal variations. Indoor temperature is greatly affected by seasonal changes for many buildings used as plant factories.

3.9.3 Element Balance in Nutrient Solution

After long cultivation periods, the elements in the nutrient solution may no longer be balanced even when an automatic fertilizer supply system is used. Certain ions in the water tank may or may not be taken up too quickly by the plants. For example, potassium (K^+) ion levels tend to decrease faster than other cations in the nutrient solution in the case of growing lettuce plants in hydroponics. Potassium ions are selectively taken by the plants through ion channels in the plant roots. Since K^+ ions are monovalent while Ca^{2+} and Mg^{2+} are divalent cations, K^+ ions are more easily absorbed by plants than Ca^{2+} and Mg^{2+} ions. Therefore, the concentration of K^+ ion decreases as cultivation continues, and it causes an imbalance in the composition of the nutrient solution.

3.9.4 pH Adjustment

This is a well-known parameter but sometimes may unintentionally be ignored by cultivators until a bigger problem arises. It mainly occurs due to the fast absorption of ammonia. One of the alternatives is to not use ammonia fertilizers and periodically renew the nutrient solution in the main tank. A regular assessment of pH is also recommended. Cations and anions are balanced in the nutrient solution in the absence of plants; therefore, the H^+ ions would not change and the pH remains stable. However, when a plant absorbs cations (especially ammonium ion, NH_4^+), it releases H^+ ions from the roots to maintain its pH balance, thus decreasing the pH of the nutrient solution. Since NH_4^+ ions are easily absorbed by the plants, it leads to a rapid drop of pH in the nutrient solution. On the other hand, when a plant absorbs anions (especially nitrate ions, NO_3^-), the plant releases bicarbonate ions (HCO_3^-) from the roots. As the H^+ from this HCO_3^- ion reacts with OH^- to become H_2O, the H^+ ions decrease, resulting in an increase in the pH of the nutrient solution. For these reasons, when cultivating plants for a long time, the pH initially decreases due to the absorption of NH_4^+ ions and then gradually increases due to the absorption of NO_3^- ions.

3.9.5 Algae

This is probably an issue for all plant factory operators. Algae need light, water and nutrients. Wherever these three essential elements coexist, algae thrive. Algae absorb light photons, consume nutrients and make the panels and cultivation beds dirty, which reduces light reflection of the panels, thus decreasing the light use efficiency of the system. It may also cause a contamination of products that accelerates rot in plant tissues. It is recommended to regularly monitor the remaining water on the panel surface and to limit light exposure to the nutrient solution. To limit algae, it is suggested to use a cover on the small gaps between floating panels and the edges, clean the panels regularly and use less water in sponges during germination and supply water from the bottom to keep the sponge surface dry.

3.9.6 Tip Burn

Tip burn is the necrosis at the margins of young developing leaves on the inner parts of vegetable plants and is generally considered a Ca deficiency-related disorder. Several external factors can affect the occurrence of tip burn, such as light, temperature, humidity and nutrient solution (Saure 1998). Most researchers agree that tip burns increase with an increased growth rate. Hence, there is always a conflict with an expectation of the fastest growth and the least occurrence of tip burn. Sometimes the cultivators have to reduce plant's relative growth rate in order to grow well-balanced plants. Incidences of tip burns can be reduced by increasing the relative humidity (>90%) during the night for 3 h, introducing air circulation on top of young leaf area and increasing Ca^{2+} concentration in the nutrient solution or applying foliar sprays of Ca salts. Another alternative is to develop plant-factory-suited varieties as mentioned above.

3.9.7 Disease, Microbe and Insects

The issues of disease, microbe and insects are the biggest problems encountered by plant factory managers. In extreme conditions of pest and/or disease incidence, the factory may have to close, which can result in both production and financial losses. Therefore, disease and insect control are vital for plant factory management. The carry in/out process must be strictly disinfected, and daily monitoring and reporting of the process must be reinforced. To reduce the risk of diseases and insects, the entrance passage can be separated to several rooms, and opening both side doors at the same time should be avoided. Several small cultivation rooms are also better than having a single cultivation room.

3.9.8 Seed Quality and Storage

Since a stable and predictable production is substantially important for commercial plant factories, a stable and uniform seed germination rate is required. A sudden drop in the germination rate can delay product delivery dates. This problem can be avoided by checking the variety, company and origin of the seeds and its proper growth storage. Normally seeds should be stored in a refrigerator at 4–5 °C under dry conditions. Some aromatic seeds contain essential oils, and their quality may drop faster than other vegetable seeds. It is of importance to pay attention to seed management in a plant factory.

3.10 Conclusion

In this chapter, a summary of current hydroponic systems, lighting systems, nutrient solution management, floor layout and set points of environmental factors is included. Issues faced by current plant factory operators and their possible solutions are also introduced. Any minor mistake may result in business failure, and, on the other hand, small improvements may lead to a great success. The information introduced above is expected to help the next generation of entrepreneurs, cultivators and researchers to move a small step forward and continue producing safe food in a highly efficient and sustainable way.

References

Kubota C (2015) Growth, development, transpiration and translocation as affected by abiotic environmental factors. Chapter 10. In: Kozai T, Niu G, Takagaki M (eds) Plant factory. Academic, London, p 155

Maneejantra N, Tsukagoshi S, Lu N et al (2016) A quantitative analysis of nutrient requirements for hydroponic spinach (*Spinacia oleracea* L.) production under artificial light in a plant factory. J Fertil Pestic 7:170. https://doi.org/10.4172/2471–2728.1000170

Niu G, Kozai T, Sabeh N (2015) Physical environmental factors and their properties. Chapter 8. In: Kozai T, Niu G, Takagaki M (eds) Plant factory. Academic, London, p 133

Saure MC (1998) Causes of the tipburn disorder in leaves of vegetables. Sci Hortic 76(3):131–147

Tsukagoshi S, Shinohara Y (2015) Nutrition and nutrient uptake in soilless culture systems. Chapter 11. In: Kozai T, Niu G, Takagaki M (eds) Plant factory. Academic, London, pp 171–172

Chapter 4
Design and Control of Smart Plant Factory

Yoshihiro Nakabo

Abstract Control system theory is applied to the design and modeling of smart plant factories. From control theory, the advantage of a closed plant factory is explained by the less disturbance in estimating the internal state of the controlled target. Several control models are explained from various viewpoints such as plant environment model, model-based control, hierarchical control, etc. Finally, three fundamental elements for designing a smart plant factory are shown.

Keywords Control system theory control model · Disturbance · Model-based control · Hierarchical control · Design elements

4.1 Introduction

4.1.1 General Control System Model

In this chapter, control system theory is applied to the design of the smart plant factory. In control system theory, controlled objects are generally represented as models with their own internal states and functions that convert their inputs to their outputs. Our aim is to obtain the desired outputs from controlled objects (the target plants) by controlling their inputs in accordance with their own specific algorithms (cultivation process of plant factories) which is called a controller. However, to do that, we must know internal states of the controlled objects, and they are not usually known directly. So, the output values are used to estimate the internal states. Controller is usually modeled by including this estimation.

Y. Nakabo (✉)
National Institute of Advanced Industrial Science and Technology (AIST),
Tsukuba, Ibaraki, Japan
e-mail: nakabo-yoshihiro@aist.go.jp

© Springer Nature Singapore Pte Ltd. 2018
T. Kozai (ed.), *Smart Plant Factory*, https://doi.org/10.1007/978-981-13-1065-2_4

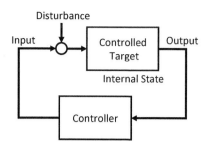

Fig. 4.1 Control system model with disturbance

4.1.2 Disturbance

If the state variables and the outputs of the controlled object vary regardless of their input, which is often seen in the reality, estimation of internal values becomes difficult, and we must consider the influence of disturbance (Fig. 4.1).

A lot of researches has been done to design a controller which encounter unexpected influences of disturbance. Here, in usual plant cultivation in open environment, the result of cultivation of plants as an output of controlled target does not necessarily become an expected value with respect to the input given by human and hence control theory. There are a lot of factors to be treated as disturbance in such case. In other words, it becomes extremely difficult to design a controller for plant cultivation in open environment by the framework of control theory.

On the other hand, in closed plant factories, the input and output of plants are strictly controlled, so there are few factors which need to be considered as disturbances. Therefore, there become possibilities of designing control systems by the control theory. To eliminate influence of disturbances, we should grasp all input variables accurately and measure the plant as output values to estimate its state variables as much as possible and also refine the target model to get a better estimation.

4.2 Control Models

4.2.1 Controlled Target of Smart Plant Factory

Now, when designing a control system, it is necessary to define what is the controlled target. In the design of a plant factory, we should notice that the controlled target must be the cultivation environment of plants rather than the plant to be cultivated as shown in Fig. 4.2.

The reason is that even if estimating internal state variables of the plants as much as possible, it is difficult to control them directly. In actual, the plant factory directly controls the cultivation environment such as temperature, humidity, light power,

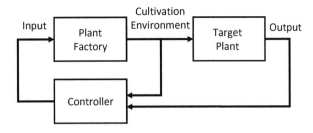

Fig. 4.2 Smart plant factory control model

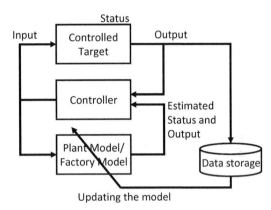

Fig. 4.3 Model-based control

cultivation solution concentration, and carbon dioxide concentration. What kind of cultivation environment gives plants how they grow is a plant growth model. The daily control of a plant factory is, for example, control of the LED, the air conditioner, the solution control device, etc. to maintain the target plant cultivation environment.

4.2.2 Model-Based Control

As mentioned above, it is not possible to directly know the state variables of the target plants. However, if we can obtain a virtual system that shows the same reaction as the target plant, i.e., if we can obtain a mathematical or computer model that can obtain the same output for a given input, we can know the internal state of the target equivalently and predict future outputs for scheduled inputs. As a result, we can use such mathematical model to adjust the input to obtain the desired output. This is the idea of model-based control which is shown in Fig. 4.3.

Obviously, an important point of model-based control is whether we can obtain a right model which shows the same behavior as the target plant. For that purpose, it is important to improve the model by feedbacking the difference between the assumed output and the actual output or by using more detailed knowledge about the target plants which is shown in Fig. 4.3 by the data storage and model update.

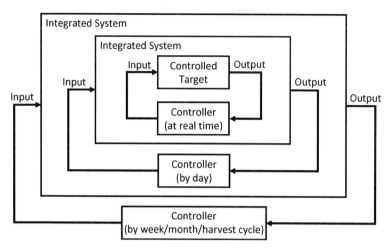

Fig. 4.4 Hierarchical control model

4.2.3 Hierarchical Control Model

If we consider the control method for the model of plants more practically, we must consider the control cycle of the controller. For example, daily temperature and humidity control by air conditioning system, adjustment of LED light, etc. need to be controlled in real time based on sensor data feedback and/or the given growth model. On the other hand, it is also important to change the habitat environment daily or weekly according to the growing condition of the plant. In addition, control models may be switched for each cultivation cycle, and different cultivation tasks and target values may be given to control.

Differences in these control cycles can be modeled with hierarchical control models which are shown in Fig. 4.4. In other words, it is a model in which systems controlled in a short cycle are handled as one integrated system, and the integrated system is newly controlled by a superordinate control system.

4.2.4 Updating Control Models as PDCA Cycle

Regarding the updating of the model described in Sect. 4.2.2, it can be considered as a PDCA (plan, do, check, act) cycle that is even higher than the hierarchical control model shown in Sect. 4.2.3. In other words, it is important to improve the entire design of the smart plant factory by repeating the execution of the four processes of PDCA.

Therefore, we design the control system as "plan" and "do" the actual cultivation, and evaluate the cultivation result by the "check" process. In other words, it is important to define the results expected to be obtained in the previous design stage. It is important to analyze the results of this evaluation and/or stored data with knowledge-based considerations or AI and to review the models, the controllers, and the control structure including the hierarchical structure and/or design of the whole plant factory as an "act" process. As a result, the PDCA cycle is executed.

4.3 Three Fundamental Design Elements of Smart Plant Factory

Next, I will change the topic from control model to design elements of a plant factory. As a control system, what are the fundamental elements for designing a plant factory is discussed.

A control system consists of models of target objects and actual control systems. Therefore, there are both models and actual equipment regarding the control as design targets of plant factories, which represents two aspects of that an abstract information model and concrete physical substances. In other words, it is necessary to design both in the real world and the virtual information world in general.

Furthermore, the space design of plant factory is another important factor. The space designs, such as cultivation racks, various piping placements, air and heat flow, space for light rays, space for human worker, and/or sizing and scaling to reach, are all important for examples. Therefore, it can be summarized as three design elements of a smart plant factory as shown in Fig. 4.5.

As shown in Fig. 4.5, the three elements mutually influence each other. Some examples of the content of each element and mutual influence are described in the figure.

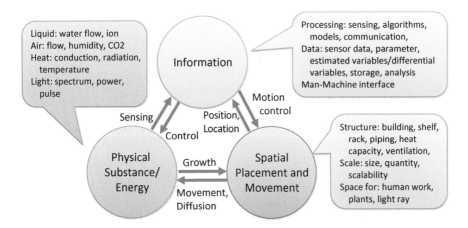

Fig. 4.5 Three design elements of smart plant factory

Chapter 5
Designing a Cultivation System Module (CSM) Considering the Cost Performance: A Step Toward Smart PFALs

Toyoki Kozai

Abstract Design factors related to the productivity of a plant factory with artificial lighting (PFAL) are discussed. The cultivation system module (CSM) is thought to be a key component of the cultivation room in the PFAL. Types of CSM are classified, and the merits and demerits of each type are discussed. The scalability of the CSM is considered a crucial design factor of the PFAL. Air movement in the CSM greatly affects the plant growth and thus is an essential design factor of the CSM. Environmental control based on phenotyping will play a significant role in a smart PFAL.

Keywords Cost performance · Cultivation system module (CSM) · Phenotyping · Productivity · Scalability

5.1 Introduction

Ideas, concepts, and design factors of the cultivation system module (CSM hereafter) are discussed in this chapter. The CSM is considered a key component of the cultivation room in a smart plant factory with artificial lighting (PFAL). The CSM is designed primarily to greatly increase the productivity and reduce the production cost and initial investment compared with the current PFALs. Designing such a CSM would be the first step toward a successful smart PFAL design and PFAL business.

A successful PFAL for commercial production is (1) well designed and built at a reasonable cost in a suitable location chosen based on qualified marketing, (2) well operated at nearly full production capacity with high cost performance under an appropriate vision and mission and reasonable short- and medium-term goals (Fig. 5.1), (3) upgradable with minimum modification and maintenance at a minimum cost, (4) well positioned in the produce value chain, (5) creating new markets

---

T. Kozai (✉)
Japan Plant Factory Association (NPO), Kashiwa, Chiba, Japan
e-mail: kozai@faculty.chiba-u.jp

© Springer Nature Singapore Pte Ltd. 2018
T. Kozai (ed.), *Smart Plant Factory*, https://doi.org/10.1007/978-981-13-1065-2_5

Fig. 5.1 Factors to be considered in the design of PFALs and the vision, mission, target and goals of PFAL R&D

that do not compete with any existing horticultural markets, (6) providing well-organized personnel development and training programs, and (7) based on appropriate marketing at a reasonable price so that almost all produce is sold out. In short, the PFAL needs to be designed, operated, and managed to maintain the above characteristics under social, technological, and global/local climate changes.

5.2 Improving the Cost Performance

The cost performance of the PFAL can be improved either by reducing the production costs or by increasing the annual sales of produce or both.

5.2.1 Reducing the Electricity and Labor Costs

Electricity and labor costs per kg (fresh weight) of produce are expected to be reduced by 50% by 2022, compared with the current PFALs, which are already making a profit as of 2017. The percentages of production costs for depreciation (for initial investment), labor (personnel expenditure), electricity, and others are, respectively, 23%, 26%, 21%, and 30% in a partially automated PFAL with LED lighting producing leaf lettuce plants in Japan (Table 5.1). In Table 5.1, the wholesale price was 0.85 USD per head (80 g) or 10.7 USD per kg. Also, percent annual sales over annual production was 95%, and percent annual production over annual production capacity was 90% (Ijichi 2018).

Table 5.1 Percentages of production cost components and profit for leaf lettuce production in a PFAL in Japan. Column A shows percentages of revenue structure components, and Column B shows percentages of production cost components (Ijichi 2018)

No.	Components	A (%)	B (%)
1	Depreciation	20	23
2	Labor	22	26
3	Electricity	19	21
4	Logistics (distribution)	6	6
5	Major supplies	6	7
6	Seeds	2	2
7	Tax, land rental	1	1
8	Others	12	14
9	Profit	12	–
J	Total	100	100

Table 5.2 Percentages of working hour components in a PFAL for leaf lettuce production in Japan (data obtained in 2016 by personal communication)

No.	Working hour components	%
1	Seeding	3
2	First transplanting	7
3	Second transplanting	21
4	Harvesting and trimming	38
5	Packaging into bags	12
6	Packing the bags in boxes	6
7	Cultivation panel cleaning	6
8	Cultivation room cleaning	7
	Total	100

As of 2017, labor hours for cultivation (seeding, transplanting, harvesting, and trimming) of another PFAL in Japan were about two times the labor hours for packaging and packing for shipping and sanitary management (cleaning, washing, and inspection) (Table 5.2). In this PFAL, transplanting, harvesting, and trimming were done manually; packing and packaging were semiautomated; and tools were used for seeding and cleaning.

Since the cultivation process is not automated in the above PFALs, a reduction in labor hours by 50% within several years would be a good numerical target to achieve through semiautomation, automation, and improvements in the production process, equipment layout, and human resource development.

The cost reduction of electricity by 50% compared to that in 2017 can be achieved within several years by smart operation of the LED lighting system, air conditioners, and other electrical equipment. The percentage of electricity consumption for lighting, air conditioning, and other electric equipment is, respectively, 75–80%, 15–20%, and 5% in case that the cultivation room is almost airtight and thermally well-insulated. "Other electric equipment" includes pumps for nutrient solution delivery, air circulation fans, sterilization units, floor and cultivation panel cleaning units, and miscellaneous. Thus, the cost reduction in lighting is important for improving the productivity per kWh of electric energy.

Around 30–40% of the electric energy consumed by LEDs is converted to photosynthetic photons (unit: mol (1 mol = 6.022×10^{23} photons)), and around

60% of photosynthetic photons are received by the plant leaves (this percentage increases with increasing the total leaf area per cultivation area). Then, only a portion of the photosynthetic photons received by plant leaves is fixed as chemical energy in carbohydrates. In a similar way, the productivity of electric energy for lighting and the productivity of photosynthetic photons (kg/mol) can be defined and be improved by improving the conversion factors of electric-light energy and light-chemical energy.

The electric energy use efficiency of air conditioners or the coefficient of performance (COP) of the air conditioners for cooling can be estimated by $(L + O)/A$ where L is the electricity consumption for the lighting, O is the electricity consumption of the electric equipment other than the lamps and air conditioners, and A is the electricity consumption of the air conditioners. The COP increases with the decrease in the air temperature outside. Cooling is necessary during the photoperiod even on winter nights to remove the heat generated by lamps and other electrical equipment.

The cultivation room needs to be airtight to minimize the loss of CO_2 and to prevent insects from entering even when the air temperature is higher inside than outside (Kozai et al. 2015). In this case, the COP is generally higher than 5. The loss of CO_2 occurs because its concentration during the photoperiod is kept at around 1000 ppm to promote photosynthesis of the plants, which means that the concentration is 600 ppm higher in the cultivation room than outside (400 ppm).

By reducing the cost of electricity and labor by 50%, the total production cost per kg of produce in Table 5.1 is reduced by 24% ($= 100$–$0.5 \times (26 + 21) - 53$) for a case in which the costs for depreciation and consumables remain the same.

5.2.2 Increasing the Annual Sales

The annual amount of sales is expected to be increased by 50% without any increase in production costs. Annual sales are approximately expressed by the product of four factors: (1) productivity per cultivation area (m^2) or cubic space (m^3), (2) price or economic value per kg of produce, (3) ratio of actual cultivation area to the total cultivation area including empty cultivation areas, and (4) amount ratio of sold produce to the total produce.

The target value of a 50% increase in annual sales is achievable mainly by adjusting the environmental control set points, selecting better cultivars, improving the cultivation system, and reducing the waste percentage of produce, based on the plan-do-check-act (PDCA) cycle. Note that the production cost and productivity of the four factors (cultivation area, electricity, labor, and consumables) tend to be interrelated (Fig. 5.2). Therefore, the cultivation room needs to be designed and operated so that the four factors have a positive effect on each other.

By reducing the production costs by 50% and increasing the annual sales by 50%, the cost performance can be doubled ($2.0 = 1.5/0.75$). In fact, several PFALs in Japan almost doubled their cost performance during 2013–2017 (personal communication). The cost performance can be further improved by installing environmental

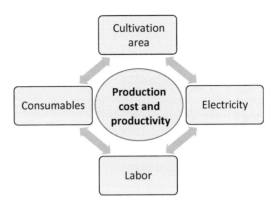

Fig. 5.2 The four factors affecting the production cost and productivity of the PFAL are generally interrelated; accordingly, the cultivation room needs to be designed and operated so that the four factors have a positive effect on each other

control software and sensors for visualization and step-by-step improvement of resource use efficiency (RUE).

Then, the market size of PFAL-grown plants will expand steadily. This expansion should lead to the creation of new markets and should not defeat the current market of greenhouse and open-field-grown produce.

5.3 Productivity, Production Costs, and Cost Performance

Before discussing the design of the cultivation system and CSM, the terms "productivity," "production cost," and "cost performance" are defined, and the current business status of PFALs in Japan is briefly described.

5.3.1 Productivity by Resource Element

Productivity in the PFAL can be calculated for the resource elements of electricity (kg/kWh; 1 kWh = 3.6 MJ (megajoules)), labor (man-hours) (kg/h), and cultivation area (kg/m^2) (Fig. 5.3, left). Amount of produce is expressed in units of kg, but it can also be expressed as the number of plants, for example. The annual productivity per cultivation space is calculated by the annual production in the cultivation room (kg y^{-1}) divided by the total cultivation (panel) area (m^2) or total cultivation air volume (m^3) in the cultivation room. As described in Sect. 5.5.2, 75–80% of the electricity (kWh) is used for lighting, and currently 30–35% of the electric energy is converted by LEDs to photosynthetic (400–700 nm) photons or photophysiological (300–800 nm) photons (mol).

Thus, we can also define the productivity in terms of electricity consumption for lighting (kg/kWh) and in terms of photosynthetic/photophysiological photons (kg/mol). We can also define the productivity based on the dry weight or the secondary metabolites of the electricity for lighting (kg (dry weight)/mol), etc.

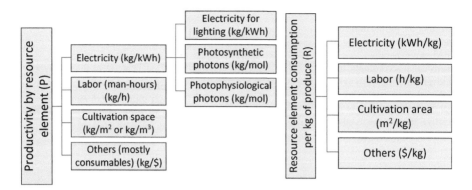

Fig. 5.3 Productivity by resource element (P) (Left) and resource element consumption per kg of produce (R) (Right). P is the inverse of R. The annual productivity per cultivation area is calculated by the annual production (kg year^{-1}) divided by the total cultivation panel area in the cultivation room (m^2). Productivity of electricity for lighting and productivity of photosynthetic photons can also be defined and estimated in a similar way. Photosynthetic photons (400-700 nm); photophysiological photons (300–800 nm)

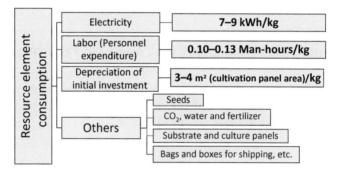

Fig. 5.4 Estimated resource element consumption per kg of fresh produce (leafy lettuce plants) in a PFAL making a profit in Japan as of 2017. The PFAL was not automated except for transplanting and culture panel cleaning

5.3.2 Resource Element Consumption per kg of Produce

Resource element consumption per kg of produce is the inverse of productivity (Fig. 5.3, right). Electric energy consumption per kg of produce is often called the "basic energy unit," which is used as an indicator of energy consumption in a production process. Typical values of resource element consumption for electricity, labor, and cultivation area are, respectively, 7–9 kWh/kg, 0.10–0.13 man-hours/kg, and 3–4 m^2 (cultivation panel area)/kg (Fig. 5.4).

Factors affecting the labor hours include (1) layout of cultivation beds, equipment, machines, tools, and consumables; (2) flow of workers, plants/produce/plant residue and tools, consumables, and used materials; and (3) cultivation system design and plant traits.

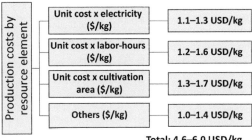

Fig. 5.5 Estimated production costs of leafy lettuce plants per kg of fresh produce by resource element in a PFAL in Japan. Costs for logistics and sales are excluded in "Others" (see Fig. 5.4 for resource element consumptions). One USD equals to 105 Japanese Yen (JPY). Unit electricity price: 15–20 JPY/kWh. Labor cost: around 1200 JPY/h for part-time worker and around 500,000 JPY/month for regular employee

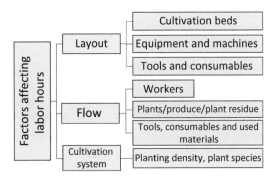

Fig. 5.6 Factors affecting the labor hours

5.3.3 Production Cost and Cost Performance

The production cost per kg of produce for each resource component can be calculated as the product of the unit economic value ($/kg) and the resource element consumption per kg of produce (Fig. 5.5, 5.6). Then, the cost performance (CP) of the PFAL can be calculated by the ratio of sales (S) to the production cost (C). The S value can be calculated as the product of the economic value (U), the volume of produce (P) (Fig. 5.7), and (1.0 − L) where L is the produce loss ratio.

C is divided into variable and fixed costs. Fixed costs include (1) depreciation for building, facilities, and equipment and (2) tax or rent for land/building, insurance and maintenance, basic salaries, and basic charges for electricity and municipal water. Variable costs include (1) seeds, CO_2, fertilizer, substrate, plastic bags, boxes, shipping, etc. and (2) variable charges for electricity, municipal water, and overtime work.

Light environmental factors affecting the biomass productivity and the economic value of produce per kg include (1) photosynthetic photon flux density (PPFD) integral or the product of the PPFD, photoperiod, and days of cultivation; (2) ratio of

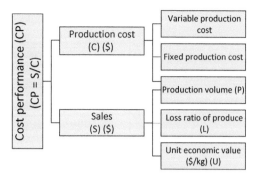

Fig. 5.7 Cost performance (CP) defined as S divided by C can also be calculated by $(U \times P \times (1 - L))$. The profit is expressed by $(S - C)$

photosynthetic photons received by the plants in mol/plant to the photosynthetic photons emitted by the lamps (mol/lamps); (3) light quality at each growth stage affecting the secondary metabolite production related to the color, shape, flavor, nutritional composition, etc.; and (4) spatial uniformity of the light environment.

5.3.4 Payback Period

Payback period refers to the number of years required to recoup the funds expended in the initial investment or to reach the break-even point (Wikipedia). Payback period N (years) can be roughly estimated by the following equation:

$$N = I/(P \times W - C)$$

where I is the initial investment per cultivation area ($/m²), W is the planned yearly production per m² (kg/(m² year), P is the planned yearly average sales price per kg of produce ($/kg), and C is the planned yearly direct production cost (excluding the depreciation cost) per cultivation area ($/(m² year)). After the start of production, the planned values of W, P, and C can be replaced with the actual values.

5.4 Concept of Cultivation System Module (CSM)

The cultivation system module (CSM) is defined as the minimum unit of the cultivation system as a key component in the cultivation room (Fig. 5.8). The CSM is placed on each tier of the cultivation rack or is stacked directly without using a cultivation rack.

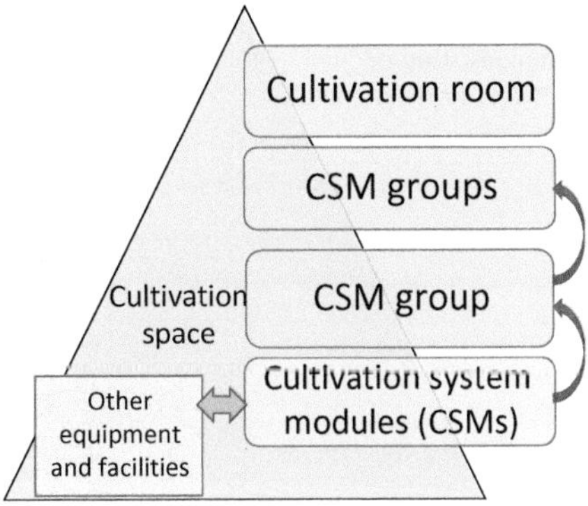

Fig. 5.8 Cultivation system module (CSM) as a key component of the cultivation room in the PFAL. Each CSM has a local control unit with a memory and input/output devices

Fig. 5.9 Spatial components of the cultivation room. Packaging and shipping are conducted outside the cultivation room

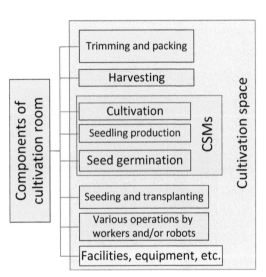

5.4.1 Components of the Cultivation Room

A typical cultivation room is divided into two spaces: one is for cultivation and the other for facilities and equipment such as the nutrient solution tank, CO_2 supply unit, and air conditioners. The space for cultivation is further divided into the space for CSMs with or without cultivation racks and the space for manual or automatic seeding, transplanting, harvesting, trimming, transporting, and packing (Fig. 5.9).

The components of the cultivation room can also be classified into hardware, firmware, and software (Fig. 5.10). The hardware consists of the facilities, equipment, and cultivation racks. The firmware consisting of a microcomputer, sensors, hardware interface, and software is implemented in each CSM connected directly or indirectly with all other CSMs.

The software components for environment control and PFAL management are shown in Fig. 5.11. The software for plant production process management and

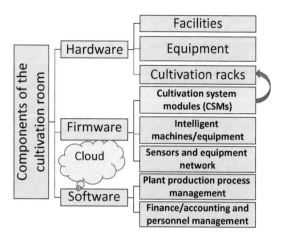

Fig. 5.10 The cultivation room consisting of hardware, firmware, and software

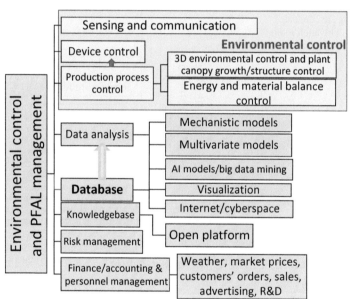

Fig. 5.11 Software components for environment control and PFAL management (component relationships are not shown in the figure)

finance/accounting and personnel management is mostly stored in the cloud server. The database contains the genomic data and the time-series data on (1) plant environments; (2) resource inputs; (3) products (resource outputs); (4) phenomes (plant phenotypic traits); (5) machine/equipment/human interventions; (6) resource use efficiencies; (7) costs, sales, and cost performance; and (8) weather, market prices, and so forth.

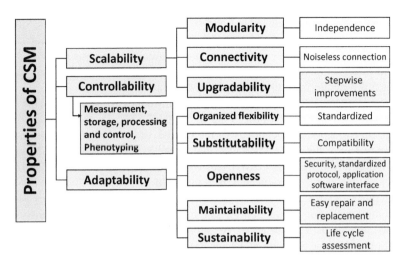

Fig. 5.12 Properties to be implemented in the cultivation system module (CSM)

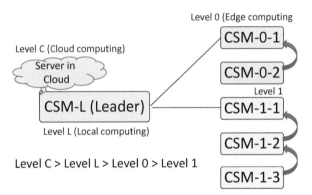

Fig. 5.13 The CSM-L acts as a leader for all CSM-0s and CSM-1s in the same group. The CSM-L is connected to a server in the cloud. More sensors and actuators are installed in CSM-L than in CSM-0 and CSM-1, and more are installed in CSM-0 than in CSM-1

5.4.2 Function and Configuration of the CSMs

The CSM is characterized by its properties of scalability (potential to be enlarged to accommodate an increase in number of CSMs and/or cultivation racks), controllability, adaptability, and so forth (Fig. 5.12). The CSM needs to be light in weight and simple in structure.

Each CSM handles the low-level measurement, control, and information processing itself. The CSMs are grouped, and one of the CSMs in each group acts as a group leader (CSM-L hereafter) for medium-level measurement, control, and information processing (Fig. 5.13) (the minimum number of CSMs in the same group is unity).

In the CSM-L, a firmware kit is installed for communicating with and supervising other CSMs (CSM-0 and CSM-1 hereafter). CSM-0 has a larger number and/or higher quality of sensors/actuators than CSM-1. CSM-1 can be changed to the CSM-L by attaching the firmware kit for CSM-L instead of that for CSM-1.

Likewise, CSM-1 can be changed to CSM-0. Namely, the firmware kit for CSM-1 is a simpler version of that for CSM-0, and that for CSM-0 is a simpler version of that for CSM-L. The physical size and system configuration are the same for all CSMs in the same group. On the other hand, the environment in a CSM group can be controlled differently from that in other CSM groups.

A server (computer) in the cloud connected with all the CSM-Ls handles the high-level control and information processing for all the CSMs. On the other hand, any CSM group can work independently of the CSM-L to some extent even when it is disconnected from the server, provided that power is supplied to the CSM group. Furthermore, any CSM-0 or CSM-1 can work independently of the CSM-L to a lesser extent when it is disconnected from the CSM-L.

With this setup of CSMs, a series of experiments can be conducted during large-scale commercial plant production for sale by using a minimum set of sensors and actuators, and big data can be collected and analyzed systematically to improve the production process step by step (Chapter 25).

5.5 Plant Production Process Measurement and Control in the CSM-L

The CSM-L consists of nine components as shown in Fig. 5.14.

In the CSM-L, seven groups of variables are measured or estimated: (1) environmental factors (Fig. 5.15); (2) supply rates of resource elements to the CSM (Fig. 5.16); (3) production rates (Fig. 5.17); (4) plant phenotypic traits (Figs. 5.18 and 5.19); (5) indices such as resource use efficiencies (RUEs) (Fig. 5.20) and parameter values of the models for plant growth, etc.; (6) signal inputs/outputs from/to equipment/actuators and sensors; and (7) productivity in terms of labor, cultivation area, and electricity (see Sect. 5.5.4). Using the CSM-L, we can provide an RUE value as a set point and control the RUE directly instead of providing a set point and controlling the environmental factors.

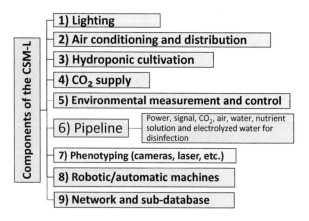

Fig. 5.14 Components installed in the cultivation system module L (CSM-L)

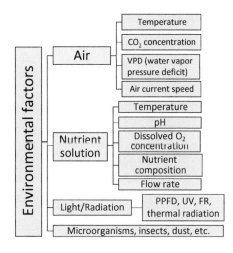

Fig. 5.15 Environmental factors measured and controlled in the CSM. PPFD, UV, and FR denote, respectively, photosynthetic photon flux density, ultraviolet, and far infrared

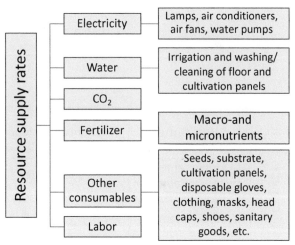

Fig. 5.16 Resource supply rates measured and controlled in the CSM-L

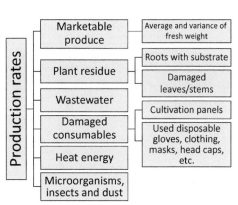

Fig. 5.17 Production rates measured and/or controlled in the CSM-L

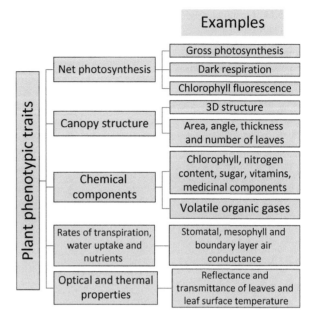

Fig. 5.18 Plant phenotypic traits to be measured in the CSM

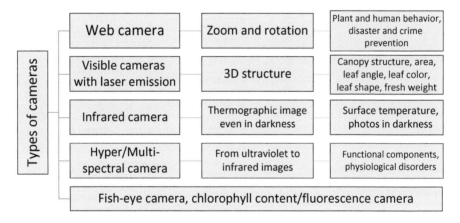

Fig. 5.19 Types of cameras for plant phenotyping

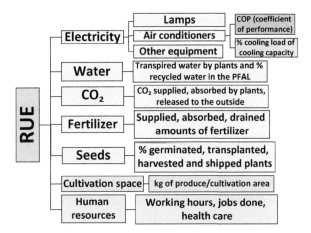

Fig. 5.20 Resource use efficiencies (RUE) in the CSM-L to be estimated

5.6 Air Movement in the Cultivation Room and the CSM

5.6.1 Air Movement Effects on Plant Growth and Other Environmental Factors

Air current speed and air flow pattern (turbulent or laminar flow and flow direction) over and within a plant canopy have a significant effect on the photosynthesis, transpiration and surface temperatures of the plants, and thus the plant growth (Yabuki 2004; Kitaya 2016). Similarly, the flow rate and flow pattern of the nutrient solution around the plant roots in the cultivation bed affect the water, dissolved O_2, and nutrient uptake of the plants. This is because the photosynthesis, transpiration, and water/nutrient/dissolved O_2 uptake of the plants are the result of the diffusion process of CO_2, water vapor, liquid water, and dissolved O_2 within and around the plants. This means that the plant growth is affected by the movement of the air and nutrient solution even under the same air temperature, CO_2 concentration, vapor pressure deficit (VPD), PPFD, and nutrient composition/strength/pH.

5.6.2 Air Movement Around the Cultivation Racks

The air movement around the cultivation racks in the cultivation room is affected by (1) the air outflow rate and its direction at the air conditioner outlets and air circulation fans and 2) the layout of the cultivation racks (height, horizontal distance between the cultivation racks, and vertical distance between the ceiling and the top of the cultivation racks). Accordingly, air movement patterns in a large cultivation room are, in general, different from those in a small cultivation room.

5.6.3 Air Movement over the Plant Canopy

The air movement over and within the plant canopy on the tier is affected by (1) the distance between the plant canopy surface and the ceiling of the tier, (2) the plant canopy leaf density (leaf area per cubic meter), (3) the PPFD at the plant canopy surface, (4) the heat generation rate by LEDs during the photoperiod, and (5) the air movement just outside the cultivation racks. Therefore, the air movement over the plant canopy indirectly affects the horizontal distribution of the air/leaf temperature, CO_2 concentration, VPD, and PPFD there.

There are many publications showing the effects of environmental factors on plant growth. However, such data has mostly been obtained at university laboratories or public/private research institutions and is often not applicable in a large-scale cultivation room of a PFAL for commercial production usually due to the difference in air movement, which is not described as an environmental condition.

5.6.4 Law of Similarity in Fluid Dynamics

Known as the law of similarity in fluid dynamics, fluid flow behaviors in a small and a large cultivation space are similar only when their values of dimensionless parameters such as the Prandtl and Reynolds numbers of the fluid are almost the same. The law of similarity is the basis for simulating large-scale phenomena in a small-scale laboratory. On the other hand, it can generally be said that the dimensionless parameter values are different in small- and large-scale production systems. Thus, it is essential to consider the law of similarity in designing the CSMs, although this kind of theoretical and experimental analyses can be done only by experts.

5.6.5 Simplest Way to Avoid the Law of Similarity Issue

The simplest way to avoid the of law of similarity issue is to develop a CSM that can be used for both laboratory-scale and large-scale PFALs, so that the experimental results obtained by using only one CSM can be used in a large-scale PFAL consisting of many CSMs. Then, the results obtained using the CSM are not influenced by the size of the cultivation rack, the number of tiers, the number of racks, and the size of the cultivation room (Fig. 5.21). If we find the optimal environmental conditions using a CSM-L in a large-scale PFAL, we can apply the optimal environment to all the CSMs in the large PFAL. In this case, plant growth should be the same or very similar in the same CSM group if the seeds are sown at the same time and in the same way.

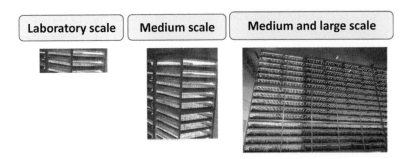

Fig. 5.21 Plant growth differs in laboratory-, medium-, and large-scale cultivation systems even when the air and nutrient solution temperature, CO_2 concentration, light environment, vapor pressure deficit (VPD), and nutrient solution composition are the same, if the flow patterns of the air and nutrient solution are different

5.7 Basic Design Concept of the CSM-L for Scalable PFALs

5.7.1 Types of CSMs

CSMs with horizontal cultivation panels/trays can be classified into four types with respect to the air flow pattern and the feasibility of RUE estimation (Table 5.3, Fig. 5.22, Fig. 5.23). Characteristics of Types A–D, not listed in Table 5.3, are given in Table 5.4. The suitable size and weight of the CSM depend on its type. In general, the CSM would be 5–12 m long, 2–5 m wide, and 0.5–5.0 m high and would weigh 20–100 kg. The CSM can then be connected and/or stacked relatively easily. The cultivation rack would be 10–30 m long, 2–5 m wide, and 2–30 m high.

In Type B-a, the horizontal air current speed and direction are controllable by changing the rotation speed and direction of the air fans at either end of the CSM, and the net photosynthesis and water uptake of the plants can be estimated if the traveling distance (fetch) of the air in the CSM is long enough to measure the significant difference in CO_2 concentration and VPD at the inlet and outlet of the CSM (around 6–10 m) (Kozai 2013). The RUEs can then be estimated. Most cultivation systems in the current PFALs belong to Type D. Photographs of Types B-b and D are given in Fig. 5.24. Figures 5.25 and 5.26 show examples of Types C-a and A, respectively. The cultivation rack with four tiers in Fig. 5.26 is in a plastic-covered room consisting of a CSM (Type A). Type A can also be used as four Type B-a CSMs if the environmental factors at the inlet and outlet are measured for each tier. Figure 5.27 shows the container for freight shipment. This container can be used as a Type A CSM if both ends are closed or as a Type B-a if either end is connected to another container. These types of CSMs do not need a cultivation rack for stacking.

To estimate the use efficiencies of electricity, the supply rates for electricity, photosynthetic photons, and water need to be measured or estimated. The net photosynthesis rate, water uptake rate, and transpiration rate of the plants are

Table 5.3 Classification of CSMs by air flow pattern and the estimation of RUEs online

Type	Controllability of air flow pattern	Estimation of RUEs
A	Air flow pattern is controllable	Easy
B-a	Longitudinal air current speed is controllable (see also Fig. 5.23, center)	Feasible
B-b	Cross-sectional air current speed is controllable (see also Fig. 5.24, left)	Difficult
C-a C-b	Air supply rate is controllable, while horizontal air flow pattern is not controllable (see also Fig. 5.25)	Very difficult
D	Air flow pattern is not controllable (see also Fig. 5.24, right)	Impossible

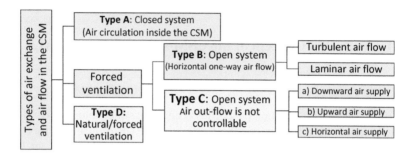

Fig. 5.22 Types of air exchange and air flow in the CSM. Types B and C are further classified into Type B-a and Type B-b and Type C-a and Type C-b (see also Figs. 5.23 and 5.24)

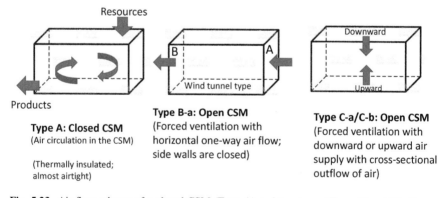

Fig. 5.23 Air flow scheme of a closed CSM (Type A) and two types (Types B and C) of open CSMs (see also Fig. 5.25). Light reflective films are pasted on the inside surface of the side walls in Types A and B

Table 5.4 Characteristics of Types A, B-a, B-b, C-a, C-b, and D (see also Figs. 5.22–5.25)

Type	No.	Characteristics
A, B-a	1	Highly light reflective film at the side walls (no openings at the sides), which creates uniform PPFD over the plant canopy and low loss of light energy through the side openings
	2	Observation of the plants through the side openings is not possible. Web cameras are necessary to observe the plants
	3	Horizontal air current speed is controllable
	4	No walkway is necessary between the CSMs, resulting in high spatial density of the CSMs in the culture room
	5	Risk of spreading small insects and pathogens across the CSMs is minimal
B-a	1	Air flow direction is reversible by using reversible fans
	2	A slight gradient is observed regarding the temperature, VPD, and CO_2 concentration along the longitudinal air current during the photoperiod, especially at a high plant density
	3	When the differences in CO_2 concentration and VPD at the inlet and outlet of the air are too small to accurately estimate the rates of net photosynthesis and transpiration of the seedlings, the differences can be increased by reducing the air current speed and/or the cross-sectional area of the CSM
B-b	1	Virtually no gradient of temperature, VPD, or CO_2 concentration during the photoperiod across the CSM
	2	Some space is necessary at both sides of the CSM for air intake and release
B-b, C-a, C-b, D	1	Rates of net photosynthesis and water uptake/transpiration and RUE cannot be estimated
C-a, C-b	1	The air is supplied either downward or upward, and the air supply rate is controllable. The air flow pattern, air temperature, CO_2 concentration, and VPD are relatively uniform along and across the CSM
D	1	The structure of the cultivation space is simple
	2	Spatial distribution of air temperature, VPD, and CO_2 concentration is uneven due to uneven spatial distribution of the air current speed and PPFD, especially at a high plant density (leaf area/m^3)

Notes: CSM (cultivation system module), PPFD (photosynthetic photon flux density), VPD (vapor pressure deficit)

estimated based on the CO_2 and water balance of the CSM (Kozai 2013; Kozai et al. 2015).

5.7.2 Types of Nutrient Flow in the Cultivation Beds

The types of nutrient solution flow in a hydroponic unit are shown in Fig. 5.28. One-way nutrient solution flow with no drainage is ideal for (1) high controllability of pH, nutrient composition, strength, and flow rate and (2) minimizing the total volume of nutrient solution, total length of piping, and pipe diameter. Actually, the nutrient film technique (NFT) system with a longitudinal nutrient solution flow is the

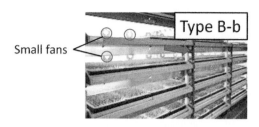

Fig. 5.24 Type B-b (left): Forced ventilation with cross-sectional horizontal air flow (small fans each with a diameter of 10 cm are installed on the back wall at an interval of 30 cm for sucking the room air from the front). Type D (right): The air in the cultivation space is exchanged with the air in the walkway at both sides by the combination of natural and forced ventilation. Air movement over the plant canopy is restricted at a high planting density. However, enhanced air exchange is necessary to promote photosynthesis and transpiration of the high-density plant canopy

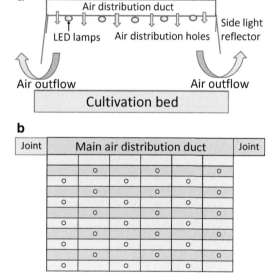

Fig. 5.25 (**a**) Cross-sectional scheme of Type C-a with downward air flow through the holes at the ceiling and air outflow from both sides. (**b**) Plan view of the ceiling of Type C-a. The open circles denote air exit holes. The ceiling panel is connected to adjacent ceiling panels at the joint ducts

most common. The NFT system with a cross-sectional nutrient solution flow would be an attractive choice in the CSM.

5.7.3 LED Lighting System

Figure 5.29 shows the factors to be considered in designing the lighting system. The factors to be considered in the selection of LEDs include (1) point, line, and surface light sources; (2) efficacy (mmol/J) and efficiency (J/J) and luminous efficacy

5 Designing a Cultivation System Module (CSM) Considering the Cost...

Fig. 5.26 The cultivation rack with four tiers is in an airtight plastic-covered room consisting of a CSM (Type A), developed by PlantX Corporation. This CSM can also be used as four Type B-a CSMs if the environmental factors at the inlet and outlet are measured for each tier

Fig. 5.27 Image of CSMs (Type A or Type B-a) that do not use a cultivation rack in the cultivation room. The distance between the CSMs is changeable. This CSM can be used for either Type A or Type B

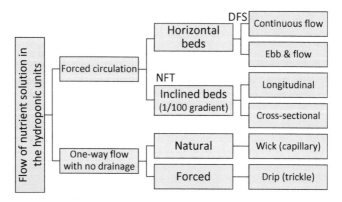

Fig. 5.28 Types of nutrient solution flow in the hydroponic unit. A hydroponic system that does not use substrate (supports) or a circulation unit for the nutrient solution (one-way) would be ideal. Also, it is desirable to have no drainage of wastewater containing unused fertilizer. NFT and DFS denote, respectively, nutrient film technique and deep flow system

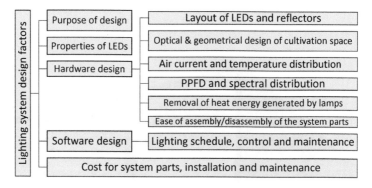

Fig. 5.29 Factors to be considered in designing the lighting system

(lumen/W); (3) spectral distribution of light; (4) angular distribution of light; (5) price per (mmol/s) by bulk purchase; (6) lifetime, waterproofness, size, shape, weight, heat sink design, etc.; (7) time-dependent light quality control function; and (8) ease of fitting/detaching.

The light environment not only affects various aspects of plant ecophysiology but also affects the cooling cost, electricity cost, and spatial distribution of air flow and temperature around the plant canopy. Smart LED lighting systems that consider all the above factors are expected to be developed.

5.7.4 Batch Production and Push/Pull Production

Batch production means that all seedlings are transplanted in the CSM at the same time and all plants are harvested at the same time several weeks later. Push/pull production means that a certain amount of plants is harvested each day from the cultivation panels/trays at one end of the cultivation rack or tier. Then, the rest of the cultivation panels with different growth stages are moved (pushed or pulled) to the end of the cultivation rack, and new cultivation panels with the seedlings are placed at the other end of the cultivation rack.

The batch production method in Types A and B-a CSMs enables the estimation of RUEs as a function of the plant growth stage. Also, the air current speed, air temperature, VPD, and CO_2 concentration can be changed as the plants grow. On the other hand, RUEs obtained with the push/pull method are the averages of different plant growth stages in Types A and B-a.

5.7.5 Automation and Robotization

Automatic machines for seeding, transplanting, transporting and labeling, packing, packaging, and culture panel washing are already being used in many PFALs for large-scale commercial production. These automatic machines can be used in large-scale PFALs with any type of CSM described in this chapter.

On the other hand, harvesting and trimming of damaged leaves are generally done manually, which accounts for nearly 40% of the total labor hours (Table 5.2). Automation or robotization of harvesting and trimming is a challenge. In this area, the introduction of technologies such as image processing, robot hands with high-resolution multipurpose cameras, and AI would be crucial.

5.7.6 Optical (Spectral) Sensing

Various types of wireless optical sensors are considered a basic technology of smart agriculture (Kameoka and Hashimoto 2015) and so is the smart PFAL with a phenotyping unit. Table 5.5 shows the major optical (spectral) sensing methods which can be used for sensing the plants in PFALs. In the near future, a nondestructive testing sensor might be developed based on infrared spectroscopy.

5.8 Concluding Remarks

The design factors of the CSM are discussed in this chapter as the first step toward designing smart PFALs. The potential productivity of the PFAL is much higher than that of the current PFALs. It is considered that the performance of the PFAL can be

Table 5.5 The major optical (spectral) sensing methods for the plants in PFALs

Electromagnetic wave	Measuring method	Measurement object	Measured substance
X-ray	Fluorescent X-ray spectroscopy	Leaf and substrate	K, Ca, P, N, Mg, Fe, Zn, Cu, and S
Ultraviolet (UV)	Fluorescent spectroscopy	Leaf	Saccharides, pigments, organic acids, disease symptom
Visible (VIS)	Color and shape analytics	Leaf and fruits	Pigments, vigor state of plants
Near infrared (NIR)	Glove-type NIR analyzer	Fruits	Sugar content and spectrum change
Mid-infrared (MIR)	FT-IR/ATR*	Leaf and fruits	Moisture content; mono-, di-, and oligo-saccharides; nitrogen; agricultural chemicals; and acids

Based on Kameoka and Hashimoto (2015)
*FT-IR, Fourier transform infrared spectroscopy; ATR, attenuated total reflection

steadily improved by using the CSM-L described in this chapter. Among the four Types A to D, it is suggested that Type A and Type B-a are suited as the CSM.

Acknowledgments This chapter is partially based on results obtained from a project commissioned by the New Energy and Industrial Technology Development Organization (NEDO). The author expresses their deep appreciation to Eri Hayashi, the project leader, of Japan Plant Factory Association. The author would also like to thank all the members of the committee for productivity improvement of the PFAL organized by Japan Plant Factory Association. Some ideas presented in this chapter were obtained during discussions at the committee meetings.

References

Ijichi H (2018) NAPA research report: chapter 3 plant factory business – current status and perspectives of plant factory business. Nomura Agri-planning Consult (in Jpn), 58–80., http://www.nomuraholdings.com/jp/company/group/napa/data/20180219.pdf (in Japanese)

Kameoka T and A Hashimoto (2015) Effective application of ICT in food and agricultural sector – optical sensing is mainly described -, IEICE transactions on communications E98-B, no. 9, pp 1741–1748

Kitaya Y (2016) Air current around single leaves and plant canopies and its effect on transpiration, photosynthesis, and plant organ temperatures, Chapter 13. In Kozai et al. (eds) LED lighting for urban agriculture. Springer, Dordrecht, pp 177–187

Kozai T (2013) Resource use efficiency of closed plant production system with artificial light: concept, estimation and application to plant factory. Proc Japan Acad Ser B 89(10):447–461

Kozai T, Niu G, Takagaki M (eds) (2015) Plant factory: an indoor vertical farming system for efficient quality food production. Academic. Amsterdam, 405 pages

Yabuki K (2004) Photosynthetic rate and dynamic environment, Kluwer Academic Publishers, Dordrecht, 126 pages

Part II
Recent Outcomes in Development and Business

Chapter 6
Business Planning on Efficiency, Productivity, and Profitability

Kaz Uraisami

Abstract Productivity is the index to measure every operation including horticulture called plant factory or controlled environment agriculture. Numerous indexes may approach "efficiency" and "productivity" issues. Some are simpler and some more complicated; some have larger magnitude and some have smaller impact on "profitability." Most of the indexes are inevitably correlated with one another, and no improvement may stand alone if the goal is to achieve the overall profitability. We invite wisdom from the business universe to analyze efficiency, productivity, and profitability and to cause the first two to help improve the last one pertaining to plant factory. This chapter introduces "business planning sheet" developed by Japan Plant Factory Association and Asahi Techno Plant. It simulates the process to engineer, procure, construct, and operate plant factories and to sell the produce thereof. It helps plant factory management to control the indexes for efficiency and productivity in order to achieve the profitability goal by distinguishing the causes from the results.

Keywords Business planning · Operational efficiency · Operational productivity · Business profitability · Sensitivity analysis · Correlation · Yield

6.1 Introduction

Plant factory business, if we may call it business, is a long chain of various components. You should engineer, procure, and construct the facility and then operate and maintain the facility and sell your produce. You should be knowledgeable and good at horticulture including but not limited to plant physiology; light-emitting diode (LED) mechanism; temperature and humidity control, i.e., vapor pressure deficit (VPD); etc. and good at things beyond them such as engineering of the facility, choosing factory materials and equipment, factory automation, process management, and people management. You should, last but not least, be

K. Uraisami (✉)
Asahi Techno Plant Co., Ltd, Okayama, Japan
e-mail: kazuya.uraisami@gkmarginal.com

experienced and committed in fresh food marketing. There shall not be too many professionals who qualify, and if any, he or she may be very expensive or doing something else. You are welcome to the real world.

To start and maintain plant factory business, those projects first need fund raising internally or externally, and you need to convince the chief financial officer of your company, fund managers, and investment or commercial bankers to provide your plant factory with the necessary fund. How can you communicate with them and convince them? Is your plant factory better than others? How soon can you return the fund or repay the loan? To answer the first question, whether your facility is more efficient, more productive, and more profitable, efficiency and productivity indexes set out a framework for comparison with peers. To answer the second, how many years the payback period is, profitability of the facility dictates annual generation of cash flow and decides the number.

This chapter does not intend to answer such questions as "what spectrum generates the highest fresh weight per lighting PPF (photosynthetic photon flux) mol" but rather to get around this extremely important horticultural question by showing that efficiency or productivity of the spectrum alone does not answer the question. You recognize that improvement of environmental control increases fresh weight per 1 m^2 cultivation tray with or without change of LED fixtures.

Before moving onto the first discussion of operational efficiency, we shall agree what plant factories are like as hydroponics in Japan and how operation goes there so that everybody compares apples with apples. Plant factory is a commercial facility under the operational principles of industrial manufacturing commonly established and is equipped usually with a highly structured interior configuration. Inside is a strictly controlled environment in terms of temperature, relative humidity, light, CO_2, and nutrient solution parameters. Operation manuals set by plant factory manager stipulate those processes for handling plants such as seeding, germination, seedling, transplanting, cropping, packing, and shipping.

We introduce how to simulate engineering, planning, construction, operation, and sales pertaining to the plant factories producing leafy greens in Japan which, with certain adjustment, may be applied to those producing other varieties as well as to the plant factories outside Japan. In specifics, this chapter assumes:

- Each shelf in the rack is 1,200 mm x 1,200 mm large and accommodates eight of 300 mm x 600 mm plastic grow tray, and the rack hosts six shelves for vertical farming.
- No medium is used for hydroponics, and instead, plastic grow tray with crop holes hosts crops supported by sponge cubic.
- First, 300 mm x 600 mm sponge sheet with 300 holes is used for seeding, germination, and 1st-stage seedling usually for a total of 5 days.
- Then cut the sponge sheet into 300 pieces, and plant seedlings on ca 25 mm x 25 mm cubic sponge pieces into plastic grow trays with 28 holes per 300 mm × 600 mm width. These 2nd-stage seedlings continue for 10 days.
- Stocks are then transplanted into 14 holes per 300 mm × 600 mm plastic grow tray so that growing size of stocks may be controlled under certain target leaf area

index (LAI), which stands for the one-sided green leaf area per unit ground surface area (leaf area/ground area, m^2/m^2).
- One more transplanting into seven holes per 300 mm x 600 mm plastic grow tray.

Baby leaf growers may exercise the same simulation by assuming how many seeds are to be put onto 300 mm × 600 mm sheets made of sponges or other alternative materials. Keep them for germination + 1st-stage seedling and 2nd-stage seedling, and at the end of 2nd-stage seedling when each crop is ca 5 g, cut them as baby leaf produce. Input number for those two processes only, and you may still simulate efficiency, productivity, and profitability on this business planning sheet.

For a comprehensive information and review of hydroponics at plant factory in Japan, please refer to Toyoki Kozai (2015), Jung Eek Son et al. (2015), and Osamu Nunomura et al. (2015).

6.2 Operational Efficiency

Japan Plant Factory Association (hereinafter, "JPFA") organized the research committee for operational efficiency of plant factory in spring 2016 consisting of some 20 scientists and business managers and held a presentation at JPFA's monthly seminar in fall 2016. A list of checkpoints to improve efficiency is concluded which covers all the aspects of facility and operation from the top management issues down to daily operation such as cleanup duties. Five major categories are:

1. Business planning
2. Housing and construction
3. Lighting fixtures
4. Air conditioning
5. Hydroponic systems

For example, housing and construction covers:

- Component of plant factory
- Floor zoning for respective operations
- Thermal insulation
- Ventilation
- Materials for walls and floors

The committee debated very seriously which floor material for hygiene area is less slippery when wet and which plastic shoes helps; what shape of joint of wall and floor, with or without roundness, reduces the time for daily cleaning; etc. and confirmed that operational efficiency shall be improved by learning from the experienced workers. Managers may implement i) visual recording if not that far as kinetics, ii) in-depth communication with and among part-time workers, and iii) job rotation of them so that, e.g., those in charge of seeding may learn how large the small difference at the earlier stage will be when reaching the cropping stage and vice versa. He or she may be more proud of the responsibilities if they know what they are doing and why and what others are doing. "Kaizen" campaign as bottom-up initiative shall also help.

When you plan improvement on the business planning sheet from 1st-year budget to 2nd year, 2nd year to 3rd, and so on, profitability may not respond squarely to respective improvement of efficiency indexes. The better, the better of course, and the worse, the worse. However, one should note:

(1) Less labor man seconds spent on one seeding due to longer experience reduces the total labor cost in proportion.
(2) Improvement of the system or process for cultivation may reduce labor man seconds more dramatically than per seed labor improvement. For example, optimized crop hole layout on the grow tray may possibly cut transplanting of the crops from three times to two, depending on produce variety, without sacrificing productivity of lighting PPF (photosynthetic photon flux) on $1m^2$ cultivation tray.
(3) If a hasty, very efficient looking, job done at 5:00 pm on Friday afternoon ends up with less yield in terms of how many crops reach the standard fresh weight within the standard cultivation period, it possibly backfires.
(4) A slow and steady job consumes more labor time but shortens the cultivation period and/or increases the yield, and therefore despite that per crop labor man hour increases, labor man hour per kg fresh weight can be decreased and profitability can be increased.

After all, it may be not because the workers get more experienced and process things faster, but because more fresh weight is obtained due to such other causes as improvement of VPD control making leaf lettuce grow faster.

Before moving from discussion of "efficiency" to the next section on "productivity," I provide you with "business planning sheet" whose copyright belongs to JPFA and Asahi Techno Plant. Initial investment and daily operation of plant factory and annual budget vs actuals are all simulated on this Excel sheet, and most of the business issues can be tried and tested thereon.

You wonder automation with robots shall or shall not replace labor works and how you may decide? Is it always to the maximum extent possible or not and if not, to what extent? Total cost of ownership analysis ("TCO") concludes the decision, and as the term dictates, cost and merits, positive or negative, shall be totaled including the running and maintenance cost of robots, relevant change of consumables, and a possible increase of labor management cost. You should of course be noted that the consumer needs may change and the production system may need change or be obsolete. Depending on how large the initial incremental investment is and how soon it is recovered, you are exposed to an additional fixed cost risk and may need plan B.

6.3 Business Planning Sheet

Below in Tables 6.1, 6.2, 6.3, 6.4, 6.5, 6.6, and 6.7, kindly find the business planning sheet whose Excel file consists of the four sheets, (1) elements, (2) initial investment, (3) budget and actuals for P/L (profit and loss statement) and cash flow, and (4) dashboard for the hypothetical case. Due to a limited size of one page as well as to the convenience of explanation, they are separated into seven tables in this book.

(1) Elements to Table 6.1 Cultivation area elements and Table 6.2 Lighting elements, etc.

(2) Initial investment to Table 6.3 Initial investment.

(3) Budget and actuals for P/L and cash flow to Table 6.4 Standard operations, Table 6.5 Standard operations P/L and cash flow, Table 6.6 Budget 5-year operations, and Table 6.7 Budget 5-year P/L and cash flow.

(4) Dashboard is to be reviewed in Sect. 6.5 Operational Productivity.

The sheet is originally denominated in Japanese yen, and for the sake of simplicity, exchange rate is fixed at US$1 = JPY100.

You will see how to plan the major terms such as size of housing, number of cultivation trays inside, PPFD (photosynthetic photon flux density) on 1 m^2 cultivation trays, number of LED fixtures necessary, initial investment in total and per m^2 cultivation area, fresh weight kilogram per m^2 cultivation area, labor man seconds per each operation, sales price per kg fresh weight produce, etc.

Plant factory managers focus eventually on the overall profitability and shall never mix the causes and the results and shall distinguish what improves profitability from what seems improved when profitability is improved. In conclusion, the three key indicators to improve profitability appear not on elements but on budget and actuals, and they are not fixed but floating over the budget period:

(i) Number of days necessary for cultivation before reaching target fresh weight (cultivation period),
(ii) Number of holes 1 m^2 cultivation area accommodates (cultivation density)
(iii) Ratio of the crops reaching target fresh weight over the number of transplanted crops (cultivation yield)

Priority for the managers to work relies on the Pareto analysis, which estimates the benefit delivered by each action and then selects a number of the most effective in the order of effectiveness of respective actions. You cut a tree one by one and, of course, the largest tree first and the second largest the second. Analyze what are the causes and what are the results, and do not mix them the other way around.

We simulate the decision process to build a plant factory in Table 6.1 where:

1. Housing for the hypothetical plant factory with 1000 m^2 width and 3.5 m height is leased to the plant factory owner.
2. Six of 400 mm height shelves are located at the rack, and 500 mm height nutrient water tank is located at the bottom of each rack.
3. Leafy green variety is to be cultivated by hydroponics.

Table 6.1 Cultivation area elements

Groups	Items		(Unit)	Numbers
Cultivation	Seeding per day		Grain	6000
	Yield: germination/seeding		Stock	90%
	Planting per day		Stock	5400
	Yield: stocks shipped/stocks picked		Stock	90%
	Shipping per day		Stock	4860
	Days for cultivation	Germination + 1st-stage seedling	Day	8
		2nd-stage seedling	Day	10
		Transplanting 1st stage	Day	8
		Transplanting 2nd stage	Day	8
Tray	Holes for cultivation	Germination + 1st-stage seedling	Hole	48,000
		2nd-stage seedling	Hole	56,921
		Transplanting 1st stage	Hole	43,200
		Transplanting 2nd stage	Hole	43,200
	Size of cultivation tray	Length	mm	600
		Width	mm	300
	Holes per tray	Transplanting 2nd stage	Hole	7
	Cultivation trays	Transplanting 1st stage	Sheet	3086
		Transplanting 2nd stage	Sheet	6172
Cultivation rack	Size of cultivation racks	Length	mm	1200
		Width	mm	1260
		Height	mm	2900
		Size	m^2	1.51
	Trays per shelf in racks	Germination + 1st-stage seedling	Sheet	8
		2nd-stage seedling	Sheet	8
		Transplanting 1st stage	Sheet	8
		Transplanting 2nd stage	Sheet	8
	Cultivation shelves	Germination + 1st-stage seedling	Shelf	20
		2nd-stage seedling	Shelf	255
		Transplanting 1st stage	Shelf	386
		Transplanting 2nd stage	Shelf	772
		Subtotal	Shelf	1433
	Shelves per cultivation racks	Germination + 1st-stage seedling	Shelf	6
		2nd-stage seedling	Shelf	6
		Transplanting 1st stage	Shelf	6
		Transplanting 2nd stage	Shelf	6
	Cultivation racks	Germination + 1st-stage seedling	Rack	4
		2nd-stage seedling	Rack	43

(continued)

6 Business Planning on Efficiency, Productivity, and Profitability

Table 6.1 (continued)

Groups	Items		(Unit)	Numbers
		Transplanting 1st stage	Rack	65
		Transplanting 2nd stage	Rack	129
		Subtotal	Rack	241
	Space b/w racks	Length	mm	500
		Width	mm	750
	Total bottom area of cultivation racks		m^2	364
	Total area of cultivation shelves		m^2	823
Room area	Picking operation, material storage, refrigerator		m^2	177
	Total size		m^2	1000

Table 6.2 Lighting elements, etc.

Groups	Items	(unit)	Numbers
Lighting	PPFD on tray	μmol/m^2/s	100
	Effective ratio	–	0.85
	Radiation	μmol/s	244,969
	Conversion factor	–	4.59
	Lighting power	W	53,370
	Energy conversion efficiency	–	0.31
	Watt per LED fixture	W	32
	LED fixture per cultivation tray	–	0.47
	LED fixture per cultivation shelf	–	4
	LED fixture total	–	5784
	Photoperiod	Hours/day	16
Air conditioning	COP (summer time)	–	2.25
	COP(other)	–	3.50
	Weighted average	–	2.56
Electricity contracted	Capacity (×1.1 with allowance)	kW	196.1
Sales	Price per kilogram fresh weight	US$/kg	12
Housing	Monthly rent (one 120th of housing price)	US$	8000

Note: COP (coefficient of performance) stands for the ratio between energy usage of the compressor and the amount of useful heat extracted from the condenser. Higher COP value 2^2 represents higher efficiency

Based on this hypothecation, elements and initial investment are fixed, and business planning simulation continues.

In more details, as described in Sect. 6.1, we assume:

① 300 mm × 600 mm sponge sheet with 300 holes are used for germination + 1st-stage seedling
② 300 mm × 600 mm plastic grow tray with 28 holes for 2nd-stage seedling.

Table 6.3 Initial investment

Items			Unit prices	Pieces	Total	Depreciation years
Housing	Interior renovation (water sink included)	Per 100 m^2	15,000	10	150,000	50
	Electricity works	Per 100 m^2	3500	10	35,000	15
	Air conditioning	Per 100 m^2	6000	10	60,000	7
	Drain water reuse system				0	7
	CO$_2$ gas systems	Per 100 m^2	1800	10	18,000	7
Rack	Cultivation racks	Per 1.2 m × 1.2 m	500	241	120,500	7
	Multipurpose racks	Per 100 m^2	1000	60	60,000	7
	Wind fans for racks	Per 100 m^2	200	400	80,000	7
	Water supply and drain systems	Per 100 m^2	7,000	10	70,000	7
Lighting	LED	Per fixture	85	5784	491,640	7
	Circuit board	Per rack	160	241	38,560	7
	Other materials	Per 6 racks	990	40	39,798	7
	Mounting frames for LED	Per rack	70	1446	101,220	7
	Reflection materials	Per rack	25	1446	36,150	7
	Control panel (incl. dimming device)	Per 6 racks	60	40	2412	7
	Dimming circuit board	Per rack	160	241	38,560	7
Cultivation	Cultivation tray	Per piece	25	12,000	300,000	7
	Cultivation pool	Per pool	100	2866	286,600	7
	Water pump	Per rack	500	241	120,500	7
Facilities	Sensors	Per set	50,000	1	50,000	7
	Air shower room	Per room	20,000	1	20,000	7
	Refrigerating room	Per room	10,000	1	10,000	7
	Cropping and shipping board	Per set	2000	1	2000	7
	Processing board	Per board	500	50	25,000	7
	Container for cropping	Per rack	10	241	2410	7

(continued)

6 Business Planning on Efficiency, Productivity, and Profitability

Table 6.3 (continued)

Items		Unit prices	Pieces	Total	Depreciation years
	Packaging machines	Per machine	5000	3	15,000	7
	Storage racks for trays	Per set	5000	2	10,000	7
	Storage racks for materials	Per set	5000	2	10,000	7
	Desks, PCs, lockers	Per set	15,000	1	15,000	7
	Total				2,208,350	

③ 300 mm × 600 mm plastic grow tray with 14 holes for posttransplanting 1st-stage cultivation.
④ 300 mm × 600 mm plastic grow tray with seven holes for posttransplanting 2nd-stage cultivation.

Each shelf in the rack is 1200 mm × 1200 mm large for eight of 300 mm × 600 mm sheets, and each rack accommodates six shelves for vertical farming. By assuming how many days each of those four grow steps described above necessitates, you may calculate how many holes and how many trays are necessary for each step and how many shelves and how many racks are. You may eventually come up with the size of the operation, how many you grow and subject to the yield assumed, and how many you crop inside the facility assumed.

If you plan to produce baby leaves, you find each crop hole accepts more than one seed, and not a hole but a sheet shall be a unit for certain number of seeds. As described in Sect. 6.1, apply grow trays for (1) germination + 1st-stage seedling and (2) 2nd-stage seedling, and at the end of 2nd-stage seedling when each crop is ca 5 g, cut them for cropping and shipping. Input number of cultivation days for (3) transplanting 1st stage and (4) transplanting 2nd stage both nil, and conclude cultivation at the end of 2nd-stage seedling. Those numbers of labor man hour for operations related to (3) and (4) are not taken into account.

You move to Table 6.2 for the purpose of DLI (daily light integral) planning where DLI stands for the number of photosynthetically active photons delivered to a specific area over a 24-h period. Set the PPFD level onto the grow tray in the unit of $\mu mol/m^2/second$ ($\mu mol\ m^{-2} s^{-1}$) and the effective rate how much thereof is applied inside the canopy (how much not escaping outside), which then lead you to how many LED fixtures you need over 1200 mm × 1200 mm cultivation area on the shelf. You need to assume (i) conversion factor, i.e., how many μmol/s of radiation be generated per 1 W of lighting power, and (ii) energy conversion efficiency (J/J), i.e., how many lighting power be generated per electricity.

Here conversion factor and energy conversion efficiency are assumed to be 4.59 and 0.31, respectively, which, as multiplied, then stands for photosynthetic photon

Table 6.4 Standard operations

Business plan						Standard operation
Production/ Sales	Seeding per day		Seed			5000
	Yield: germination/seeding		%			90.0%
	Transplanting per day		Stock			4500
	Yield: stocks shipped as one in one/stocks cropped		%			70%
	Yield: stocks shipped as two in one/stocks cropped		%			20%
	Yield: stocks not shipped/stocks cropped		%			10%
	Weighted yield: stocks shipped/stocks cropped		%			80%
	Stock packs shipped		Stock			3600
	Cultivation shelves total – cultivation shelves in operation					12
	Fresh weight per pack		g			90
	Fresh weight shipped per day		kg			324
	Price per kilogram		US$			12.0
	Annual sales		US$			1,399,680
	Total number of days in operation		Day			360
Cultivation Systems	Number of days necessary	Germination + 1st-stage seedling	Day			8
		2nd-stage seedling	Day			10
		Transplanting 1st stage	Day			10
		Transplanting 2nd stage	Day			10
	Number of planting holes necessary	Number of planting holes necessary	Hole			300
		Number of planting holes necessary	Hole			28
		Number of planting holes necessary	Hole			14
		Transplanting 2nd stage	Hole			7
	Number of cultivation trays necessary	Germination + 1st-stage seedling	Sheet			133
		2nd-stage seedling	Sheet			1694
			Sheet			3214

(continued)

Table 6.4 (continued)

Business plan							Standard operation
		Transplanting 1st stage					
		Transplanting 2nd stage	Sheet				6429
	Number of cultivation shelves necessary	Germination + 1st-stage seedling	Shelf				17
		2nd-stage seedling	Shelf				212
		Transplanting 1st stage	Shelf				402
		Transplanting 2nd stage	Shelf				804
	Total number of shelves necessary		Shelf				1434
	Toal number of shelves available in Plant Factory		Shelf				1446
Lighting	PPFD on trays		µmol/s/m²				100
	Photoperiod		Hours				16
	COP		Numbers				2.56
	Electricity for water pumps and fans		kW				20
	Monthly electricity consumption		kWh	Contract capacity	kWh	139,671
	Annual electricity cost	Capacity payment(US$)	14.45		196	33,996
		Metered payment (US$)	Summer price	0.14	US$/kWh	222,697
			COP	2.25	Pumps and fans	
			Non-summer price	0.13	US$/kWh	
			COP	3.50	Pumps and fans	
Labor	Germination + 1st-stage seedling		Second/stock				1
	2nd-stage seedling		Second/stock				3
	Transplanting 1st stage		Second/stock				3
							3

(continued)

Table 6.4 (continued)

Business plan							Standard operation
	Transplanting 2nd stage			Second/stock			
	Cropping			Second/stock			13
	Shipping (packing, storing in freezing room)			Second/stock			13
	Cleaning, etc.			Hours for CEA			8
	Total			Total hours			50
	Hourly rate			US$			8.50
	Number of part-time staff (6 hours/day)			Head count			9
	Labor cost total			US$			165,240
	Welfare and commuting	Vs hourly rate		20%			33,048
	Labor cost grand total						198,288
Consumables	Seeds			Cents/stock			1.5
	Seeding sponge			Cents/stock			1.0
	Nutrient hydro			Cents/stock			1.0
	CO_2			Cents/stock			1.0
	Packaging materials			Cents/stock			7.0
	Transportation			Cents/stock			10.0
	Waste cost (roots, sponge)			Cents/stock			0.7
	Others			Cents/stock			5.0
	Total			US$/stock			375,192

Table 6.5 Standard operations P/L and cash flow

P/L and cash flow		(Unit: US $)			Standard operation
Gross sales					1,399,680
Fixed cost	Depreciation			Fixed amount	3000
				Fixed amount	291,383
	Maintenance and renovation		3%		66,251
	Building monthly rent (US$)		8000	Monthly	96,000
	Electricity and water (capacity)			Monthly	33,996
	Subtotal				490,630
Variable cost	Electricity and water (capacity)				222,697
	Labor cost				198,288
	Consumables				154,872
	Subtotal				575,857
Cost over sales	Electricity and water (capacity)				18%
	Labor cost				14%
	Consumables				11%
Cost per stock	Electricity and water (capacity)	(Cents/stock)			20
	Labor cost	(Cents/stock)			15
	Consumables	(Cents/stock)			12
Gross profit					333,193
SGA					292,320
Operating profit					40,873
EBITDA					335,256
Cash on hand beginning of year	Cash funded for the project	3,000,000			0
Cash on hand end of year					335,256

Note:
P/L profit and loss statement, *SGA* selling and general administrative expenses, *EBITDA* earnings before interest, taxes, depreciation, and amortization

Table 6.6 Budget 5-year operations

Business plan				Year 1 plan	Year 3 plan	Year 5 plan
Production/sales	Seeding per day		Seed	4950	4850	5000
	Yield: germination/seeding		%	85.0%	90.0%	90.0%
	Transplanting per day		Stock	4208	4365	4500
	Yield: stocks shipped as one in one/stocks cropped		%	50%	65%	70%
	Yield: stocks shipped as two in one/stocks cropped		%	25%	20%	20%
	Yield: stocks not shipped/stocks cropped		%	25%	15%	10%
	Weighted yield: stocks shipped/stocks cropped		%	63%	75%	80%
	Stock packs shipped		Stock	2630	3,274	3,600
	Cultivation shelves total − cultivation shelves in operation			17	14	12
	Fresh weight per pack		g	80	80	90
	Fresh weight shipped per day		kg	210.4	261.9	324.0
	Price per kilogram		US$	12.0	13.0	13.0
	Annual sales		US$	908,820	1,225,692	1,516,320
	Total number of days in operation		Day	360	360	360
Cultivation systems	Number of days necessary	Germination + 1st-stage seedling	Day	8	8	8
		2nd-stage seedling	Day	14	12	10
		Transplanting 1st stage	Day	10	10	10
		Transplanting 2nd stage	Day	10	10	10
	Number of planting holes necessary	Germination + 1st-stage seedling	Hole	300	300	300
		2nd-stage seedling	Hole	28	28	28
		Transplanting 1st stage	Hole	14	14	14
		Transplanting 2nd stage	Hole	7	7	7
			Sheet	132	129	133

6 Business Planning on Efficiency, Productivity, and Profitability

Number of cultivation trays necessary	Germination + 1st-stage seedling	Sheet					
	2nd-stage seedling	Sheet			2282	1972	1694
	Transplanting 1st stage	Sheet			3005	3118	3214
	Transplanting 2nd stage	Sheet			6011	6236	6429
Number of cultivation shelves necessary	Germination + 1st-stage seedling	Shelf			17	16	17
	2nd-stage seedling	Shelf			285	247	212
	Transplanting 1st stage	Shelf			376	390	402
	Transplanting 2nd stage	Shelf			751	780	804
Total number of shelves necessary		Shelf			1429	1432	1434
Total number of shelves available in plant factory		Shelf			1446	1446	1446
Lighting	PPFD on trays	$\mu mol/s/m^2$			100	100	100
	Photoperiod	Hours			16	16	16
	COP	Numbers			2.56	2.56	2.56
	Electricity for water pumps and fans	kW			20	20	20
	Monthly electricity consumption	kWh	Contract capacity	196	139,245	139,512	139,679
	Annual electricity cost	Capacity payment(US$)	14.45	US$/kWh	33,996	33,996	33,996
		Metered payment (US$)	Summer price	0.14	222,019	222,444	222,710
		COP	2.25	Pumps and fans			
		Non-summer price	0.13	US$/kWh			
		COP	3.50	Pumps and fans			

(continued)

Table 6.6 (continued)

Business plan				Year 1 plan	Year 3 plan	Year 5 plan
Labor	Germination + 1st-stage seedling		Second/stock	2	1	1
	2nd-stage seedling		Second/stock	5	4	3
	Transplanting 1st stage		Second/stock	5	4	3
	Transplanting 2nd stage		Second/stock	5	4	3
	Cropping		Second/stock	20	15	13
	Shipping (packing, storing in freezing room)		Second/stock	15	15	13
	Cleaning, etc.		Hours for CEA	8	8	8
	Total		Total hours	63	56	50
	Hourly rate		US$	8.50	8.50	8.50
	Number of part-time staff (6 hours/day)		Head count	11	10	9
	Labor cost total		US$	201,960	183,600	165,240
	Welfare and commuting	Vs hourly rate	20%	40,392	36,720	33,048
	Labor cost grand total			242,352	220,320	198,288
Consumables	Seeds		Cents/stock	1.5	1.5	1.5
	Seeding sponge		Cents/stock	1.0	1.0	1.0
	Nutrient hydro		Cents/stock	1.5	1.0	1.0
	CO_2		Cents/stock	1.5	1.5	1.0
	Packaging materials		Cents/stock	7.0	7.0	7.0
	Transportation		Cents/stock	10.0	10.0	10.0
	Waste cost (roots, sponge)		Cents/stock	0.7	0.7	0.7
	Others		Cents/stock	5.0	5.0	5.0
	Total		US$/stock	312,908	354,831	375,192

6 Business Planning on Efficiency, Productivity, and Profitability

Table 6.7 Budget 5-year P/L and cash flow

P/L and cash flow				(Unit: US$)		Year 1 plan	Year 3 plan	Year 5 plan
Gross sales						908,820	1,225,692	1,516,320
Fixed cost		Depreciation		Fixed amount		3000	3000	3000
		Maintenance and renovation		Fixed amount		291,383	291,383	291,383
		Building monthly rent (US$)		3%		66,251	66,251	66,251
		Electricity and water (capacity)		8,000	Monthly	96,000	96,000	96,000
		Subtotal			Monthly	33,996	33,996	33,996
						490,630	490,630	490,630
Variable cost		Electricity and water (capacity)				222,019	222,444	222,710
		Labor cost				242,352	220,320	198,288
		Consumables				151,971	154,477	154,872
		Subtotal				616,342	597,242	575,870
Cost over sales		Electricity and water (capacity)				28%	21%	17%
		Labor cost				27%	18%	13%
		Consumables				17%	13%	10%
Cost per stock		Electricity and water (capacity)	(Cents/stock)			27	22	20
		Labor cost	(Cents/stock)			26	19	15
		Consumables	(Cents/stock)			16	13	12
Gross profit						−198,152	137,820	449,820
SGA						232,937	272,354	292,320
Operating profit						−431,089	−134,533	157,500
EBITDA						−136,705	159,850	451,883
Cash on hand beginning of year		Cash funded for the project	3,000,000			791,650	679,291	1,282,024
Cash on hand end of year						654,945	839,141	1,733,907

efficacy of 1.42 µmol per J, probably an average number of LED fixtures available in the market today. More advanced or efficient LED product may be available for less cost in the market in the near future and may generate higher µmol per J; however, this is not the subject of this chapter.

You assume an ex-factory price of the produce and the rent for housing and complete elements and then move to Table 6.3 Initial investment.

Unit prices used here may be more or less than you expect in your home markets and most likely more expensive than you need to pay when this book is published thanks to a fast development of more efficient vertical firming systems for less throughout or somewhere in the world. Number of years for depreciation is based on the tax life rules applicable in Japan and may not be so in other countries. I recommend, in the later section, a management accounting approach to the useful life of LED fixtures considering that the price of LED fixtures and other material are dropping faster than they depreciate over the tax law useful life years.

Upon completing the initial investment budget, you may then simulate your operations in Table 6.4 Standard operations and find out how the efficiency and productivity indexes correlate to the profitability and what the key indexes for the overall profitability in Table 6.5 Standard operations P/L (profit and loss statement) and cash flow. In this context, "standard operation" shall be an estimated future level of operations.

In Table 6.4, you should set those indicators practical, if not conservative, that are not fixed at the time of the initial planning:

1. Number of days for cultivation in four stages, i.e., germination + 1st-stage seedling, 2nd-stage seedling, posttransplanting 1st stage, and posttransplanting 2nd stage
2. Yield of the number of germination over the seeds and that of the number of the crops having reached the target fresh weight over the number of transplanted crops
3. Labor man hour consumption of each step of operations per crop, i.e., germination, transplanting, cropping, packing, cleaning, etc.
4. Cost of respective consumable items, i.e., seeds, sponges, nutrient hydro, CO_2, packaging materials, etc.

Transportation cost is assumed as one of those consumable items on the sheet but on P/L and cash flow is reported as sales and general administration cost. It occupies a sizable portion of the cost and, as it varies substantially depending on who and where to sell at what size, takes a different set of management skills to control.

After setting the standard operation, you then move to the same exercise for 5-year budget planning on Table 6.6 Budget 5-year operations and Table 6.7 Budget 5-year P/L and cash flow. Take into consideration any possible improvement over the 5-year period very seriously.

Due to a limited size of the page, only Year 1, 3, and 5 are shown. Developing the budget for 5 years leads you to an estimate of how many years are necessary to receive a full payback of what is invested. In Table 6.7, recovery of initial investment is to be completed in 7.8 years as you may recover US$452,000 each in year 6, 7, and

8 while in the first 5 years, US$1,734,000 is recovered of US$3 million initially invested.

If you so wish, please contact with Kaz Uraisami at kazuya.uraisami@gkmarginal.com or JPFA at info@npoplantfacotry.org for further information or a softcopy of business planning sheet. We may introduce you to attend the English language seminars to be held by JPFA at Kashiwa-no-ha, Kashiwa, Chiba, Japan.

6.4 How to Analyze and What to Find from the Sheet

You use "business planning sheet" to build, analyze, and manage the business plan of yours, preferably before you build plant factories, as follows:

(1) Grasp all the components necessary to construct the facility and all the materials to operate and maintain it, and plan the sales price of the crops. Business planning sheet lists up virtually all the pieces you should put together in this horticulture jigsaw puzzle, even though you may be more detailed or be simpler if you so wish.
(2) Try good numbers and bad numbers for higher or lower probability when filling out the index and indicator column on the sheet, and find out the outcome. Payback period, how many years you need to receive your investment back, varies substantially depending on the numbers assumed and convinces you, your boss, or your fund providers to or not to implement the investment or to amend the plan.

Dream may not always come true and you need to remain very practical and flexible. Do not fill in such numbers as $100 \mu mol/m^2/s$ PPFD on the cultivation tray, with a very high efficacy of 2.5 µmol/J, with a perfect design for 100% effective ratio inside canopy with perfect uniformity on the tray without any reflection board, and LED is purchased for US$20 per fixture with guarantee of 100,000 hours. Even this perfect system shall not complete the cultivation from seeding to cropping in 20 days at 100% yield.

Series of trial and errors on the sheet supported by academic or scientific knowledge and practical information in depth help reach what is a practically perfect combination of the indexes. It is your target but not standard, and "standard operation" is set as what you have a good chance to reach within 5-year planning period.
(3) For example, which of the following three lighting plans optimizes the initial investment per 1 m^2 cultivation tray and is the most efficient?:

   (i) 100 $\mu mol/m^2/s$ PPFD for 16-hour photoperiod
   (ii) 200 $\mu mol/m^2/s$ PPFD for 8-hour photoperiod
   (iii) 200 $\mu mol/m^2/s$ PPFD for 16-hour photoperiod

If the question is which is the lowest, the answer is 100 µmol/m²/s PPFD for 16-hour photoperiod. If the question is which produces the most fresh weight per 1 kWh electricity consumed, we need more information, "how many days each lighting choice takes before cropping," and as long as DLI and the number of cultivation days are inversely proportional to each other, the answer is the same.

If the yield of produce over seedling, i.e., how many crops can come out as produce on 1 m² cultivation tray from seedling, does not change and the cultivation period is reversely proportional to DLI, 200 µmol/m²/s PPFD for 16-hour photoperiod achieves by far the highest profitability. Shorter cultivation period improves profitability substantially.

(4) Through the process of trial and error described in (2), you find not all the indexes are the causes and not all have the same impact on profitability. Some are the causes and some are the results or the phenomena. Start with Pareto analysis of the causes and cut a tree one by one, the larger trees being the first.

The next section reviews the numerous indexes for operational productivity.

6.5 Operational Productivity

Following the conclusion of the 2016 Operational Efficiency Research Committee, JPFA in spring 2017 convened again some 20 scientists and business managers to review operational productivity. New committee has started with debating what indexes operational productivity shall be measured on. They must be clear, precise, and universal in definition and in unit. Table 6.8 lists those indexes proposed as possible "major productivity indexes," and Table 6.9 adds more:

Initial investment per 1 m² cultivation area is US$1019/m², and initial investment per kg daily fresh weight production is US$6816/kg, which means cultivation area necessary for daily 1kg fresh weight production is 6.69 m². If cultivation area necessary for daily 1kg fresh weight production does not change, the first two numbers are simply proportional to each other while initial investment per kg daily fresh weight production and daily fresh weight production per US1$ initial investment, which is not listed here, are inverse to each other.

Electricity consumed for lighting and water pumps and not including heat pumps, necessary for 1 kg fresh weight, is 10.34 kWh/kg, and fresh weight per 1 kWh electricity consumed for lighting and water pumps is 0.10 kg/kWh. Obviously these two numbers are inverse to each other as well.

Of the pairs of inverse numbers, the former is how small the cost is per production and the latter is how large production is for the cost. The same is the case with (i) lighting PPF consumed per kg fresh weight (45.73 mol/kg) vs fresh weight per 1mol lighting PPF (0.02 kg/mol) and (ii) man hour per kg fresh weight (0.154 MH/kg) vs fresh weight per man hour (6.49 kg/MH). You may choose either of a pair of inverse numbers that better fits your purpose. In general, production per cost indexes

Table 6.8 Major productivity indexes

Items	Index	(Unit)	Standard operation
Cultivation area per 1 m^2	Initial investment	US$/m^2	1019
	LED fixtures	US$/m^2	310
	Other materials	US$/m^2	709
Cultivation area per 1 m^2 per day	Fresh weight produced	kg/m^2/d	0.148
	Daily light integral (DLI)	mol/m^2/d	5.76
	Cost for DLI	US$/m^2/d	43.8
	Electricity consumed by LED fixtures	kWh/m^2/d	1.53
	Cost for electricity consumed	Cents/m^2/d	23.5
	Economic depreciation of LED fixtures	Cents/m^2/d	20.3
	Cost for labor	Cents/m^2/d	23

Table 6.9 Other productivity indexes

Items	Index	(Unit)	Standard operation
Initial investment	Per 1 kg daily production	US$/kg	6816
	Per one daily stock crop (80g)	US$/80g	613
Cultivation area	Area necessary for 1 kg daily production	m^2/kg	6.75
	Daily stock crop production per 1 m^2 area	crop/m^2	1.65
Electricity	Electricity necessary for one stock crop	kWh/crop	1.29
	Lighting and water pumps only	kWh/crop	0.93
	Electricity necessary for 1 kg fresh weight	kWh/kg	14.37
	Lighting and water pumps only	kWh/kg	10.34
	Fresh weight per 1 kWh electricity consumed	kg/kWh	0.07
	Lighting and water pumps only	kg/kWh	0.10
	Lighting PPF consumed per 1kg fresh weight	mol/kg	45.73
	Fresh weight per 1 mol light PPF	kg/mol	0.02
Labor	Man hour for one crop	MH/crop	0.014
	MS for man second	MS/crop	50
	Man hour for 1kg fresh weight	MH/kg	0.154
	MS for man second	MS/kg	554
	Fresh weight per 1 MH	kg/MH	6.49

Table 6.10 Economic depreciation and managing cost of LED fixtures

Additional assumption			
	LED fixture guaranteed hours	35,000	
	LED fixture replacement hours	20,000	
	Light hours per day	16	
	LED fixtures months before replacement	42	
Economic cost of LED fixtures		(Unit)	Standard operation
	1 LED fixture for 1-hour light period	Cent/LED/h	0.43
	Per 1 kWh electricity consumed by LED fixtures	Cents/kWh	13.28
	Per 1mol/m^2/hour	Cents/mol/m^2/h	2.62
	Per 1 kg fresh weight	Cents/kg	121.39
	Per 1 stock crop	Cents/crop	10.93

help in planning the operation, and cost per production indexes help in managing the operation.

One more table, Table 6.10, is added here to propose accounting for the cost of LED fixtures as a matter of productivity indexes. LED fixtures account for one third of the initial investment, and it is a good news and bad news situation that their purchase cost is going south. I propose plant factory managers exercise a firm and most updated grip choosing and operating LED fixtures. Not accounting for tax useful life but down to 90% illuminance period shall be used to calculate the photoperiod hourly cost. In Table 6.10, 20,000 h is used as the period higher than 90% of the product specified lighting PPF is expected, and if you use the fixture 16 hours a day, it lasts 42 months above 90%. If you remain profitable with hourly depreciation of LED fixtures with this schedule, you have a choice of replacing the subject LED fixtures and use them as sunk cost for such other purposes as supplement lighting for the canopies with higher LAI.

Photosynthetic efficacy of respective spectrum is not the major issue for this chapter; however, as reported by Mr. Qingwu (William) Meng on July 27, 2017, on URBANAG News, we today recognize that spectrum combination is a floating target half a century after Dr. Keith McCree created a classic photosynthesis curve (McCree 1971). Spectrum of 2 red and 1 blue combination is now exposed to a challenge that not a leaf by leaf measurement but a grow canopy wide measurement is necessary in order to optimize the spectrum for cultivation. A group of researchers at Utah State University (Snowden et al. 2016) finds that a light spectrum with up to 30% green light is generally as good as red and blue light for plant biomass gain, and while the upper leaves of a plant absorb most red and blue light, they transmit more green light to lower leaves for photosynthesis.

To manage lighting PPF cost, I propose we look simultaneously both at depreciation cost of LED fixtures and hourly cost of electricity. In the business planning sheet, electricity necessary for 1kg fresh weight is 10.34kWh/kg if not including the cost for air conditioning, and as kWh electricity cost is set at 15.36 cents, the electricity cost is US$1.59 per 1kg fresh weight. The cost of LED fixtures to produce 1kg fresh weight is, as the cost for 1kWh electricity to be consumed by LED fixtures is 13.28 cents, US$1.37. Therefore, the total lighting PPF cost per kg fresh weight is US$2.96/kg. This is a more practical approach to the total lighting PPF cost than looking at LED fixtures only as the source of initial investment cost and depreciation cost for accounting or for tax reporting.

You pick up several productivity indexes as the key to monitor and to control the operation, and we move to the next section, "How to manage and achieve business profitability?" Indexes inform you whether things are going as planned or not. Kindly be noted that the indexes are often the results and not the causes. If electricity necessary for 1 kg fresh weight reduces or fresh weight per 1kWh electricity consumed increases, it is either because:

(i) Improvement of LED fixtures: lighting PPF per 1 kWh of electricity consumed increases, and thanks to more lighting PPF, fresh weight per 1 kWh electricity consumed increases.
(ii) Improvement of environmental control: more fresh weight from the same amount of lighting PPF while electricity consumed to generate lighting PPF does not decline.

To the best of my knowledge as one of plant factory managers, (ii) above happens more often with larger magnitude than (i).

6.6 How to Achieve Business Profitability

In the next two sections, we review PPFD productivity vs profitability and man hour productivity vs profitability. As already presented a few times, productivity indexes are powerful tools to compare plant factories with one another; however, they are not always the causes of higher profitability but the results of other indicators improved. Improving those indicators needs a package of skills and know-hows both from horticultural and management perspectives.

To better control environment of plant factory, you shall command plant physiology, LEDs, sensing technology for many environmental elements including temperature and humidity for vapor pressure deficit, factory automation, pest control, and many more things beyond. You should also be, last but not least, experienced and committed in people management and fresh food marketing. There shall not be too many professionals who qualify, and if any, he or she may well be doing something else. You are welcome to the real world.

People management may be measured by operational efficiency indexes such as how many seconds one seeding takes, one crop transplanting takes, etc. One hundred sponge sheets for daily operation, 300 mm x 600 mm, each for 300 seeds, may be different from 100 times of one sponge sheet. Keeping up the good job everyday by the same staff or by rotation may be even more different. You need to manage the people as the team and manage the operation as the process.

If you plan to sell your produce to the retailers, B2B2C (business to business to consumer) business model shall be established. Popular punch lines are:

- Locally grown
- Hydroponically grown
- Clean produce
- Greenhouse grown
- Sustainably grown
- Uses less water than soil-grown produce
- No synthetic pesticides

Depending on the countries, regions, and areas, combination may change. If you look to the grocery stores of vegetable packages (salad mix) or delicatessen for sandwiches, you need a good B2B2B2C model. You may of course choose to be a direct retailer, B2C, by operating your own stores or *marche* as "locally grown," and the choice of punch lines is more important. Your tactics depend on how large your operation is, how close to the larger consumer markets, how seasonal your produce is, who you compete with, etc.

Coming back from people management, marketing, etc. to operation productivity indexes, we look at correlation between productivity and profitability in the next two sections.

6.7 PPF Productivity and Profitability

Two often used productivity indexes are (i) fresh weight per 1kWh electricity consumed for lighting and (ii) fresh weight per man hour (6.49 kg/MH). While fresh weight per 1 kWh electricity consumed for lighting correlates with PPFD, fresh weight per man hour does with labor man hour (or seconds) for respective operation such as seeding, transplanting, etc. We start with PPF productivity and profitability.

First, I examine the assumption that 200 $\mu mol/m^2/s$ PPFD for 16-h photoperiod shortens the number of days necessary from seeding to cropping by ca 50% compared with 100 $\mu mol/m^2/s$ PPFD for 16-h photoperiod and how twice as much PPFD may or may not improve productivity. Four cases are exhibited on business planning sheet:

(1) 100 $\mu mol/m^2/s$ PPFD for 16-h photoperiod for 38-day cultivation period
(2) 200 $\mu mol/m^2/s$ PPFD for 16-h photoperiod for 21-day cultivation period
(3) 200 $\mu mol/m^2/s$ PPFD for 16-h photoperiod for 27-day cultivation period
(4) 200 $\mu mol/m^2/s$ PPFD for 16-h photoperiod for 27-day cultivation period with 5% lower yield to cropping

6 Business Planning on Efficiency, Productivity, and Profitability

Table 6.11 PPFD vs operational productivity compares the cases from (1) through (4) described above, and Table 6.12 PPFD vs cost of LED fixtures reviews how the total of LED fixtures cost and lighting PPF cost varies in cases from (1) through (4).

You find:

(1) Initial investment per 1 m^2 cultivation area increases if PPFD is set 200 µmol/m^2/s simply because the LED cost almost doubles, but not quite doubles because those equipped on the rack, such as circuit board, mounting frames, etc., does not change regardless of LED fixtures per rack.
(2) Initial investment per one daily crop or per kg fresh weight reduces dramatically if the number of cultivation days declines half even though DLI almost doubles and so is electricity consumption by LED fixtures per 1 m^2 cultivation tray. Keep in mind that fresh weight produced per 1mol lighting PPF remains the same. What this means is that PPFD productivity does not improve but turnover rate improves substantially and so does the overall profitability. Payback period is down from 7.8 years for 100 µmol/m^2/s PPFD for 16-hour photoperiod in 38-day cultivation period to 4.4 years for 200 µmol/m^2/s PPFD for 16-hour photoperiod in 21-day cultivation period.
(3) Horticulture discussion may center on what intensity of PPFD is optimal for leafy lettuce, baby herbs, variety by variety, but this is not the major topic of this chapter. We assume here that shortening of cultivation period depends on DLI increase and twice as much DLI may reduce cultivation period by half or less than half. How will productivity and profitability change if 200 µmol/m^2/s PPFD is not 200% efficient nor productive and reduces cultivation days only by one third. We may also assume that stronger PPFD rather generates tip burn (physiological disorder observed as partial change of color to brown on leaf tip and edges) and ends up with lower yield. This is considered in (3) 200µmol/m^2/s PPFD for 16-hour photoperiod for 27-day cultivation period and (4) 200 µmol/m^2/s PPFD for 16-hour photoperiod for 27-day cultivation period with 5% lower yield to cropping. In these two cases, fresh weight produced per 1mol lighting PPF substantially declines from 100 µmol/m^2/s PPFD for 16-hour photoperiod. Conclusion is still that, in spite of the lower productivity in terms of fresh weight produced per 1mol lighting PPF, the overall profitability improves and provides you with the less years for payback.
(4) Does this mean that a number of cultivation days, in other words, turnover rate, are more important than productivity index of fresh weight produced per 1mol lighting PPF? I would rather perceive *not* and that following three indicators shall rather determine the productivity index of fresh weight per 1mol lighting PPF, (i) cultivation period, (ii) cultivation density, and (iii) cultivation yield. When you hear of the news that your competitor produces more fresh weight per 1mol lighting PPF, please do not jump into a wrong conclusion that they apply the spectrum different from yours which is more efficient for photosynthesis.
(5) You may translate (i) and (ii) in (4) above that integration of each cultivation day multiplied by respective width which cultivation hole occupies on the grow tray on that single day is the key indicator for the turnover rate. A steady growth, or

Table 6.11 PPFD vs operational productivity

Items	Index	(Unit)	100μmol/m²/s photoperiod 16 hours/day	200μmol/m²/s photoperiod 16 hours/day (cultivation days 50% reduced)	200μmol/m²/s photoperiod 16 hours/day (cultivation days 1/3 reduced)	200μmol/m²/s 16 hours/day (cultivation days 1/3 reduced) (yield lowered)
Cultivation area per 1m²	Initial investment	US$/m²	1019	1246	1246	1246
	LED fixtures	US$/m²	310	537	537	537
	Other materials	US$/m²	709	709	709	709
per one daily crop	Initial investment (80g crop)	US$/80g	613	379	528	545
Cultivation area per 1m² per day	Fresh weight produced	kg/m²/d	0.148	0.293	0.210	0.204
	Daily light integral (DLI)	mol/m²/d	5.76	11.52	11.52	11.52
	Cost for DLI	US$/m²/d	438.0	814.0	814.0	814.0
	Electricity consumed by LED fixtures	kWh/m²/d	1.53	2.84	2.84	2.84
	Cost for electricity consumed	Cents/m²/d	23.5	43.9	43.9	43.9
	Economic depreciation of LED fixtures	Cents/m²/d	20.3	37.6	37.6	37.6
	Cost for labor	Cents/m²/d	23.3	42.4	31.5	31.2

per 1 mol PPF			0.026	0.025	0.018	0.018
Yield	Fresh weight produced	kg/mol				
	Crop 80g or larger	%	70	70	70	65
	2 crops in 1 package 80g or larger	%	20	20	20	25
Number of cultivation days	Germination + 1st-stage seedling	Days	8	6	6	6
	2nd-stage seedling	Days	10	5	7	7
	1st-stage transplanting	Days	10	5	7	7
	2nd-stage transplanting	Days	10	5	7	7
Annual revenue		US$ 000	1516	3002	2153	2086
Gross profit		US$ 000	450	1392	648	654
Operating income		US$ 000	158	884	263	279
EBITDA		US$ 000	452	1178	628	573
Payback period		Years	7.8	4.4	6.99	7.59

(continued)

Table 6.11 (continued)

Items	Index	(Unit)	100μmol/m²/s photoperiod 16 hours/day	200μmol/m²/s photoperiod 16 hours/day (cultivation days 50% reduced)	200μmol/m²/s photoperiod 16 hours/day (cultivation days 1/3 reduced)	200μmol/m²/s 16 hours/day (cultivation days 1/3 reduced) (yield lowered)
Initial investment	Per 1kg daily production	US$/kg	6816	4209	5869	6058
	Per one daily stock crop (80g)	US$/80g	613	379	528	545
Cultivation area	Area necessary for 1kg daily production	m²/kg	6.75	3.41	4.75	4.91
	Daily stock crop production per 1m² area	Crop/m²	1.65	3.26	2.34	2.27
Electricity	Electricity necessary for one stock crop	kWh/crop	1.29	1.21	1.68	1.74
	Lighting and water pumps only	kWh/crop	0.93	0.87	1.21	1.25
	Electricity necessary for 1kg fresh weight	kWh/kg	14.37	13.42	18.70	19.30
	Lighting and water pumps only	kWh/kg	10.34	9.65	13.45	13.88
	Fresh weight per 1kWh electricity consumed	kg/kWh	0.07	0.07	0.05	0.05
	Lighting and water pumps only	kg/kWh	0.10	0.10	0.07	0.07

			45.73	46.19	64.41	66.48
	Lighting PPF consumed per 1kg fresh weight	mol/kg				
	Fresh weight per 1 mol Light PPF	kg/mol	0.022	0.022	0.016	0.015
Labor	Man hour for one crop	MH/crop	0.014	0.013	0.013	0.014
	MS for man second	MS/crop	50	46	48	49
	Man hour for 1kg fresh weight	MH/kg	0.154	0.142	0.147	0.150
	MS for man second	MS/kg	554	510	528	540
	Fresh weight per 1 MH	kg/MH	6.49	7.05	6.82	6.66

Table 6.12 PPFD vs cost of LED fixtures

LED fixture economic depreciation assumed						
	LED fixture guaranteed hours	35,000				
	LED fixture replacement hours	20,000				
	Light hours per day	16				
	LED fixtures months before replacement	42				
Economic cost of LED fixtures		(Unit)	100 μmol/m²/s photoperiod 16 hours/day	200 μmol/m²/s photoperiod 16 hours/day (cultivation days 50% reduced)	200 μmol/m²/s photoperiod 16 h/day (cultivation days 1/3 reduced)	200 μmol/m²/s 16 h/day (cultivation days 1/3 reduced) (yield lowered)
	1 LED fixture for 1-hour light period	Cent/LED/h	0.43	0.43	0.43	0.43
	Per 1 kWh electricity consumed by LED fixtures	Cents/kWh	13.28	13.28	13.28	13.28
	Per 1 mol/m²/hour	Cents/mol/m²/h	2.62	2.62	2.62	2.62
	Per 1 kg fresh weight	Cents/kg	121.39	122.62	170.98	176.49
	Per 1 stock crop	Cents/crop	10.93	11.04	15.39	15.88

even a slower growth, at the earlier stage may be more than compensated for by faster growth at the later stage from a turn over perspective.
(6) You shall control and maintain very carefully lighting PPF density over LAI of the canopy at the optimal level, and hence, you may need more LED fixtures at the later stage when LAI becomes higher over the given width of grow tray. Fine-tune PPFD and the spectrum for the respective varieties of leafy greens at respective cultivation stages.

6.8 Man Hour Productivity and Profitability

In this section, I examine the assumption that labor time for seeding, transplanting from seeding sponge to the first grow tray (28 holes on 300 mm × 600 mm) to the second (14 holes on 300 mm × 600 mm) and to the third (7 holes on 300 mm × 600 mm), declines from 10 seconds to 6 seconds of man hour over the 5-year planning period and that the total labor time declines by half. Lower yield caused by the shorter labor time is also reviewed. Table 6.13 Man hour vs operational productivity compares the four cases in details.

You find:

(1) Six seconds down to 10 seconds for seeding, transplanting, etc. per crop do not improve productivity nor profitability dramatically even though it does not seem to be an easy goal for the on-site managers.
(2) Once the yield starts to lower, damage therefrom more than offsets the lower labor cost.
(3) In Japan the labor cost is lower compared with the USA or EU countries, but the employment practice demands stability for part-time workers as well, and in the USA, the part-time labor cost is higher. In either case, the gross labor cost does not shrink substantially responding to efficiency improvement and stability, and quality of the labor shall be more important for plant factory operations.

6.9 Correlation of Efficiency and Productivity to Profitability

Correlation of efficiency and productivity to profitability is reviewed using the four cases in Sect. 6.7. Business planning sheet answers such a question as "how much decrease of yield due to tip burn offsets the positive impact from stronger PPFD due to shorter cultivation period?" If we accept the assumption in the Sect. 6.7, the answer is (comparing (i) and (ii)):

Table 6.13 Man hour vs operational productivity

Items	Index	(Unit)	10 seconds per crop (seeding, transplanting ×3)	6 seconds per crop (seeding, transplantingx3)	6 seconds per crop (seeding, transplantingx3) yield lowered	Labor time total 50% reduced
Cultivation area per 1m^2	Initial investment	US$/m^2	1019	1019	1019	1019
	LED fixtures	US$/m^2	310	310	310	310
	Other materials	US$/m^2	709	709	709	709
Per one daily crop	Initial investment (80g crop)	US$/80g	610	610	633	610
Cultivation area per 1m^2 per day	Fresh weight produced	kg/m^2/d	0.148	0.148	0.144	0.148
	Daily light integral (DLI)	mol/m^2/d	5.76	5.76	5.76	5.76
	Cost for DLI	US$/m^2/d	438.0	438.0	438.0	438.0
	Electricity consumed by LED fixtures	kWh/m^2/d	1.53	1.53	1.53	1.53
	Cost for electricity consumed	Cents/m^2/d	23.5	23.5	23.5	23.5
	Economic depreciation of LED fixtures	Cents/m^2/d	20.3	20.3	20.3	20.3
	Cost for labor	Cents/m^2/d	23.3	20.9	20.8	14.6
Per 1 mol PPF	Fresh weight produced	kg/mol	0.026	0.026	0.025	0.026
Yield	Crop 80g or larger	%	70	70	65	70
	Crop 80g or larger	%	70	70	65	70

(continued)

Table 6.13 (continued)

Items	Index	(Unit)	10 seconds per crop (seeding, transplanting ×3)	6 seconds per crop (seeding, transplanting×3)	6 seconds per crop (seeding, transplantingx3) yield lowered	Labor time total 50% reduced
Number of cultivation days	Germination + 1st-stage seedling	Days	8	8	8	8
	2nd-stage seedling	Days	10	10	10	10
	1st-stage transplanting	Days	10	10	10	10
	2nd-stage transplanting	Days	10	10	10	10
Annual revenue		US$ 000	1516	1516	1469	1516
Gross profit		US$ 000	450	472	427	516
Operating income		US$ 000	158	180	141	224
EBITDA		US$ 000	452	474	436	518
Payback period		Years	7.8	7.4	7.98	6.68

(i) Number of cultivation days under 100 µmol/m^2/s set as 8 days for germination + 1st-stage seedling, as 10 days for 2nd-stage seedling, as 10 days for transplanting 1st stage, and as 10 days for transplanting 2nd stage generating 80% yield.
(ii) Under 200 µmol/m^2/s set as 6 days for germination + 1st-stage seedling, as 5 days for 2nd-stage seedling, as 5 days for transplanting 1st stage, and as 5 days for transplanting 2nd stage generating 50% yield (i.e., down from 80% to 50%),

both (i) and (ii) may achieve approximately the same profitability.

(iii) Under 200 µmol/m^2/s set as 6 days for germination + 1st-stage seedling, as 7 days for 2nd-stage seedling, as 7 days for transplanting 1st stage, and as 5 days for transplanting 2nd stage generating 70% yield

which may also produce the same result.

This means, if the cultivation days at your plant factory is reduced almost by half, even if your yield is lower by 30%, you are approximately breakeven. In other words, if you cut the cultivation days by almost half, any improvement of yield higher than 50% back to 80% by regulating tip burn shall bring in additional profitability. If the facility is equipped with LED dimming system, managers may opt to control how fast they grow the produce and how much they regulate tip burn so that they may reach the optimized correlation package of profitability to efficiency and productivity.

This applies not only to PPFD productivity vs profitability but also to man hour per each operation as well as to, e.g., such a complicated calculation of more expensive LED with specialized spectrum vs profitability. In virtually all cases, three key factors control productivity, (1) cultivation period, (2) cultivation density on tray, and (3) cultivation yield, and cultivation period is more important at the stage that cultivation density declines when each hole for crops takes the larger space.

This chapter does not cover how to maximize the speed of plant growth; however, experience of mine as one of plant factory managers convinces that slower yet steadier growth at the earlier stage sets the destiny of the produce. If your budget affords 50 more µmol/m^2/s on grow tray either at the earlier or at the later stage, you should definitely choose the earlier as:

(i) Lighting PPF cost per crop is smaller due to higher density.
(ii) Stronger PPFD at the earlier stage sets the lifetime speed of cultivation.
(iii) Stronger PPFD at the earlier stage is less likely to cause tip burn and that at the later stage is more likely to.

6.10 Sensitivity and Risk Analysis for Funding Raising

The last section discusses how to complete capital funding for plant factories. Business planning sheet calculates the payback period, i.e., how many years are necessary to recover the investment back. We should convince the financiers to provide your plant factory with the necessary fund for engineering, procurement and construction, and working capital for a start-up period of operations before annual cash flow turns positive.

Efficiency and productivity indexes set by the three key indicators and other factors may get better or worse than planned from time to time. Commonly shared business wisdom is that:

(1) Establish "standard" numbers for the indexes and the indicators to be used in the business planning sheet, and assign the probable numbers for each year in the planning period.
(2) Analyze how much better or worse those numbers fluctuate over the period at 90% probability for the worst case scenario.
(3) Depending on where funding comes from, propose the terms and conditions for share allocation and/or the loans, and discuss them to reach agreement with and satisfaction of the fund providers.

For example, the bankers may ask you whether you may reduce the number of days or, more precisely, the cumulative width per day that one crop occupies on the cultivation tray from seeding through cropping and whether you may increase the daily fresh weight produce per $1m^2$ cultivation tray. Depending on what your roles are, your answers may be geared to such productivity indexes as lighting PPF consumed per 1kg fresh weight or man hour for 1kg fresh weight. You emphasize how good or how experienced you are, and the bankers may ask how other plant factories are producing and what the average number is. They may almost always ask you how far the worst case scenario deteriorates, to or not to make you furious.

At this point of time, track record and information available for plant factories are few and limited, and it is difficult to convince the fund providers what the worst case scenario is. They may demand the lower share issue price for the higher allocation or the lower loan-to-value ratio for the higher debt service coverage ratio. Standard numbers for indexes, sensitivity analysis, and risk scenario approach are yet to be developed, and we need the industry-wide research and cooperation for sure.

6.11 Conclusion

Efficiency, productivity, and profitability shall be translated from one to another in order to facilitate collective understanding of the value of the plant factories by and among horticulture specialists, operating managers, marketing experts, and financial professionals. If there be no money, no risk taking is possible, and if there be no risk taking, no return is possible.

Repeated simulation convinces you that Pareto analysis does not start with what you usually think is the largest cause and improvement of the largest does not come from what you think is the most important. Financial specialists demand not the scientific data from the horticultural standpoint but the confidence from the statistical approach. Standard numbers for indexes, sensitivity analysis, and risk scenario approach are yet to be developed, and we need the industry-wide research and cooperation.

If this approach ever prevails and proves practical, it shall involve horticulture scientists, plant factory operators, manufacturers of the equipment and materials used, housing builders, etc. all together. Global cooperation shall start sooner rather than later as the standard numbers for the indexes for respective plant factories in respective countries are diversified while agriculture in the controlled environment is not a luxury but a survival concern of the whole biosphere. Those who are growing vegetables in plant factories, and those who are going to, shall analyze efficiency, productivity, and profitability and learn not only how to review the results but also how to improve the causes. This set of information and knowledge, when communicated and shared among the operators and with the financiers, shall certainly cultivate the global wisdom.

References

Jung Eek Son, Hak Jin Kim, Tae In Ahn (2015) Hydroponic Sys- tems. Chapter 17 of Plant factory an indoor vertical farming system for efficient quality food production. Academic, Amsterdam

Kozai T (2015) Plant production process, floor plan, and layout of PFAL, Chapter 16 of "Plant factory an indoor vertical farming system for efficient quality food production. Academic, Amsterdam

McCree KJ (1971) Significance of enhancement for calculations based on the action spectrum for photosynthesis. Plant Physiol 49:704–706

Osamu Nunomura, Toyoki Kozai, Kimiko Shinozaki, Takahiro Oshio (2015) Seeding, Seedling Production and Transplanting. Chapter 18 of Plant factory an indoor vertical farming system for efficient quality food production. Academic, Amsterdam

Snowden MC, Cope KR, Bugbee B (2016) Sensitivity of seven diverse species to blue and green light: interactions with photon flux. PLoS ONE 11(10):e0163121. https://doi.org/10.1371/journal.pone.0163121

Chapter 7
Renewable Energy Makes Plant Factory "Smart"

Kaz Uraisami

Abstract The plant converts the light energy into the chemical energy and so does the solar panel the light energy into the electric energy. Is it then possible at all that the solar panel converts light into electricity, which is then converted back into light by artificial lighting devices and then in turn into chemical by the plant? We review, as an experiment, how the power systems provide the plant factories with photovoltaic (PV) renewable energy generated nearby and supplied off-grid (i.e., stand-alone power system or mini-grids typically to provide a smaller community with electricity). Photovoltaics stand for converting light into electricity using semiconducting materials that exhibit the photovoltaic effect. A typical photovoltaic system employs solar panels as the largest price component.

The number of "solar sharing" facilities has increased in Japan which support solar panels over the cultivation field like wisteria trellis without blocking more than certain percentage of sunshine. However, not only sharing the light between power generation and agriculture, but also directly combining power generation with horticulture has become practical thanks to price commoditization of the solar panels and the lithium-ion batteries. Plant factory or the greenhouses with supplemental lighting by the light-emitting diode (LED) may well be the crossing point of these three innovations in the near future.

Keywords Photovoltaic · Renewable energy · Battery storage · Off-grid

7.1 Introduction: Global Price Movement as Commodities

Global price of the solar panels declined from US$10 to US$1 in the 20-year period from 30 years ago to 10 years ago and then has become one third of what it cost 10 years ago as shown in Fig. 7.1. It is devastating to those manufacturers in the USA or in the EU countries who have been on the frontline developing technology for

---

K. Uraisami (✉)
Asahi Techno Plant Co., Ltd, Okayama, Japan
e-mail: kazuya.uraisami@gkmarginal.com

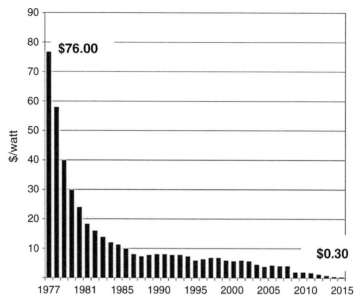

Fig. 7.1 Price history of silicon photovoltaic (PV) cells. (https://en.wikipedia.org/wiki/File:Price_history_of_silicon_PV_cells_since_1977.svg#filehistory as of May 2 2015)

photovoltaic power generation. On the other hand, such a drastic price reduction of the solar panels as commodities mostly lead by those manufactures in China has contributed to a massive explosion of the total size of power generation by photovoltaic (PV) plants in the world. Naturally the total system price of the power plant is not determined by the solar panels only and may vary from one country to another as construction and compliance with regulations are very local and are not exportable nor importable. Reportedly, the solar power system for 1 megawatt costs no more than one million US dollars nowadays in many countries and when the cost of civil engineering at the given site is more than reasonable, as low as three quarters thereof.

To make the system even more attractive is the price destruction of storage batteries, which has recently started to follow the price trend of solar panels. Solar power plants shall owe a sincere gratitude to the automobile industry that needs mass production of batteries for electric vehicles (EVs) (Fig. 7.2).

According to the recent report by McKinsey & Company, technology development of lithium-ion batteries has led to the reduction of energy storage cost for commercial customers to an attractive level. A greater uptick in EV adoption in terms of mass of economy shall only accelerate this trend. Major players are now scaling up their lithium-ion manufacturing capacity in order to meet the demand, and those in China are once again converting the advanced batteries to the commodities.

Batteries may now play the more central role in the photovoltaic energy markets departing from niche uses as grid balancing. Solar power system for 1 megawatt with

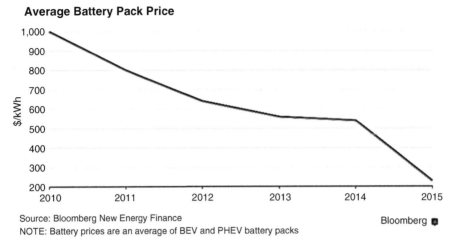

Fig. 7.2 Average battery pack price. (Bloomberg.com https://www.bloomberg.com/news/articles/2016-10-11/battery-cost-plunge-seen-changing-automakers-most-in-100-years October 11, 2016 by Reed Landberg)

3 days' consumption worth of storage system costs no more than five quarters of one million US dollars. For the first time, solar may become a stable alternative to conventional power generator fuel.

7.2 Smart Energy for Smart Plant Factory

Plant factory becomes smarter if the power for artificial lighting, heat pumps, etc. is supplied by clean CO_2-free energy, which has become possible at the competitive electricity price. One of the two challenges solar power has been facing is stability of generation due to day time and night time and weather changes whether forecasted or not. Battery storage system has alleviated this burden, once unconquerable, by storing the power and balancing the output, and a tremendous price reduction of batteries has blessed this solution with reality. The other challenge is the possible location of plant factories pertaining to the power grid. Solar power curtailment remains significant as an issue as long as the system needs to be connected with the grid, and if not, i.e., works independently on an off-grid basis, the issue may automatically disappear.

Total cost of 1 megawatt solar power plant used to be three million US dollars, of course without battery storage capacity, and is now reduced to one million US dollars or even less with a few exceptions including Japan. The system consists of:

(i) Solar modules
(ii) Solar module racks
(iii) Power conditioners

(iv) High-voltage cabling systems
(v) Civil works
(vi) Electric works
(vii) Others

In Japan, both civil and electric works cost high and so do others partially due to the "feed-in tariff (FIT)" system still operating, which imposes more regulatory works on the engineering, procurement, and construction companies in exchange for FIT purchase price.

Off-grid operation does not mean you may do without power conditioners and high-voltage cabling systems and add no extra cost reduction. However, off-grid means you are free from limitation or curtailment by the grid lines, and you should consider daily and seasonal change of temperature of the location, hiring the staff, shipping your produce to the consumer market including one for locally grown, and the last but certainly not the least, where the sun shines fine.

You convert the photon flux of the sunshine including those not optimal photosynthetic to the electric power and then convert it again, through LED or fluorescent light, to the photosynthetic photon flux in the form of artificial lighting. The second of the two conversions is necessary in any event to supply artificial lighting for plant factory, and you may be proud that the first is friendlier to the environment if without CO_2 emission. Price destruction of the solar systems shall make "smart energy smart plant factory" practical and then, as a result, plant factory more affordable to local communities.

7.3 Experimental Calculation

If you acquire 45,000 square meters of land and construct the plant factory on the lot of 1000 square meters thereof filling out the most of the rest with solar panels, you may supply 100% of the electricity necessary for artificial lighting and heat pumps, etc. inside the plant factory from the independent photovoltaic plant system of your own. While many things are hypothetical, the experimental calculation is as follows:

(i) Plant factory on 1000 square meter lot has 240 cultivation racks (1200 mm × 1200 mm wide and 3000 mm high) with the total of cultivation rack bottom in width of ca 360 square meters. Each rack has six cultivation shelves (300 mm high), i.e., 1440 total cultivation shelves for the total 2160 square meters in width. If you demand PPFD of 200 $\mu mol/m^2/s$ at 85% effective ratio over the cultivation shelf, you need 5800 of 32 W LED fixtures, and if you run them for 16-hour photoperiod, including heat pumps, etc. at weighted average coefficient of performance (COP) rate of 2.6, you need to secure the contract capacity of ca 390 kW from the power company for your plant factory. If you operate the plant factory 360 days a year, annual consumption of electricity is 3370 MWh.

(ii) In Japan, 1 megawatt photovoltaic plant generates 1100 MWh as average, and while the number varies throughout the world, I assume 1000 square meter lot plant factory with 3 meters in height may remain lit by 3 megawatt photovoltaic generation system.
(iii) Depending on the shape of the location site, etc., 3 megawatt photovoltaic generation system often needs 30,000 to 45,000 square meters of land.
(iv) Capacity of the battery storage is much more difficult and trickier to calculate. An extreme assumption that no sunshine available for 24 hours x 3 days may demand storage capacity of 28 MWh.

Summing up the experimental calculation above, in order to launch plant factory of 1000 square meters lot fully hosted by the solar power and battery storage system for the electricity, you need to secure ca 30,000 to 45,000 square meters of land for 3 MW photovoltaic system and 28 MWh storage system. Cost of 3 MW system shall be justified from the standpoint of total cost of ownership if local electricity price is 7 cents to 10 cents per kWh and 1 MW system is three quarters to 1 million US dollars. Cost of battery storage for 1 MWh is or may soon come down to US $200,000, and 3 days' worth of 3 MWh is ca US$2 million. The latter is more difficult to justify partly because one needs to review and find out how bad or not so bad if supply of artificial lighting is temporarily reduced with or without minimum battery storage size to maintain the heat pump operation 24 hours a day regardless of the external environment.

7.4 Conclusion

PFAL stands for "plant factory with artificial lighting" and by definition cannot survive without the electricity for artificial lighting. Whether it is generated by burning fossil fuels or converted from the power of falling water, shining sun, blowing wind, or flowing tide, the cost is the most important consideration; however, one may contend that plant factory is smarter if the overall social cost is minimized and the social value is optimized. Cost includes emission of CO_2 and value includes a stable supply of food in spite of inconvenience of uncontrolled environment outside. The days may come very soon that the sun generates photosynthetic reaction on the leaves outside or does so on the solar panels outside and then inside via LEDs on the leaves. Stability of vegetable cultivation shall offset the cost of photovoltaic power systems without emitting CO_2 gas.

Chapter 8
Total Indoor Farming Concepts for Large-Scale Production

Marc Kreuger, Lianne Meeuws, and Gertjan Meeuws

Abstract Indoor farming can contribute to the world food production in the coming decades. Indoor growing would greatly reduce the need for water and pesticides to grow crops and enable the production of safe, clean, nutritious, and affordable food. However, to realize the yields needed, some of the plant's requirements must be met. For this a climate system was designed where light, temperature, and evaporation are controlled independently. In the modular system, a laminar airflow controls evaporation, independent of light levels and without impact of infrared light. Together with crop models, calculations and predictions of yields can be made, which are necessary to define commercial success in advance.

The Plant Balance Model allows to further increase yields and quality by fine-tuning the specific effects of temperature, evaporation, light, and crop management.

In the modular indoor farm, a large variety of crops can be grown. It can consist of a single-layer growing area dedicated for vine crops like tomato, cucumber, and pepper. Alternatively, a multilayer area can produce herbs, lettuce, and other small crops.

Keywords Plant balance · Tomato · Vine crops · Indoor farming

8.1 Introduction

The growing population and the need to provide enough fresh, tasty, healthy, and affordable food urge agriculture to make a dramatic shift in efficiency. Moving from field to greenhouses, including nutrition and artificial lighting, in the last decades marked a great improvement in food production. However, this is not enough to feed the entire population, certainly in the coming decades with the population rising to nine billion and water becoming the limiting factor in many agricultural areas. Additionally, the current supply chain is too inefficient; too many losses occur during production, transportation, storage, and due to a mismatch between supply

---

M. Kreuger (✉) · L. Meeuws · G. Meeuws
Seven Steps To Heaven B.V, Zwaanstraat 31U, 5651 CA, Eindhoven, The Netherlands

© Springer Nature Singapore Pte Ltd. 2018
T. Kozai (ed.), *Smart Plant Factory*, https://doi.org/10.1007/978-981-13-1065-2_8

and demand. And even more, today's crops were selected to increase yields and for being tolerant to abiotic stress and resistant against pathogens. Specific benefits related to feeding a world population in terms of nutritional levels are hardly part of the current breeding programs.

Indoor farming can be regarded as the next generation of agriculture and may very well be one of the few solutions to make that shift. Growing crops without daylight enables higher yields, quality, and nutrition. Food can be produced everywhere with low water usage, without pesticides, and at reasonable prices. The reduction in footprint will be dramatic, especially on those places where water, food, and resources are scarce. The impact on infrastructure and environment will be limited, securing a sustainable food supply.

In this chapter is described how a different view on plant growth, combined with integrated technology, offers the opportunity for a scalable indoor farming solution. The creation of a climate, perfectly fit for plants, will secure the high yields that are necessary for both food production and profitability. Only when both are realized can indoor farming deliver on the promise many people feel it has.

8.2 What Makes Plants Grow

Plant production is driven by light, temperature, and evaporation, generally described as the climate. Outdoors these are intimately linked as the sun provides light as well as heat. Conventional greenhouse growers are in a constant battle with the climate, as this is a continuous uncertainty. Good growers handle the climate very well and know how to control its impact on their crop. The ability to grow plants in an indoor farm offers many more control mechanisms than in a greenhouse. The climate in an indoor farm is very stable, with no need to anticipate to the outside situation. Growers can therefore focus more on crop management and maintenance of the technical operation and equipment.

From the plants' points of view, nature can be a horrible place. It can be too hot or too cold; there is too much or too little light; there can be night, too little or too much water, no nutrients, wind, insects, grazing animals, etc. If it were up to plants, they would want a totally different environment: a stable temperature and flow of air and water and enough nutrients and light. They most likely would not prefer sunlight, as 50% of that light is heat that plants need to get rid of through evaporation. This is how leaves keep their temperature constant and is most prevalent in the upper layers of leaves that catch most of the sunlight (Crawford et al. 2012; Medrano et al. 2005). The temperature control mechanism is key to regulate growth and development but makes outdoor plants use a lot of water. As water uptake by the roots is a passive process following nutrient uptake, the balance between uptake and evaporation greatly impacts plant habitat. Low evaporation with high root pressure will result in higher water content. Taller and elongated plants are generally not favorable in agriculture.

In field and greenhouse evaporation of plants is mostly driven by vapor pressure deficit (Turner et al. 1984; Seversike et al. 2013; Yang et al. 2012). This is the difference between the water content in the air, at a certain temperature, and the saturated water vapor pressure. This is strongly depending on temperature and is calculated and visualized in the Mollier diagram. As leaves heat up in sunlight (Tyree and Wilmot 1990), the temperature in the stomata increases, and as a result, the vapor pressure deficit and consequently evaporation increase as well. Diffusion of water to the outside of the leaf, transpiration, will increase and in turn result in cooling of the leaf. So, in sunlight radiation intensity, photosynthesis and evaporation go hand in hand.

8.3 Evaporation of Plants Indoors

The control of evaporation is essentially different in an indoor farm compared to a greenhouse and field. Air quality is therefore critical for optimal growth indoors where evaporation is mostly driven by airflow and vapor pressure deficit (Turner et al. 1984; Seversike et al. 2013; Yang et al. 2012). Mixing air in a closed environment to control air quality will result in differences in temperature and humidity. This will lead to differences in growth and evaporation rates of a growing crop. Consequently, the uniformity within the growing area will decline. Additionally, crop quality may be affected on certain locations in the growing area. Either way, control levels are declining.

From the viewpoint of the plant, the indoor climate design can be radically different from the usual outdoors or greenhouse climate. The understanding of what makes plants grow in an indoor farm is of great importance. The indoor use of light-emitting diodes (LEDs) as light source has a great impact on the plant's responses. The restricted use of only a few wavelengths to target specific receptors and the low levels of infrared result in a totally different light perception by plants. Plants can do with only blue and red light as this is absorbed by chlorophyll, phytochromes, cryptochromes, and other receptors (Kong and Okajima 2016). These are most of the receptors that direct photosynthesis and developmental responses. Growth under blue and red light only has a very broad optimal ratio; more than 15% blue light in general works well (Hogewoning et al. 2010).

Ultimately evaporation and photosynthesis should be controlled independently. For these two things are needed; the light source needs to drive photosynthesis only and not evaporation. Evaporation needs to be controlled by vapor deficit and air speed alone and preferably not by infrared radiation. The first can be achieved by using very efficient LEDs and not placing them close to the crop. Efficient LEDs with a high µmole/Joule value can be placed several meters from the crop. At that distance, virtually no infrared is left at the plant level while sufficient photosynthetically active radiation (PAR) levels are maintained. Crops placed in close vicinity to the LEDs, like in a multilayer system, experience relatively larger amounts of infrared. This will stimulate evaporation and will fundamentally be linked to the

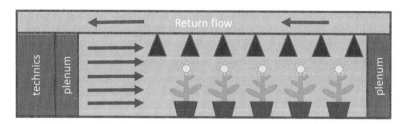

Fig. 8.1 Schematic presentation of an indoor farm with a laminar air flow. Blue arrows indicate the direction of the airflow. Purple triangles represent lighting equipment. The left plenum is the air inlet; the right plenum is the exit. In the technical area is the air treatment system. It also contains the control units and fertigation systems

photosynthesis levels. With increasing efficiencies of LEDs, a decline in infrared light in indoor farms is evident and will continue in the coming years. As a result, leaf heating is declining as well making the control of evaporation increasingly important. In other words, if evaporation drops and the logistic processes of water and nutrient supply are lost, photosynthesis will be inhibited as well.

Following this strategy, a specific climate and associated technology was created to serve best the needs of the plants. The specific construction results in a high refresh rate of approximately $30\ h^{-1}$. These conditions maintain sufficient evaporation without creating wind stress as a slow moving laminar flow of air moves through the unit. This is realized by an air in and outlet consisting of a plenum space where air is moving through multiple holes. As a result, a large volume of air is conditioned every minute, allowing the incoming and outgoing air to be almost identical to the set point, without creating a high air speed. The uniform distribution of air quality allows every leaf in every position to evaporate equally. The vapor pressure deficit in combination with air speed will determine the loss of water from the plants. Higher deficit and higher air speed both increase evaporation.

In such an environment, light, temperature, and evaporation can be controlled independently which allows full control of plant development. A schematic representation of such a system is shown in Fig. 8.1 but can be scaled to almost any size. A similar principle can be applied in a multilayer system. Dimensions will change then as the heat load per square meter floor area will increase.

8.4 Do We Really Need Far-Red Light in Indoor Farming?

Plants sense different wavelengths of light with several receptors. Each of these receptors results in a specific response of the plant (Kong and Okajima 2016). Purified chlorophyll absorbs red and blue light to drive photosynthesis. Phototropins and cryptochromes absorb blue light. They are involved in circadian rhythms, phototropism, inhibition of hypocotyl elongation, stomatal opening, and various other responses. Phytochromes absorb red (R) and far-red (FR) light and are involved in the shade avoidance response, flower induction, germination, and de-etiolation (Castillon et al. 2007; Cerdan and Chory 2003; Demotes-Mainard et al. 2016; Possart et al. 2014). Outdoors plants experience far-red light every evening (Kasperbauer 1987) when a peak in the ratio FR/R occurs a few minutes before darkness. Plants also use far-red light to sense their surroundings (Jaillias and Chory 2010). In shade, the ratio FR/R is much higher than in bright sunlight. In a response to escape from the lack of light, plants elongate or flower early. Auxin production is induced causing (vascular) tissue to elongate. As a result, plants get taller, reaching the light earlier, but are forced to put more energy in stems. This change in dry matter partitioning will have its impact on the harvest index as a significant higher portion of dry matter must be put in stems (Kasperbauer 1987). Leaves also expand more in far-red light. In general, epidermal cells are larger, but the cell number remains fairly constant. As a result, stomata get diluted over the leaf surface (Chitwood et al. 2015). Whether this has an impact on evaporation is not known.

In some crops, early flowering is induced by high FR/R ratios and/or by end of day far-red light (Cerdan and Chory 2003; Kim et al. 2008). In general, there are many routes and mechanisms that lead to flowering (Simpson et al. 1999), and most plants have the genetic capability to use at least some of them. In most areas plants had to adapt to seasons and local environments. Far too often seasons limit growth, flowering, and seed set as these must be properly timed to secure the next generation. Responses to day length variation (e.g., long-day, day-neutral, and short-day plants) are critical for maintenance of the species. Next to regular seasonality, flower induction can be achieved in multiple ways. "Stress-induced flowering" is considered to be an off-season flowering mechanism induced by "adverse" climate conditions (Roitsch, 1999; Takeno 2016). Obviously, these can be mimicked indoors if the triggers are known. Alternatively, "luxury-induced flowering" is something that almost exclusively can be achieved indoors where plants can be forced to follow different developmental paths. In the described design, the outdoor climate is not meant to be copied, but a new, highly controlled and stable climate is created.

A common factor in flower induction is the formation of the FT protein (Corbesier et al. 2007), also known as the florigen. This is generally accepted as the molecule that triggers the formation of flowers. Additionally, it was shown that a sugar compound, trehalose-6-phosphate (Van Dijken et al. 2004; Wahl et al. 2013), is involved and essential in the process. This compound alone, when present in the shoot meristem, can induce flowering independent of all other signals (Wahl et al. 2013). The same compound is also produced only when sugar is present in abundance (Lastdrager et al. 2014; Stitt and Zeeman 2012). Trehalose-6-phosphate acts as a "sugar sensor" to tell the organism energy is not a limiting factor. Since flowering demands a lot of energy, the sensor may be the lock on the door to ensure plants flower only when enough energy is available. It also shows that at least one alternative route to flowering is driven by sugar and sugar-related compounds, which happens to be under control in an indoor farm environment. The possibility of making plants flower at will may not be too far away since the creation of plants with excess sugar is very well possible.

Fig. 8.2 Growing tomato plants in an indoor farm with laminar air flow and LEDs on the ceiling

During fruit set in tomato, far-red light inhibits the synthesis of lycopene (Llorente et al. 2016a; Llorente et al. 2016b). This "self-shading" of fruits ensures lycopene and carotenoids are produced at the right time. Additionally, postharvest treatment of ripe fruits with far-red light reduces lycopene levels (Gupta et al. 2014). Tomato production with only blue and red light may therefore lead to fruits with altered, higher levels of lycopene and carotenoids, though this needs to be proven.

Grown under LED lights, biomass production can be stimulated by red and blue light only, provided the balance between the two is not too far off (Massa et al. 2008; Hogewoning et al. 2010). Growth, from germination till fruit set, therefore should be possible with only the two colors. The absence of far-red light in indoor farms prevents the perception of shadow by plants. Leaves below other leaves will simply receive less light and will not experience shadow. Additionally, flower induction can be achieved in other ways as well, so the need for far-red light is questionable. An example of such culture is shown in Fig. 8.2 where tomato plants are grown under blue and red light only and produce normal fruits.

8.5 The Plant Balance

An indoor farm takes away all the constraints for plants that can be found outdoors. Plants receive what they need in optimal quality and quantity. With their leaves, light, nutrients, water, and CO_2, plants generate energy in the form of carbohydrates, sugar. The sugar is used for basically three things. First in line is maintenance; the energy needed to keep the organs functional and alive. Second in line is growth; any surplus of energy can be used for the formation of new organs (roots, stems, leaves, flowers) and ultimately seeds. This is the delicate balance between source and sinks. Last in line is the secondary metabolism; any energy left over will be used for compounds normally not produced or at very low levels. These are compounds not vital for the plant and often produced in times of stress. Many can have a role in defense against pathogens or adverse conditions. Others are used to attract, like flower colors, fragrances, and tasty compounds. In indoor farms, however, stress is absent, and these compounds can be regarded as luxury compounds.

In field conditions light and temperature are often connected, as 50% of the sunlight is infrared light. More light is mostly resulting in a higher (leaf)temperature with all consequences. Indoors these can be separated as LEDs have much less infrared light. Therefore, the rate of dry matter production through photosynthesis can be uncoupled from the metabolic rate. In practice, this means that dry matter production and content are under control. The ultimate control of the surplus of energy available for the plants allows us to control growth, development, and metabolism. Consequently, essentially different plants can be produced, consistently and scalable. Compared to field conditions and to a lesser extent greenhouses, the energy production in an indoor farm can, in principle, be much higher. This is a key difference between field/greenhouse and indoor farming simply because so much more energy, sugar, can be made (see Fig. 8.3).

In 1989 Gertjan and Lianne Meeuws started Buro Meeuws, a consultancy and research firm in horticulture. They were pioneers in the field of hydroponics,

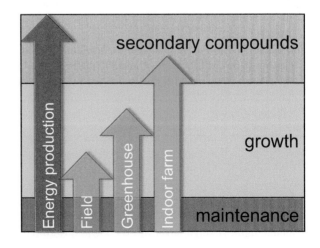

Fig. 8.3 Schematic representation of energy production by plants in the field, greenhouse, or indoor farm. Energy is needed for maintenance, growth, and secondary compounds, in that order

Fig. 8.4 The plant balance model

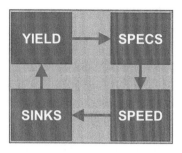

assimilation lights, and robotics. They have consulted with over 200 leading horticultural companies all over the world. In the 1990s, they worked on a big data project, collecting millions of data over 5 years from 500 measuring fields from different crops and their habitats. The goal of the project was to develop and validate algorithms that help explain and predict the growth of plants; this resulted in the Plant Balance Model.

The Plant Balance Model allows us to calculate biomass production, based on plant characteristics and the environment. Additionally, it enables the prediction of future productivity and/or changes resulting from adaptions in the climate. The model, as shown below, starts at the top left (Fig. 8.4):

Yield is the maximum amount of biomass that can be produced of a certain product, depending on the actual technical equipment of the indoor farm. This is often expressed as dry matter per square meter per year. It is based on the level of photosynthesis and the harvest index (the percentage of produced biomass that we use).

Specs are the actual or desired specifications of the produce grown. It consists of the desired weight of the product and its dry matter content.

Speed is related to temperature and the time it takes to complete a cycle. For instance, how long it takes to produce three leaves and a truss for tomato. This is driven by Growing Degree Units (GDUs) indicating that a raise in temperature will lead to an acceleration of growth. For most crops GDUs are known for each developmental stage.

Sinks is related to the number of sinks needed to produce the desired number and quality of products or fruits. Linked to the previous three items, this can be used to calculate the plant or stem density for instance. It enables the calculation and prediction of yields when the climate parameters like temperature, light intensity, etc. are known.

In theory, this means dry matter production per square meter of growing area can be calculated. Based on a set of assumptions and data gathered over the years (G. Meeuws, L. Meeuws, unpublished data), the maximum amount of dry matter production is 30 $kg.m^{-2}.year^{-1}$. One of the assumptions is that 300 $\mu mol.m^{-2}.s^{-1}$ of only blue and red light is sufficient for this. In practice, however, this is hardly achieved for a number of reasons (Table 8.1). First, the leaves must absorb the available light. A single layer of leaves will absorb 65% of the light, the next layer

8 Total Indoor Farming Concepts for Large-Scale Production

Table 8.1 breakdown of dry matter production per year and reductions due to daylength, harvest index, and planting events

	Unit	DM (kg.m^{-2}.year^{-1})
24/24 hrs. of light	300 µmol.m^{-2}.s^{-1}	30
Leaf area index = 3	100% light interception	30
Daylength/photoperiod	20 hours	25
Harvest index	70%	18
Planting events, 1x per year	42/52 weeks	14

This is an example of a calculation. For specific crops data will be different. DM is dry matter production

again 65% of the transmitted 35%. Three layers of leaves over a certain area equals a leaf area index (LAI) of 3 and will absorb most of the light (Atwell et al. 1999). As plants grow their LAI will increase, as a result the average LAI may be different. Second the daylength or photoperiod may not be 24 hours. In the example, a daylength of 20 hours is used which is commonly used for most crops. Third is the harvest index. This is the percentage of the product biomass of the total biomass. For example, in tomato this may be 70%, but for lettuce, it can be close to 100%. Finally planting events may result in periods where there is no harvest, in the example 10 weeks. The actual dry matter production of the produce comes down to 15 kg*m^{-2}*year^{-1} in this example. To convert dry matter to food by the addition of water, the dry matter percentage of the crop is a key variable. A high dry matter percentage will lead to less fresh weight produce. A low dry matter percentage will increase yield as simply more water is added to the same amount of dry matter. When calculating and predicting yields starts with dry matter as described, the performance of the indoor farm can be calculated. In the end, the added water determines dry matter percentage and fresh weight produced.

8.6 Large-Scale Production Facility

The first indoor farm dedicated to grow vine crops like tomato, pepper, and cucumber containing the laminar air flow system was realized in Cincinnati, Ohio (OH), United States (USA). It consists of a single-layer growing area of 360 m^2 equipped with the nutrient film technique (NFT) gutters as growing system (see Fig. 8.5). A second similar-sized multilayer area for herbs, lettuce, and other small crops is part of the same building. The two growing areas are separated by a technical aisle containing all air treatment equipment, fertigation units, heat pumps, and control units. The farm is owned and operated by 80 Acres Urban Agriculture limited liability company (LLC).

Fig. 8.5 The 80 Acres Urban Agriculture LLC indoor farm in Cincinnati, OH, USA

8.7 Future Developments

The creation of an indoor farm as described will enable the growth of any crop. The Plant Balance Model will calculate and predict yields. Obviously yields must be adequate to justify any investment. All these are needed to make indoor farming viable on any but certainly a larger scale. A contribution to the world food security and production needs to be commercially viable but also scalable.

As developments in LED technology, climate control, and plant science continue, they come together more and more. At the interface, new insights and clever solutions will increase yields even more. Additionally, insight in nutritional values in relation to health and medicine can now become a common platform. The unprecedented control of plant quality offers new opportunities to extend above simply producing food. The combination of technology, genetics, and specific climates can give the term superfood a new meaning.

References

Atwell B, Kriedeman P, Turnbull C (1999) Plants in action. Macmillan Education Australia Pty Ltd, Melbourne

Castillon A, Shen H, Huq E (2007) Phytochrome interacting factors: central players in phytochrome-mediated light signaling networks. Trends Plant Sci 12:515–523

Cerdan P, Chory J (2003) Regulation of flowering time by light quality. Nature 423:881–885

Chitwood D, Kumar R, Ranjan A et al (2015) Light-induced indeterminacy alters shade-avoiding tomato leaf morphology. Pl Physiol 169:2030–2047

Corbesier L, Vincent C, Jang S et al (2007) FT protein movement contributes to long-distance signaling in floral induction of arabidopsis. Science 316:1030–1033

Crawford A, McLachlan D, Hetherington A et al (2012) High temperature exposure increases plant cooling capacity. Curr Biol 22(10):R396

Demotes-Mainard S, Peron T, Corot A et al (2016) Plant responses to red and far-red lights, applications in horticulture. Environ Exp Bot 121:4–21

Gupta S, Sharma S, Santisree P et al (2014) Complex and shifting interactions of phytochromes regulate fruit development in tomato. Pl Cell & Environ 37:1688–1702

Hogewoning S, Trouwborst G, Maljaars H et al (2010) Blue light dose–responses of leaf photosynthesis, morphology, and chemical composition of *Cucumis sativus* grown under different combinations of red and blue light. J Exp Bot 61:3107–3117

Jaillias Y, Chory J (2010) Unraveling the paradoxes of plant hormone signaling integration. Nat Struct Mol Biol 17:642–645

Kasperbauer M (1987) Far-red light reflection from green leaves and effects on phytochrome-mediated assimilate partitioning under field conditions. Pl Phys 85:350–354

Kim S, Yu X, Michaels S (2008) Regulation of *CONSTANS* and *FLOWERING LOCUS T* expression in response to changing light quality. Pl Physiol 148:269–279

Kong S-G, Okajima K (2016) Diverse photoreceptors and light responses in plants. J Plant Res 129:111–114

Lastdrager J, Hanson J, Smeekens S (2014) Sugar signals and the control of plant growth and development. J Exp Bot 65(3):799–807

Llorente B, D'Andrea L, Rodriguez-Concepcion M (2016a) Evolutionary recycling of light signaling components in fleshy fruits: New insights on the role of pigments to monitor ripening. Front Pl Sc 7, 263

Llorente B, D'Andrea L, Ruiz-Sola M et al (2016b) b. Tomato fruit carotenoid biosynthesis is adjusted to actual ripening progression by a light-dependent mechanism. Plant J 85:107–119

Massa G, Kim H, Wheeler R et al (2008) Plant productivity in response to LED lighting. Hortscience 43(7):1951–1956

Medrano E, Lorenzo P, Sanchez-Guerrero M et al (2005) Evaluation and modelling of greenhouse cucumber-crop transpiration under high and low radiation conditions. Sci Hortic 105:163–175

Possart A, Fleck C, Hiltbrunner A (2014) Shedding (far-red) light on phytochrome mechanisms and responses in land plants. Plant Sci 217–218:36–46

Roitsch (1999) Source-sink regulation by sugar and stress. Curr Op Pl Biol 2:198–206

Seversike T, Sermons S, Sinclair T et al (2013) Temperature interactions with transpiration response to vapor pressure deficit among cultivated and wild soybean genotypes. Physiol Plant 148:62–73

Simpson G, Gendall A, Dean C (1999) When to switch to flowering. Ann Rev Dev Biol 99:519–550

Stitt M, Zeeman S (2012) Starch turnover: pathways, regulation and role in growth. Curr Op Pl Biol 15:282–292

Takeno (2016) Stress-induced flowering: the third category of flowering response. J Exp Bot 67(17):4925–4934

Turner N, Schulze E, Gollan T (1984) The responses of stomata and leaf gas exchange to vapour pressure deficits and soil water content - I. Species comparisons at high soil water contents. Oecologia 63:338–342

Tyree M, Wilmot T (1990) Errors in the calculation of evaporation and leaf conductance in steady-state porometry; the importance of accurate measurement of leaf temperature. Can J For Res 20:1031–1035

Van Dijken A, Schliemann H, Smeekens S (2004) Arabidopsis trehalose-6-phosphate synthase 1 is essential for normal vegetative growth and transition to flowering. Plant Phys 135:969–977

Wahl V, Ponnu J, Schlereth A et al (2013) Regulation of flowering by trehalose-6-phosphate signaling in *Arabidopsis thaliana*. Science 339:704–707

Yang Z, Sinclair T, Zhu M et al (2012) Temperature effect on transpiration response of maize plants to vapour pressure deficit. Environm Exp Bot 78:157–162

Chapter 9
SAIBAIX: Production Process Management System

Shunsuke Sakaguchi

Abstract In order to increase commercial-use PFAL (plant factory with artificial lighting) profitability, it is essential to improve area productivity and RUE (resource-use efficiency). Described here are functions demanded of production process management systems aiming at the achievement of the same, as well as examples of adoption.

Keywords Production process management system · Plant growth · Index value · Productivity

9.1 Introduction

The major merit of a PFAL is the ability to increase the growth rate of plants by creating the ideal environment for plant growth. In doing so, a PFAL can achieve a productivity in terms of area (land) more than 100 times that of open ground (fields). Its second merit is the capacity for stable production of high-quality plants irrespective of the impact of weather conditions.

In order to realize a highly productive factory, as shown in Fig. 9.1, not only are actuators to appropriately control cultivation environment required but also cultivation equipment equipped with sensors to evaluate plant growth and CP (cost performance). However, in reality, in the great majority of PFALs operating in the red, there is a lack of the cultivation environment control performance required to increase productivity to levels that would put them in the black. This is the main reason why only around 25% of PFALs operate in the black in Japan (Kozai et al. 2015). There are many variables that have an impact on productivity, including cultivation environment conditions and damage to plants during work. Furthermore, these variables are themselves mutually interconnected. Accordingly, it is too complicated to understand all of those relationships with human intuition alone,

---

S. Sakaguchi (✉)
PlantX Corp, Kashiwa, Chiba, Japan
e-mail: sakaguchi@plantx.co.jp

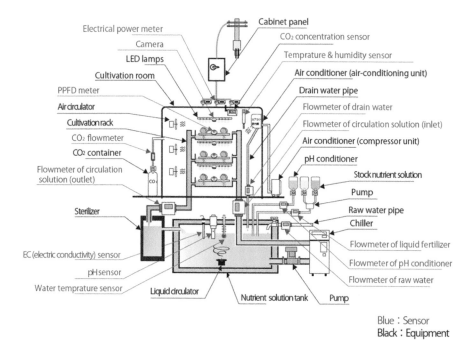

Fig. 9.1 Example of cultivation equipment including actuators and sensors. *PPFD* photosynthetic photon flux density

and it is for the abovementioned reasons that, in order to maximize the merits of PFALs, a production management system that can maximize PFAL productivity by quantitatively managing plant growth is essential.

9.2 System Description

A PFAL production management system requires plant growth, production process, and sales management functions. This is because PFAL profitability is greatly impacted by a PFAL's production output, product sales prices, and sales yield compared to its cultivation area, electric power consumption, and labor costs.

9.2.1 Management of Plant Growth

The first step in production management is a quantitative assessment of what is taking place inside the factory. The following are examples of what should be quantitatively assessed in a PFAL:

1. Cultivation environment state variables (air temperature, VPD (water vapor pressure deficit), CO_2 concentration, leaf-surface PPFD (photosynthetic photon flux density), nutrient solution ion concentration, etc.)
2. Plant growth rate and related rate variables (fresh weight increase rate, leaf surface-area increase rate, net photosynthesis rate, nutrient uptake rate, etc. Here, rate variables are variables including time units such as kg/sec and me/kg/sec).
3. Plant quality (size, shape, color, texture, presence/absence of physiological disorders, etc.).
4. Resource supply rate and RUE.

In particular, the creation of value in PFALs depends on the growth of the plants themselves, so circumstances in which plant growth rates are not managed cannot be described as productivity management. Furthermore, RUE is greatly impacted by plant growth rates, so there is a need for multifaceted measurement and evaluation of plant growth.

9.2.2 Management of Production Process

In a PFAL, what is desired is a shortening of the production lead time through an increased plant growth rate, along with a simultaneous reduction in the hands-on human workload that includes transplanting and harvesting. Not only does unnecessary automation of work alone not enable a shortening of the production process, there conversely is also the risk it may lower CP. Accordingly, a function that shortens the production process both effectively and efficiently is required.

9.2.3 Sales Management

Generally, in order to increase the profitability of a factory, even where productivity is stable, it is essential to link factory production control and inventory control in order to adjust production output in response to fluctuations in demand. Production factories in the manufacturing industry can adjust production by stopping production lines temporarily in response to fluctuations in demand.

However, with a PFAL, the production process cannot be stopped temporarily until seeds that have been sown are harvested as plants. Accordingly, mistakes in production adjustment mean product shortages or disposal, which in one way or another cause an operating loss. For that reason, in order to minimize harvest shortages or disposal, a function for the adjustment of factory growth management and sales management is required.

By applying these advanced techniques to PFAL production management, strict production management to rival that of the production factories in the manufacturing industry will be possible. Together with Toyoki Kozai (Professor Emeritus, Chiba University), PLANTX Corp. has developed SAIBAIX, which is a dedicated PFAL production management system and is providing the same for commercial-use PFALs. In the following sections, the above functions using SAIBAIX will be explained and interspersing examples analyzed.

9.3 Measurement of State and Rate Variables

In order to increase PFAL productivity, it is important to manage not only the setting values for the various main cultivation environment factors but also the plant growth rate. No matter how precisely setting values for state variables such as temperature, VPD, CO_2 concentration, and nutrient solution ion concentration are maintained in a cultivation room, if those setting values do not create conditions to increase the growth rate for plants, then productivity will not improve. The plant growth rate can be estimated online from rate variables such as the net photosynthetic rate, the nutrient solution ion uptake rate, and the water uptake rate (Kozai 2013; Kozai et al. 2015). Furthermore, using these rate variables enables the almost-real-time estimation of the RUE and CP for resources including electricity, CO_2, and water, which will be discussed below. Next, an example of estimating net photosynthetic rate and uptake rate will be shown.

Plants grow through photosynthesis, so the net photosynthetic rate is a rate variable that directly represents plant growth. A plant's net photosynthetic rate can be estimated from the CO_2 concentration and CO_2 supply volume (Kozai 2013). Figure 9.2 shows an example of the net photosynthetic rate in a cultivation room. The diagram shows the net photosynthetic rate for all plants being cultivated in the cultivation room. Monitoring the net photosynthetic rate enables noninvasive estimation of the growth rate. Furthermore, by constantly comparing the steady-state net photosynthetic rate and the current values, growth abnormalities can be detected online.

It is difficult to notice unforeseen changes in the cultivation environment such as a toxic-substance outbreak in the nutrient solution or a drop in PPFD due to equipment failure by just monitoring state variables alone. Plants respond quickly to such changes in environment, so additional monitoring of variables that indicate plant growth will enable the detection of growth abnormalities before any declines appear in harvest weights.

Functional plants such as mineral-rich vegetables or medicinal plants can be cultivated in PFALs. In order to cultivate these plants, it is necessary to manage not only plant growth but also the functional components of the plants. To that end, it is important to manage the nutrient solution ion uptake rate. Figure 9.3 shows the ion uptake rate when cultivating lettuce. The diagram is the ion uptake rate in a nutrient solution system controlled to maintain a uniform EC (electrical conductance). As described in Chapter 24, even if EC values are uniform, the balance of the ion concentration in the nutrient solution changes according to the uptake of fertilizer

9 SAIBAIX: Production Process Management System

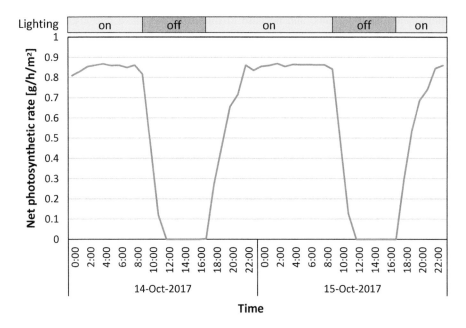

Fig. 9.2 Average net photosynthetic rate per unit area when cultivating lettuce. Bands at the top indicate ON/OFF for cultivation rack lighting

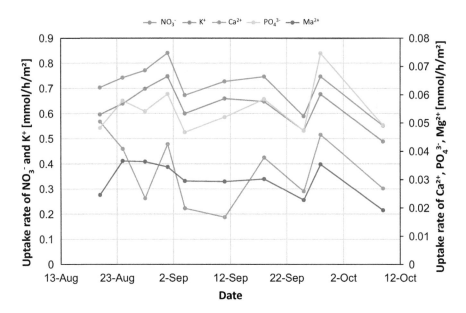

Fig. 9.3 Ion uptake rate per unit area when cultivating lettuce. Electrical conductance has been kept constant in this example. Fluctuations in uptake rate are considered to be caused by changes in solution ion concentration and fluctuations in plant bodyweight due to harvesting

components by the plants, and because of this, the ion uptake rate also changes. Clarifying this relationship between the ion uptake rate and the rate of change of components inside the plant bodies and appropriately adjusting the ion concentration allows plants to have the desired components.

Growth management of plants using the rate variables described above will not only simply enable early detection of abnormalities, it will also deliver operational advantages such as reductions in product development times, so that the importance will probably only increase in the future.

9.4 Online Estimation of Resource-Use Efficiency and Cost Performance

The first step in improving the profitability of a factory involves accurately ascertaining problems through a quantitative evaluation of productivity and CP. In general, production factories are always measuring and managing items related to productivity, such as component costs, manufacturing costs, yields, and manufacturing lead times. In the same way, in PFALs it is necessary to constantly manage the cost for resources, such as seeds, CO_2, water, and fertilizer, and electric energy costs, harvest yields, and plant growth rates. PFALs are different from open ground (fields), using only electric energy in a closed space, so the energy input/output relationship can be measured and evaluated almost in real time. RUE and CP of PFAL are defined in detail for each resource, such as lighting, CO_2, or fertilizer (Kozai 2013). SAIBAIX can manage such RUE and CP.

Figure 9.4 shows the monthly electricity consumption and electricity CP. Electric power costs are one of the largest cost factors for PFALs, and their reduction has a great effect. Of the electric energy, energy for lighting composes the largest percentage (Fig. 9.5). Accordingly, the amount of electricity consumption per month in Fig. 9.4 is mainly proportional to the factory's operating ratio, in other words, the percentage of lighting equipment turned on. In the figure, despite January 2015 having almost the same amount of electricity consumption as the previous month, the weight of the harvest has increased. This can be considered as being due to a temporary improvement in cultivation environment factors other than lighting, such as air temperature or nutrient solution. In this way, productivity can sometimes be increased without raising costs by improving the cultivation environment. Furthermore, where there exist plural conditions that deliver the same harvest weights, the condition(s) that increase CP can be selected. Additionally, methods to reduce electric power costs are introduced in Chapter 21 in Kozai et al. (2015).

Figure 9.6 shows the daily amount of drainage water from heat pumps. One of the greatest advantages of PFALs is their high WUE (water use efficiency). So long as there are no large divergences from target ion concentrations or problems such as proliferations of disease-causing bacteria, then the nutrient solution can be continued to be used. Moisture transpired by plants is collected as drainage water from the

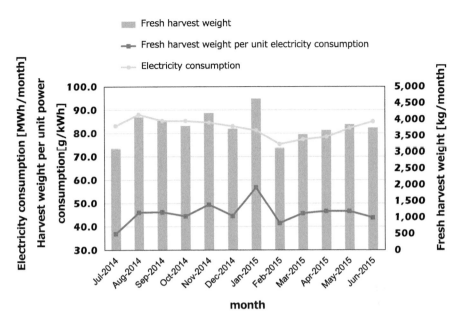

Fig. 9.4 Monthly production, electricity consumption, and electricity use efficiency (Courtesy of JA Touzai Shirakawa "Miryoku Manten Yasai no Ie")

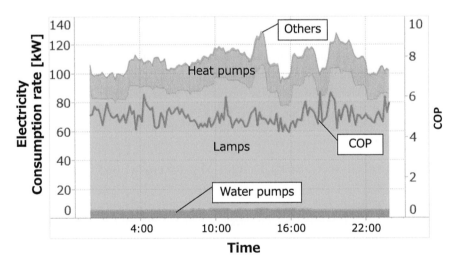

Fig. 9.5 Example of shift in electric power consumption over 1 day. The COP (coefficient of performance for air-cooling heat pumps) = (lighting power + pump power + other electric power)/air-cooling heat pump power (Measured at JA Touzai Shirakawa "Miryoku Manten Yasai no Ie")

air-conditioners in the cultivation room, and when recycled, of the raw water supplied to the cultivation system, hardly any is lost, with the exception of the water included in the shipped plants. As a result, compared with a maximum of 2%

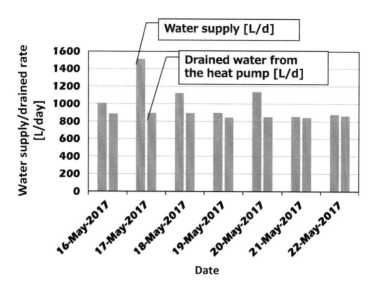

Fig. 9.6 Relationship between amount of water supplied for cultivation and amount of heat-pump drain water collected. The amount supplied changes depending on the presence/absence of work inside the cultivation chamber such as harvesting or cultivation solution renewal. The drain water collection rate for days with no work inside the cultivation chamber (May 21 and 22) was more than 98%. Drain water is not reused in this example, but when it is, the water supplied is roughly the same as the amount of water contained in the harvested plants. (Raise Co. Ltd., Chiba University Factory)

for the WUE for a greenhouse, the WUE for a PFAL can be kept at a level of 80% and sometimes more than 90% (Kozai 2013).

In this way, by monitoring and improving RUE and CP, productivity can quickly be improved and continuously maintained at high levels.

9.5 Noninvasive (Camera Image) Measurement of Plants

The structure, color, texture, presence/absence of physiological disorders (e.g., yellowed leaves, tipburn), etc., for produced plants are important information when evaluating product values. Even today in the agricultural sphere, much of this visual information is subjectively evaluated by humans. However, not only is there individual difference in evaluations by humans, in terms of cost and feasibility; it is also not realistic to carry out 100 % inspections of large-scale PFALs. For this reason, the development of a quantitative evaluation method using image data is essential. The automation of evaluation will be the key to maintaining high-quality and stable production, which is the greatest strength of PFALs.

The below are examples of main evaluation items using images:

1. Seed submersion state when planting
2. Germination rate

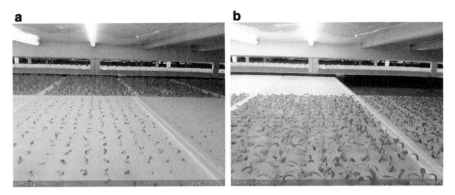

Fig. 9.7 Growing lettuce seedlings. (**a**) 4th day after seed sowing. (**b**) 10th day after seed sowing. (Raise Co. Ltd., Chiba University Factory)

Fig. 9.8 Images of lettuce seedlings before and after transplanting. (**a**) Before transplanting (23rd day after seed sowing. 144 heads/m^2). (**b**) After transplanting (24th day after seed sowing. 33 heads/m^2). (Raise Co. Ltd., Chiba University Factory)

3. Seedling growth state (leaf unfolding/separation, leaf area, number of leaves, etc.) (Fig. 9.7)
4. State before/after transplanting (planting density, seedling damage, seedling collapse, etc.) (Fig. 9.8)
5. Growth rate
6. Condition at harvest (size, color, texture, presence/absence of physiological disorders, etc.)
7. 3D structure of plant (height, thickness of leaves, number of leaves, leaf area)

In particular, as shown in Chapters 2 and 26, plant 3D structure information will only increase in importance in the future because of phenotyping. The application of image data analysis in PFALs has only just begun but will probably see rapid implementation moving forward together with progress in AI (artificial intelligence) technology and phenotyping technology.

9.6 Visualization for Production Process Management

Generally, to efficiently perform production in a "factory," it is necessary to (1) increase production quantity, (2) improve quality, and (3) shorten process times (Nakao et al. 2002). As described above, PFAL productivity is mainly determined by the growth of plants themselves. However, at the same time, the PFAL production process includes many processes involving human participation, such as transplanting, harvesting, packaging, and shipping handling. Not only do these ancillary processes represent roughly 1/4 of production costs, there is much work handling the plants themselves, so this directly impacts plant growth and product prices. Accordingly, if these processes are not carried out appropriately and efficiently, it is difficult to improve profitability.

In order to reduce individual difference in quality and speed in work carried out by humans, it is necessary to establish clear procedures and evaluation criteria for each process and enable anybody to carry out work that satisfies a fixed standard. An example of a method to realize this is DPD (decision-based process design) (Nakao et al. 2002). DPD is a technique that uses the so-called tacit-knowledge decision-making processes of workers to improve processes. Specifically, first, the decision-making processes tacitly undertaken by a highly skilled worker in each process are analyzed in detail. Then, the know-how which has become clear in that is further optimized before being standardized. Then, by incorporating that standardized know-how into process structure instead of individual knowledge or skill, a fixed standard of work quality is guaranteed. This method has been applied in the production factories of more than 100 companies, including mold manufacturing factories in the manufacturing industry. Depending on the situation, results have seen lead time reduced by 95% or more.

After using DPD process analysis methods to analyze the processes in commercial large-scale PFALs, they were segmented into roughly 200 steps, including processes relating to cultivation covering seed sowing to harvest, processes relating to packaging and shipping, and processes for materials replenishment and equipment maintenance (Fig. 9.9). As a result of work-process optimization based on the DPD analysis, the workload per harvest weight had been reduced by more than half.

The near future will probably see the realization of highly productive PFALs due to the overall systematization of production processes, including worker work designation functions, individual plant traceability functions, and linking with various kinds of automated machines.

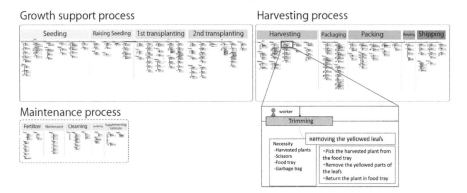

Fig. 9.9 Process chart for large-scale commercial-use PFAL, from seed sowing to shipping. The enlarged diagram at the top-right shows part of the harvest process. The entire process is comprised of roughly 200 processes, from seedlings to shipping. (Created based on the production process of a PFAL of Raise Co. Ltd. at Chiba University) (Jointly prepared with Stone Soup Inc.)

9.7 Modeling, Multivariate Analysis, and Big Data Mining

As described above, in PFALs it is now possible to obtain many datasets, including cultivation environment information, plant growth information, the operational status of workers and automated machines, and resource and energy use efficiency. Until now, data relating to the causal relationship between environment and growth in agricultural cultivation was mostly only obtainable few times a year at best. However, trials are continually underway in PFALs, so new data will accumulate every day. In order to extract and effectively use desired information from such vast data, sophisticated data analysis techniques will be necessary. Fortunately, in recent years there has been remarkable practical progress in data analysis techniques. With the spread of things like open-source machine learning software frameworks, there continue to be large-scale reductions in the costs involved in their introduction.

Figure 9.10 shows the results of an estimate of harvest weight from air temperature and water temperature using multivariate analysis. As shown in the diagram, it can be understood that rough trends can be estimated even from estimates that use simple methods and only a few variables. A factory operating in a steady state will be able to take fast countermeasures against any plant growth rate abnormalities by constantly monitoring differences between steady-state and current values in variables that indicate plant growth, such as the net photosynthetic rate described earlier. Similarly, with respect to plant components, leaf texture, etc., by analyzing the relationship between variables relating to cultivation environment and plant growth using nonlinear mathematical models and machine learning methods, the production of plants with desired characteristics will probably be possible.

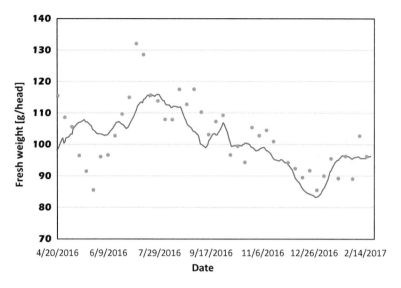

Fig. 9.10 Estimation result of harvest fresh weight using air temperature and water temperature. The red line shows estimated values, while the dots show actual measured values. Environmental factors are considered to have a large impact where large estimation errors are present. (Raise Co. Ltd., Chiba University Factory)

9.8 Commercial Application Examples

SAIBAX has been introduced in Japan's major large-scale commercial-use PFAL and an educational small-scale PFAL (Japan's first example of introduction for educational use). This paragraph will introduce examples of SAIBAIX use as well as its effects.

In 2014, JA Touzai Shirakawa became Japan's first agricultural cooperative to commence operation of a PFAL. SAIBAIX was temporarily installed in the PFAL, and a trial optimizing air-conditioning operation to homogenize cultivation environment conditions was carried out. In the trial, the operating devices of the 10-plus air-conditioning units installed inside the factory were adjusted to take into account the airflow inside the cultivation room. As a result, not only was there a large-scale reduction in the electricity consumption for air-conditioning, but the average air temperature inside the cultivation room also moved closer to the set value while air temperature distribution simultaneously improved. (For details, refer to Kozai et al. (2015), Chapter 22.)

Established in 2015, Raise (Raise Co. Ltd.) commenced operation of a large-scale PFAL inside Chiba University (Fig. 9.11). This PFAL was operated in 2012 by MIRAI as the world's first commercial-use large-scale PFAL and enables harvesting of 3000 lettuces per day. Raise took over the facility and equipment from MIRAI and recommenced operation. Before commencing operation, SAIBAIX was installed, and there was a continual effort to improve productivity. As a result, only 4 months

Fig. 9.11 Large-scale commercial-use PFAL and SAIBAX dashboard (LCD display on left side of photo). Cultivation trials using LED lighting equipment from various manufacturers are being carried out in the cultivation chamber. (Chiba University Kashiwa no Ha Campus, Japan)

after the recommencement of operations, the average harvest yield was doubled for the same number of cultivation days compared to before introduction of the system (Fig. 9.12). These results show the effectiveness of SAIBAIX. Specifically, they show that, even in the same production facility, there is the potential to increase productivity by improving cultivation environment conditions.

SAIBAIX for educational purposes was introduced when the educational small-scale PFAL was established at a Junior High School in Chiba prefecture, Japan (Figs. 9.13 and 9.14). It displays a variety of measured values, including the PFAL's internal/external air temperature, humidity, VPD, PPFD, CO_2 concentration, water temperature, EC, pH, electricity consumption, and plant growth videos, and enables learning of what those values mean. Not only does SAIBAIX allow students to understand PFAL mechanisms, it provides the opportunity for them to become interested in a wild range of science and technology fields, including plant physiology, chemistry, and electrical engineering, and in doing so contributes to the nurturing of the human resources who will be responsible for the next generation.

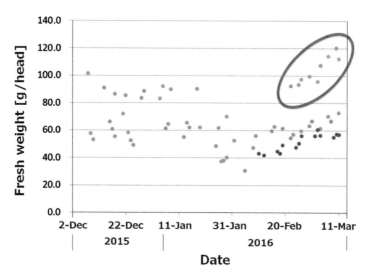

Fig. 9.12 Example of productivity improvement in a large-scale commercial-use PFAL. Different color dots show different cultivation conditions. Product is Frill Lettuce. As shown by the dots in the red circle, harvest weight has more than doubled due to cultivation conditions. (Raise Co. Ltd., Chiba University Factory)

Fig. 9.13 A small-scale educational PFAL and SAIBAX dashboard (Top-right LCD display) in a junior high school. (Chiba prefecture, Japan. Photo taken by the author)

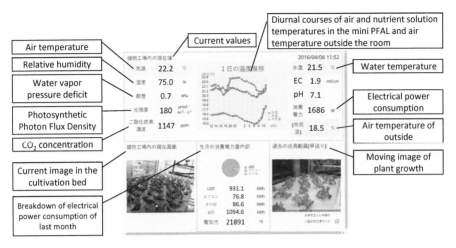

Fig. 9.14 Example of SAIBAX display at a junior high school

9.9 Concluding Remarks

Production management systems that can manage the growth of plants are essential in order to increase the profitability of commercial-use PFALs. A diverse range of advanced functions is sought in production management systems, including functions to monitor cultivation environment conditions, but also functions to monitor productivity and RUE, functions to optimize production processes, and ordering management based on harvest predictions. This chapter described a production management system equipped with such features and also examples of its application. In the future, together with advances in things like genetic engineering, phenomics research, and AI technology, production management systems will evolve even further and will probably become indispensable in the cultivation of plants with desired shapes and functional components.

References

Kozai T (2013) Resource use efficiency of closed plant production system with artificial light: concept, estimation and application to plant factory. Proc Jpn Acad, Ser B 89:447–461

Kozai T, Niu G, Takagaki M (eds) (2015) Plant factory: an indoor vertical farming system for efficient quality food production. Academic, Amsterdam, p 405

Nakao M, Yamada M, Kuwabara M, Otubo M, Hatamura Y (2002) Decision-based process design for shortening the lead time for mold design and production. CIRP Ann Manufact Technol 51 (1):127–130

Chapter 10
Air Distribution and Its Uniformity

Ying Zhang and Murat Kacira

Abstract Air distribution system in plant factories with artificial lighting (PFALs) is responsible for the air exchange and replacement to create desired growing conditions for plants. Combined effects of multitiers, heat from supplemental lighting, improper air conditioning, and air distribution system design can lead to environmental nonuniformity in PFALs. The principle for the design of air distribution system is to understand the physics of wind and how crops response to wind. This chapter describes how wind affects the photosynthesis and transpiration processes of crops by briefly explaining the theory of leaf boundary layer and boundary layer resistance. Then an example application of improving air movement to prevent plant physiological disorder (e.g., tipburn in lettuces) is introduced. For the design of air distribution system, the overall control with mixing ventilation systems and the localized control with cooling fans and perforated air tubes are described. Finally, several indices for assessment of ventilation performance are defined such as air exchange effectiveness, local mean age of air, efficiency of heat removal, and coefficient of variation.

Keywords Boundary layer · Boundary layer resistance · Air movement · Air distribution system · Cooling fan · Perforated air tube · Air exchange effectiveness · Local mean age of air · Efficiency of heat removal · Coefficient of variation · Simulation · Computational fluid dynamics

10.1 Introduction

The operational costs and resource-use efficiency in multitier-based plant factory systems can be improved by appropriate production-system design modifications for key technologies and control strategies while considering the crop-specific minimum

---

Y. Zhang · M. Kacira (✉)
Agricultural and Biosystems Engineering, The University of Arizona, Tucson, AZ, USA
e-mail: yingzhang@email.arizona.edu; mkacira@email.arizona.edu

environmental requirements such as light, air temperature, air velocity and flow pattern, CO_2, and uniformity of these variables.

Majority of the plant factories with artificial lighting have been constructed inside a pre-existing warehouse building with multilayer production shelves. It is observed that the main focus with the existing system designs has been using the internal building space to achieve higher biomass production without considering detailed engineering design fundamentals for air conditioning systems, uniformity of the environment, efficient delivery of CO_2, shelf spacing, smart lighting system and shelf designs, and interaction of crop and surrounding climate in terms of heat and mass transfer processes. Therefore, lack of detailed engineering analysis in the system design can lead to inefficient use of resources (i.e., energy, CO_2, water), nonuniform environment, higher system costs, and limit production quality, yield, and profitability.

Due to the limited and uneven air circulation inside each shelf and large production domain, the environment in multitiered plant factories with artificial lighting may not be uniform. This can limit the production quality, yield, and speed. Thus, the resource consumption is increased as the growing period is extended. It is necessary to properly design crop production, air conditioning, and air distribution systems in plant factories under sole source lighting for providing desired airflow patterns, boundary layer thickness, sufficient air current speed for optimal heat and gas exchanges, improving uniformity of the environment, and efficient delivery of CO_2.

The importance of proper air distribution in conditioned space is often underestimated. Air movement around crops in PFALs has a major impact on crop growth and physiology. Determining an optimized multitier system design requires exhaustive onsite studies and experimentation, labor, and time to analyze various configurations, design variables, and operational strategies to address the challenges indicated above. Thus, using computer modeling and simulation-based approach is a more advantageous and time-, cost-, and labor-efficient way once a validated model is developed to determine key design features and variables in detail and to recommend design optimization leading to improved resource-use efficiency. In this chapter, the mechanics of interaction between crop and surrounding climate and other important factors are discussed, some of the air distribution system alternatives are considered and evaluated, and localized climate control concept is proposed, with results and illustrations based on computer modeling-based approach and analysis on airflow uniformity in plant factory system.

10.1.1 Leaf Boundary Layer and Leaf Boundary Layer Resistance

Leaf boundary layer (LBL) is a thin layer of still air adhering to the leaf surface generated by air friction. Airflow within the boundary layer can be laminar,

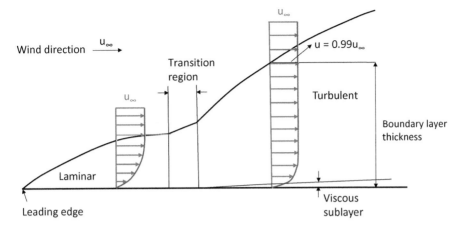

Fig. 10.1 Schematic illustration of airflowing over a smooth flat plane, indicating the transition from laminar to turbulent flow

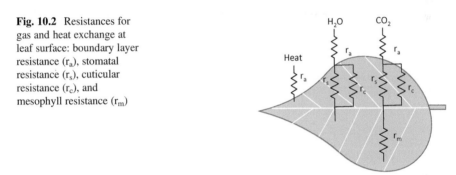

Fig. 10.2 Resistances for gas and heat exchange at leaf surface: boundary layer resistance (r_a), stomatal resistance (r_s), cuticular resistance (r_c), and mesophyll resistance (r_m)

turbulent, or transitional, which depends on the turbulence of the impinging airstream and characteristics of the leaf (Van Gardingen and Grace 1991). Figure 10.1 illustrates airflowing over the top of a flat leaf, showing the transition from laminar to turbulent flow. The arrows indicate the relative speed and the direction of airflow. Within laminar boundary layer or laminar sublayer, air movement is parallel to the leaf surface. Heat and gas transfer occurs by molecular diffusion in the laminar boundary layer. Turbulent flow is characterized by unsteady eddying motions in all sizes. The eddies assist heat and gas transfer in turbulent boundary layers. In nature, wind profile is not laminar. The flapping motion of leaves with the wind, the roughness of leaf surfaces, the veins, and serrations all affect the development of airflow and lead to a turbulent boundary layer (Van Gardingen and Grace 1991).

The exchange of heat and mass between crop and the atmosphere must overcome various resistances at leaves (Fig. 10.2). Photosynthesis can be considered a process of CO_2 diffusion from the air to chloroplast. The resistances to CO_2 diffusion are from mesophyll, cuticular, stomata, and boundary layer (Gaastra 1959). Transpiration is the transport of water and minerals from roots to leaves. Water vapor must diffuse through LBL and then be removed by moving air. The resistances in the diffusion pathway of water vapor are cuticular, stomata, and LBL, whereas only the leaf boundary layer resistance (LBLR) plays a role for heat transfer. Air movement plays a vital role affecting the heat and mass transfer between crop and its surrounding by changing the boundary layer thickness and therefore positively or negatively affects crop growth.

LBLR is the resistance in the pathways for energy and gas fluxes to and from leaf surface, which is directly related to the leaf boundary layer thickness (LBLT). The LBLT is the normal distance to leaf immediately from leaf surface to a point where the flow velocity closes to the free stream value. It is influenced by the characteristics of leaves (such as leaf size, shape, and roughness) and air movement. The average LBLT next to a flat leaf can be defined as the approximate equation below (Nobel 2009):

$$\delta^{bl} = 4.0\sqrt{\frac{l}{v}}$$

where l is the mean length of leaf in the downwind direction in m, v is the ambient wind speed in m s^{-1}, and δ^{bl} is the average LBLT in mm. The factor 4.0 (mm s$^{-0.5}$) and the exponent 0.5 vary with different leaf shapes and sizes. Generally, the average LBLT is directly proportional to the mean length of the leaf in the downwind direction and inversely proportional to the ambient wind speed.

10.1.2 Effects of Air Current Speed and Air Current Direction on Photosynthesis or/and Transpiration

Insufficient air circulation decreases photosynthesis (Pn) and transpiration (Tr) by suppressing the gas and water diffusion in the leaf boundary layer and thus limits plant growth and development (Yabuki 2013). Kitaya (2005) assessed the effect of air current speed on Pn and Tr of a seedlings canopy and single leaves of cucumber. For the seedlings canopy, Pn and Tr increased, respectively, by 1.2 and 2.8 times when the air current speed was increased from 0.02 to 1.3 m s^{-1}. Similarly, Pn and Tr of the single leaves increased, respectively, by 1.7 and 2.1 times when the air current speed was increased from 0.005 to 0.8 m s^{-1}. The results showed that enhancing the wind current speed has a positive effect on photosynthesis and transpiration. It was reported that as air current speed increases from 0.005 to 0.1 m s^{-1}, the decrease of LBLR is proportional approximately to the minus 0.37 power of the air current speed. Wind affects transpiration not only by reducing LBLT but also by removing the moist air close to the surfaces of leaves. When the moist air at boundary layer is replaced with drier air, the water potential gradient between stomata and ambient environment increases, and the transpiration rate is enhanced.

Air current direction has been shown to have a great influence on the transpiration of plants. Kitaya et al. (2000) studied the effect of vertical and horizontal air currents on transpiration rate for a model plant canopy. The speed of the downward air current was controlled in the range of 0.1 to 0.3 m s^{-1}. The results showed that the evaporation rates were 2 and 2.7 times greater in the vertical airflow than in the horizontal airflow at air current speeds of 0.15 and 0.25 m s^{-1}, respectively. Compared to a horizontal airflow, vertical airflow can effectively decrease LBLT at canopy surface and thus increase the diffusive rate of water vapor at boundary layer. Forced air movement with vertically downward air currents was recommended for a closed plant culture system with a large amount of plants at a high density.

10.1.3 Effects of Air Current Speed and Air Current Direction on Tipburn Prevention

In order to produce high-quality crops continuously, proper growing conditions for the crop must be maintained in the production space. In an indoor production system, due to the limited and uneven air circulation across a shelf and large production domain, air temperature over the crop canopy can deviate by several degrees from that of A/C unit set points, and the environment also may not be uniform thereby limiting production quality, yield, and rate. The appropriate air current speeds for enhancing gas exchanges by leaves were more than 0.3 m s^{-1} in the vicinity of the leaves (Kitaya et al. 1998). Lack of vertical airflow and limited capacity to create proper boundary layer dynamics, especially when the shelf height (head space) is limited, may result in crop physiological disorders (i.e., nutrient deficiency-induced tipburn with lettuce).

Tipburn is considered as a symptom of calcium deficiency-related disorder, and it is characterized by browning margins in lettuces. Calcium is an essential plant nutrient for strengthening plant cell walls. Calcium uptake from the roots to the leaves of plants is passive and is driven by transpiration process. The tipburn symptom may occur at inner and newly developing leaves with low transpiration rate due to the stagnant air at boundary layer even under high transpiration demand conditions despite plenty of supplies of calcium available at root zones. This defect affects the appearance of lettuces and limits its market value.

Goto and Takakura (1992) demonstrated that creating vertical airflow toward the lettuce crop canopy was effective to prevent tipburn. Kitaya et al. (2000) indicated that forced air movement with vertically downward air currents is essential in a closed crop culture system with high cropping densities and the air velocity should be at least 0.3 m s^{-1} just above the canopy boundary layer. Compared to a horizontal airflow, vertical airflow can effectively decrease the thickness of boundary layer at canopy surface and thus increase the diffusive rate of water vapor at boundary layer. Forced air movement with vertically downward air currents was recommended for a closed plant culture system with a large amount of plants at a high density. Shibata et al. (1995) investigated the effect of forced airflow on the growth and occurrence of tipburn in butterhead-type lettuce grown in a plant factory. A 2^3 factorial experimental design was created with three factors (vertical airflow, horizontal airflow, and no airflow) and two levels for each factor (60% of relative humidity and 80% of relative humidity). Airflow was supplied at the velocity of 0.7 m s^{-1} with either vertical or horizontal direction from air supply systems and achieved the velocity of around 0.5 m s^{-1} at the site of cultivation bed. The study showed that the vertical airflow could effectively prevent the tipburn of lettuces. No tipburn occurred up to 40th day after sowing under vertical airflow conditions. The tipburn symptom was observed without airflow and horizontal airflow at 30th day after sowing. Lee et al. (2013) studied the occurrence of tipburn symptom of two tipburn-sensitive cultivars under four different horizontal airflow rates in an indoor plant factory. One of the reasons for tipburn occurrences at the center is the effect of leaf enclosure. Airflow was generated by three air circulating fans in a horizontal line along the side of the beds in the experiment. It was found that a stable horizontal airflow about 0.3 m s^{-1} significantly reduced the incidence of tipburn and however tipburn was still detected in the inner leaves near harvest at the center of cultivation bed. Compared to the control group without having air supply, 65% and 55% of tipburn were reduced separately for two cultivars under the air current speed of 0.28 m s^{-1}.

10.2 Air Ventilation/Distribution System in PFALs

10.2.1 Air Movement

Air movement inside and through PFALs can be driven by three forces: wind pressure, buoyancy, and mechanical force. Air infiltration/exfiltration is the unintentional inward or outward movement of air through cracks in the building envelope. It is due to the pressure differences between inside and outside. It depends on wind speed, wind direction, and the airtightness of the building envelope. The level of airtightness in a PFAL can be high with 0.01–0.02 of air change per hour. That means most of the time, the flow of air caused by infiltration/exfiltration in a PFAL is negligible.

Buoyance is the driving force of fluid movement because of the density difference of fluid. It is the pressure generated by molecular collisions in all directions and depends on the kinetic energy of fluid molecules. Denser or cooler fluids have less kinetic energy, and thus less pressure is generated by the fluids. In a PFAL without ventilation, by the force of gravity, cool air with higher density falls, and hot air with lower density rises. This creates an upward buoyant force and the flow of air forms. This also causes spatial temperature gradients in a PFAL.

Air movement is PFALs is mainly driven by mechanical ventilation using fans and air ducts. Typically, air handling units are connected to ductwork. Supply air is distributed by air distribution system to the ventilation space to create a uniform climate of temperature, humidity, CO_2, and air motion in production shelves. The design of air ventilation system in PFALs is to properly choose the type, location, and size of the supply air inlet and the return air outlet according to the room geometry, internal heat source, and desired environmental conditions. The characteristics of airflow inside a PFAL are the results of the combination of buoyancy effect and mechanical forces.

10.2.2 Air Distribution System

10.2.2.1 Overall Control

Mixing ventilation system is widely used in PFALs to provide air circulation for overall control (OC). The OC here is defined as the general control of air distribution in major areas. The principle is to supply ventilation air at a high velocity (high Reynolds number) to mix and dilute the entire room air to provide mixing and air quality equalization. The inlets usually located in the upper parts of the room supplying air in a jet type (ceiling or wall at high level) (Schiavon 2009). The requirement for outlets is to avoid short-circuiting of supply air. Three examples of air ventilation system are shown in Fig. 10.3. With a jet-type flow, the motion is mainly governed by the initial momentum of the supply air. Therefore, airflow pattern is greatly affected by the location of the inlet whist minimally by the outlet. Airflow pattern is a key factor affecting temperature gradients in the building (Randall and Battams 1979).

Side Wall Supply and Extract

In ventilation community, Archimedes number (Ar) is widely used to characterize the direction of the flow (Berckmans et al. 1993). It can be expressed in a general form as (Awbi 2008):

$$Ar = \frac{g\beta \Delta T L}{U^2}$$

where g is the gravity acceleration (m s^{-2}), β is the thermal expansion coefficient (calculated at the mean temperature between the inlet air and the air close to the wall), ΔT is the temperature difference between inlet air and the coldest (or hottest) wall of the enclosure ($^\circ$C), L is the characteristic dimension of (m), and U is the average velocity at the inlet grid. The equation reveals the relative importance of buoyant and inertia forces by combing the supply air velocity and room temperature difference. For air distribution systems with air supplied from side wall (Fig. 10.3a, b), the jet deflection from the horizontal depends on Ar. As Ar increases, the jet deflection from the horizontal increases (Randall and Battams 1979; Berckmans et al. 1993; Cao et al. 2014).

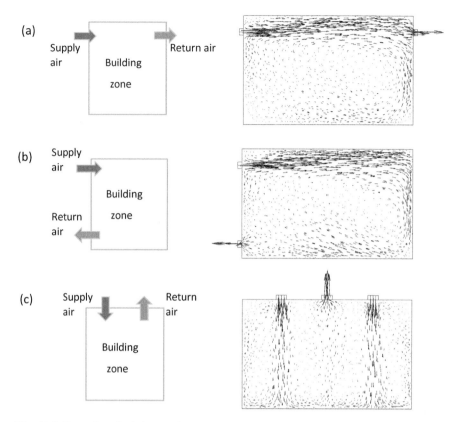

Fig. 10.3 Examples of mixing ventilation systems (left) and the corresponding airflow pattern (right): (**a**) opposite side walls supply (high) and extract (high), (**b**) one side wall supply (high) and extract (low), and (**c**) ceiling supply and extract

Ceiling Supply and Extract

The ceiling-based ventilation (Fig. 10.3c) is suitable for large space (Awbi 2015). For this type of mixing ventilation system, a potential flow can be observed along the supply air jet but no clear connective flow around the exhaust outlet (Kikuchi et al. 2003). In a PFAL, a hot aisle/cold aisle air delivery system with alternating inlets with outlets above the aisles can effectively remove the excessive heat form lighting and with uniform air temperature distribution (Fig. 10.4) (Zhang and Kacira 2017). However, for the air velocity distribution, a spatial variation of air current speed can be seen in this air distribution system, with strong airflow closed to the floor and weak airflow in the top levels of production systems. It is common to have a low average air current speed (below 0.3 m s^{-1}) at crop canopy surface.

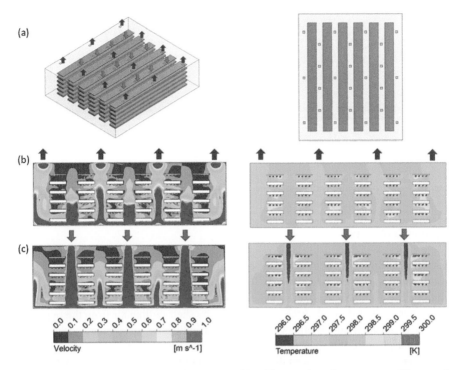

Fig. 10.4 An example of air ventilation system with ceiling supply and extract vents (Zhang and Kacira 2017): (**a**) the layout of the supply and extract vents, (**b**) the air velocity and temperature distribution at outlet vents, and (**c**) the air velocity and temperature distribution at inlet vents. The analysis and results on air velocity and temperature distributions were obtained using computational fluid dynamics modeling

10.2.2.2 Localized Control

Although mixing ventilation system can help minimize the variation of gas concentration, humidity, and temperature through the room in some degrees (Awbi 2015), the climate uniformity and air movement at crop canopy cannot be insured. With combined effects of multitiers, heat from supplemental lighting, and buoyancy forces, nonuniform air temperature and inadequate air movement at crop canopy can be found even along a single production shelf. These will lead to nonuniform crop growth and crop disorders. Besides, to control the entire space with OC is not efficient in energy. Localized control (LC) is the control strategy to use equipment to enhance air circulation just at crop canopy in each shelf. It can help to improve the climate uniformity, accelerate air movement at crop canopy, and enhance resource-use efficiency.

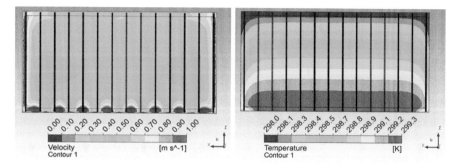

Fig. 10.5 Top view of air current speed and air temperature distributions on a horizontal plane in the shelf

Airflow/Cooling Fan

Installing airflow or cooling fan at the end of the shelf or along the length of the shelf is the simplest way as a LC. Figure 10.5 shows the top view of a production shelf equipped with fluorescent lamps and exhaust fans. The exhaust fans are installed at the center of the side wall along the length of the shelf pulling air from the opposite side wall through the shelf. A uniform distribution of air current speed on a horizontal plane in the shelf can be observed. However, the temperature of intake air is increasing by mixing with the hot air closed to the lamps as air passing through the shelf. The air temperature difference from one side to the other side depends on the temperature of intake air, the pathway of the airflow, the sensitive heat released from lamps, and the height of the shelf. Installing fans on the end of the long shelf is not recommended. If we consider the top surface of crop canopy as a big flat leaf, according to the approximate equation for LBLT described in previous section, LBLT increases with the mean length of leaf in the downwind direction. Therefore, the air circulation at the center of the shelf cannot be improved with a large LBLR.

Perforated Air Tube

For LC, using fans to circulate air downward to provide vertical airflow to crop canopy is not practical. It will lead to a mixing of hot air around lighting with the cool air at leaf surface, increasing the ambient air temperature at leaf surface. Perforated air tube can be used to deliver conditioned air to crop canopy or help air circulation around crop canopy to achieve desired climate uniformity and adequate air movement. The design of a perforated air tube is associated with the size and layout of the production system. The number, size, shape, and spacing of the discharge holes influence the static pressure in the air tube and decide the dissipate rate of jets of air from each hole (Saunders and Albright 1984; Wells and Amos 1994). The aperture ratio for a perforated pipe is defined as the total hole area to dust area. Wells and Amos (1994) reported that an aperture ratio greater than 1.5 will

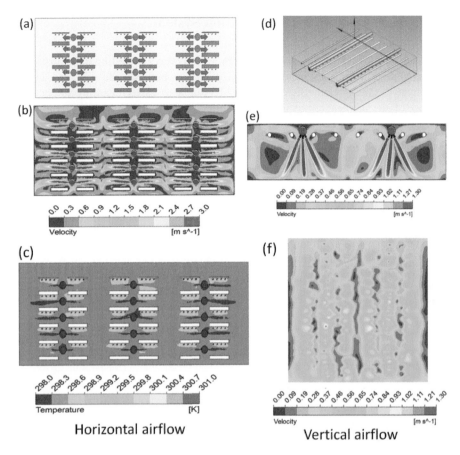

Fig. 10.6 Examples of the application of perforated air tube for the localized control of air distribution system to provide horizontal airflow (**a–c**) (Zhang and Kacira 2017) and vertical airflow (**d–f**) in production shelves (Zhang et al. 2016): (**a**) the layout of perforated air tubes (horizontal airflow), (**b**) front view of air velocity distribution (horizontal airflow), (**c**) front view of air temperature distribution (horizontal airflow), (**d**) the layout of perforated air tubes (vertical airflow), (**e**) front view of air jets (vertical airflow), and (**f**) top view of air velocity distribution (vertical airflow)

cause nonuniform duct discharges and an aperture ratio of around one will give the best compromise between uniform discharge and avoiding high inlet pressure. Figure 10.6 shows two examples with perforated air tubes to provide horizontal airflow (left) and vertical airflow (right) to production shelves (Zhang et al. 2016; Zhang and Kacira 2017).

10.3 Assessment of Air Distribution System

Ventilation performance can be evaluated by various ways according to the tasks of the ventilation. Efficiency compares the difference between the real and ideal performances. Ventilation efficiency in PFALs can be assessed regarding air change rate, age of air, and heat removal. Climate uniformity is also an important parameter to assess the performance of the air ventilation/distribution system.

10.3.1 Air Exchange Effectiveness

Air exchange effectiveness is the efficiency of the ventilation to change the air in the ventilated space with fresh air. Local (a point in a room) air change effectiveness is half of the ratio of nominal time constant (τ_n) and the room local mean age of air ($\overline{\tau_p}$) (Cao et al. 2014):

$$\varepsilon_a = \frac{\tau_n}{2\overline{\tau_p}}$$

where τ_n equals to the reciprocal of the air change rate (ACH = Q/V). ACH is equal to the ratio of air supply rate (Q) to the room volume (V).

10.3.2 Local Mean Age of Air

The local age of air is the time taken since the fresh air entered the room or building to reach a designated point (Awbi 2008). The local mean age of air is defined as (Cao et al. 2014):

$$\tau_p = \frac{1}{C(0)} \int_0^\infty C_p(t)dt$$

where $C(0)$ is the initial concentration of the tracer gas and C_p is the gas concentration at a certain point in the room at time t. It's common to use mean air velocity to reveal the lack of ventilation within a space. The local MAA distribution is also shown a sensitive parameter, which can be used to detect the stagnant zones, where have the combined accumulation of heat and moisture (Chanteloup and Mirade 2009).

10.3.3 Efficiency of Heat Removal

The ventilation efficiency for heat removal can be expressed as (Cao et al. 2014):

$$\varepsilon_t = \frac{T_R - T_S}{T_P - T_S}$$

where ε_t is the efficiency of heat removal, T_R is the temperature of the exhaust air, T_S is the temperature of the supply air, and T_P is the temperature in the occupied zone. For PFALs, T_P can be the average temperature around crop canopy.

10.3.4 Coefficient of Variation

Coefficient of variation (CV), also known as the relative standard deviation (RSD), is a statistical measurement that describes the spread of data respect to the mean:

$$c_v = \frac{\sigma}{\mu}$$

where σ is the standard deviation and μ is the mean. It allows to compare the variates whose scales of measurement are not comparable. It can be used to analyzing the climate uniformity in PFALs, such as air temperature, CO_2 concentration, and humidity. Small RSD means less variation in the evaluated variable and higher uniformity. However, since the standard deviation is divided by the mean, with a mean less than unity will lead to a high SCD and often meaningless (Chanteloup and Mirade 2009). For this situation, CV should be carefully used and some further clarification for data interpretation is needed.

References

Awbi HB (2008) Ventilation systems: design and performance. Taylor & Francis, New York
Awbi HB (2015) Ventilation and air distribution systems in buildings. Front Mech Eng 1:1–4. https://doi.org/10.3389/fmech.2015.00004
Berckmans D, Randall JM, Van Thielen D, Goedseels V (1993) Validity of the Archimedes Number in ventilation Commercial livestock building. J Agric Eng Res 56:239–251
Cao G, Awbi H, Yao R et al (2014) A review of the performance of different ventilation and airflow distribution systems in buildings. Build Environ 73:171–186. https://doi.org/10.1016/j.buildenv.2013.12.009
Chanteloup V, Mirade PS (2009) Computational fluid dynamics (CFD) modelling of local mean age of air distribution in forced-ventilation food plants. J Food Eng 90:90–103. https://doi.org/10.1016/j.jfoodeng.2008.06.014
Gaastra P (1959) Photosynthesis of crop plants as influenced by light, carbon dioxide, temperature, and stomatal diffusion resistance. Overdruk 59:1–68

Goto E, Takakura T (1992) Promotion of calcium accumulation in inner leaves by air supply for prevention of lettuce tipburn. Trans ASAE 35:641–645

Kikuchi S, Ito K, Kobayashi N (2003) Numerical analysis of ventilation effectiveness in occupied zones for various industrial ventilation systems. In: Proceedings of 7th international symposium on ventilation for contaminant control, pp 103–108

Kitaya Y (2005) Importance of air movement for promoting gas and heat exchanges between plants and atmosphere under controlled environments. In: Omasa K, Nouchi I, De Kok LJ (eds) Plant responses to air pollution and global change. Springer Japan, Tokyo, pp 185–193

Kitaya Y, Shibuya T, Kozai T, Kubota C (1998) Effects of light intensity and air velocity on air temperature, water vapor pressure and CO_2 concentration inside a crops stand under an artificial lighting condition. Life Support Biosph Sci 5:199–203

Kitaya Y, Tsuruyama J, Kawai M et al (2000) Effects of air current on transpiration and net photosynthetic rates of plants in a closed plant production system. Transpl Prod 21st Century 83–90. https://doi.org/10.1007/978-94-015-9371-7_13

Lee JG, Choi CS, Jang YA et al (2013) Effects of air temperature and air flow rate control on the tipburn occurrence of leaf lettuce in a closed-type plant factory system. Hortic Environ Biotechnol 54:303–310. https://doi.org/10.1007/s13580-013-0031-0

Nobel PS (2009) Temperature and energy budgets. Physicochem Environ Plant Physiol 318–363. doi: https://doi.org/10.1016/B978-0-12-374143-1.00007-7

Randall JM, Battams VA (1979) Stability criteria for airflow patterns in livestock buildings. J Agric Eng Res 24:361–374. https://doi.org/10.1016/0021-8634(79)90078-7

Saunders DD, Albright LD (1984) Airflow from perforated polyethylene tubes. Am Soc Agric Eng 84:1144–1149

Schiavon S (2009) Energy saving with personalized ventilation and cooling fan. PhD thesis

Shibata T, Iwao K, Takano T (1995) Effect of vertical air flowing on lettuce growing in a plant factory. Acta Hortic:175–182. https://doi.org/10.17660/ActaHortic.1995.399.20

Van Gardingen P, Grace J (1991) Plants and wind. Adv Bot Res 18:189–253. https://doi.org/10.1016/S0065-2296(08)60023-3

Wells CM, Amos ND (1994) Design of air distribution systems for closed greenhouses. In: Acta horticulturae. International Society for Horticultural Science (ISHS), Leuven, pp 93–104

Yabuki K (2013) Photosynthetic rate and dynamic environment. Springer, Dordrecht

Zhang Y, Kacira M (2017) Analysis of environmental uniformity in a plant factory using CFD analysis. In: Acta Hortic 1037:1027–1034

Zhang Y, Kacira M, An L (2016) A CFD study on improving air flow uniformity in indoor plant factory system. Biosyst Eng 147:193–205. https://doi.org/10.1016/j.biosystemseng.2016.04.012

Part III
Re-considerations of Photosynthesis, LEDs, Units and Terminology

Chapter 11
Reconsidering the Fundamental Characteristics of Photosynthesis and LEDs

Toyoki Kozai and Masayuki Nozue

Abstract Fundamental characteristics of photosynthesis – action spectrum and quantum yield – are reconsidered from the viewpoint of PFALs with light-emitting diodes (LED) lighting. The effects of green light on photosynthesis are reconsidered in view of the improved cost performance of white LEDs. The differences in action spectrum and quantum yield between a single leaf and a dense plant canopy are discussed. Also discussed are the characteristics of LED lamps that are necessary for the design and operation of LED lighting systems and which are to be presented as a product label.

Keywords Action spectrum · Energy per photon · Green light · Multiple light reflection · Quantum yield · White LED

11.1 Introduction

Fundamental characteristics of photosynthesis and those of LEDs for use in the design and operation of LED lighting systems in plant factories with artificial lighting (PFALs) are discussed in this chapter. Methods of LED lighting and the control of environmental factors other than light for PFALs are somewhat different from those for greenhouses. The efficient design and operation of an LED lighting system that has a large degree of freedom requires a thorough understanding of the fundamental characteristics of photosynthesis and LEDs.

Useful information on LEDs in horticulture not described in this chapter was obtained from a comprehensive and informative 87-page review paper by Mitchell

---

T. Kozai (✉)
Japan Plant Factory Association (NPO), Kashiwa, Chiba, Japan
e-mail: kozai@faculty.chiba-u.jp

M. Nozue
Research Center for Advanced Plant Factory, Shinshu University, Nagano, Japan

Faculty of Textile Science and Technology, Shinshu University, Nagano, Japan
e-mail: msnozue@shinshu-u.ac.jp

et al. (2015) and the book *LED Lighting for Urban Agriculture* by Kozai et al. (2016).

11.2 Absorption Spectra of Chlorophyll a/b and Carotenoids

Figure 11.1 shows the absorption spectra of isolated chlorophyll a and b and β-carotene, for which the peak wavelengths are, respectively, 430 and 667 nm, 455 and 642 nm, and 448 and 482 nm. β-Carotene is one of the carotenoids. Other carotenoids show different peak wavelengths ranging roughly from 350 to 500 nm (Lichtenthaler and Buschmann 2001). It should be noted that the absorption spectra are for isolated pigments solubilized in a solvent, neither for single leaves nor for plant canopy. From this graph, it is sometimes misunderstood that the plants utilize only blue and red light for photosynthesis and that green light is not effective for photosynthesis.

11.3 Action Spectrum and Quantum Yield Spectrum of a Single Leaf

11.3.1 Action Spectrum (Gross Photosynthetic Rate per Photosynthetically Active Irradiance)

Figure 11.2 shows the action spectrum and quantum (photosynthetic photon) yield spectrum of a single leaf on a photon basis (McCree 1972). The action spectrum

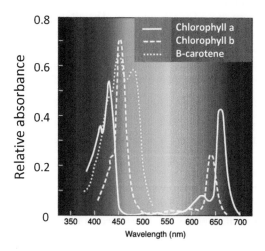

Fig. 11.1 Absorptance spectra of isolated chlorophyll a/b and β-carotene. Peak wavelengths of chlorophyll a and b and β-carotene are, respectively, 430 and 667 nm, 455 and 642 nm, and 448 and 482 nm. β-Carotene is one of the carotenoids. Other kinds of carotenoids show different peak wavelengths ranging roughly from 350 to 500 nm. (After H. K. Lichtenthaler and C. Buschmann, 2001)

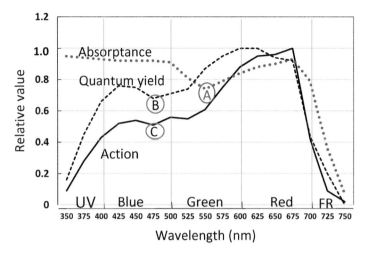

Fig. 11.2 Averaged spectra of absorptance, quantum (photosynthetic photon) yield, and action (gross photosynthetic rate) of horizontal single leaf for 22 growth chamber plant species (Reproduced from numerical data given in McCree 1972). UV and FR denote, respectively, ultraviolet and far-red

means the gross photosynthetic rate per incident photosynthetically active radiation (W m^{-2}) as a function of the wavelength. In the figure, the action spectrum shows a weak peak at 425–450 nm, a gradual increase between 525 and 550 nm, and a relatively sharp increase between 550 and 675 nm. Note that the average gross photosynthetic rate under the blue waveband (400–500 nm) is lower than that under the green waveband (500–600 nm). Blue light (400–500 nm) is absorbed by accessory pigments including the carotenoids, and the absorbed energy is transferred via excited electrons to chlorophylls a and b and finally to chlorophyll a of photosynthetic reaction centers with some energy loss.

As indicated by the circled A, B, and C in Fig. 11.2, a drop in absorptance occurs at the wavelength of 550 nm (green light), while no drop occurs at 550 nm with respect to the quantum yield and gross photosynthetic rate. Instead, a small drop occurs in the action and quantum yield spectra at the wavelength of 475 nm, which is probably due to the absorption by carotenoids.

It is also noted that the gross photosynthetic rate is positive at the wavelength ranges of 350–400 and 700–750 nm, although the wavelength range for photosynthetically active radiation (PAR) is defined as 400–700 nm.

11.3.2 Quantum Yield

Quantum yield, Q_p, is defined as the gross photosynthetic rate, P_g, divided by the photosynthetic photon absorption rate (μmol m^{-2} s^{-1}). In a case where P_g is expressed in units of μmol (CO_2) m^{-2} s^{-1}, Q_p is expressed by $P_g/(Q_a \times L_a)$, where Q_a is the photosynthetic photon flux density at the leaf (μmol m^{-2} s^{-1}) and L_a ($0 < L_a < 1$) is the absorptance of the leaf.

The quantum yield with respect to energy, Q_e, is calculated by $P_g/(E_p \times Q_a \times L_a)$, where E_p is the energy per photon. Thus, Q_e increases with decreasing energy per photon and absorption. P_g can also be expressed in units of g (dry weight) m^{-2} s^{-1}. When considering the electricity cost for lighting and electric energy use efficiency, Q_e is more important than Q_p.

11.3.3 Quantum Yield for Green Light

The quantum yield at 550–599 nm (green light) is higher than that of blue light (400–499 nm), although the absorptance of green light (0.74–0.81) is 0.11–0.18 lower than that of blue light (0.92–0.93). This drop in green light area is dependent on the thickness and chlorophyll concentration of the leaf tested (Garbrielsen 1948). Green light is reasonably efficient for photosynthesis. This fact is especially important when a significant portion of the light generated by lamps is green, such as with white LEDs (see 6.3 White LEDs). The relatively lower quantum yields in the blue and green wavebands, compared to the red waveband, would be attributable to absorption by the non-photosynthetic pigments (such as flavonoids) (Gabrielsen 1948; McCree 1972) and heat dissipation by the carotenoids (Horton et al. 1996).

11.3.4 Energy per Photon

The energy per photon, E, can be calculated by $E = h \times c/\lambda$, where h is the Planck constant (6.626×10^{-34} Js), c is the speed of light in vacuum (3×10^8 m s^{-1}), and λ is the photon's wavelength (m). As an example, for a photon with a wavelength of 450×10^{-9} m (blue), E would be 4.4×10^{-19} J, and for a photon with a wavelength of 700×10^{-9} m (red), E would be 2.8×10^{-19} J. Namely, the energy (J) per photon or mol (6.022×10^{23}) at 450 nm is 1.56 (= 700/450 = 4.4/2.8) times the energy per photon or mol at 700 nm.

Thus, red photon emission from LEDs is more electric energy efficient than blue photon emission when the conversion factor from electric energy (J) to the number of photons in moles is approximately the same regardless of the wavelength.

11.3.5 Action Spectrum in Comparison with Quantum Yield Spectrum

As described in the previous section, the action spectrum or gross photosynthesis rate under green light (500–599 nm) is as high as or higher than that under blue light (400–500 nm). Also, the quantum yield under green light (530–600 nm) is higher than that under blue light (400–500 nm). Similar action spectra determined for 33 species of plant had been also obtained by Inada (1976). Thus, it is necessary to reconsider the positive effect of green light for photosynthesis.

From the viewpoint of lighting system design, the action spectrum is more important and practical than the quantum yield spectrum. The action spectrum shows the utilization efficiency of photosynthetic photons provided by the lighting system for photosynthesis, whereas the quantum yield spectrum shows the utilization efficiency of photosynthetic photons absorbed by the leaf for photosynthesis. It is relatively easy to improve the efficiency of a lighting system by modifying its design and operation, while it is difficult to improve the efficiency of the photosynthetic characteristics of plants.

11.3.6 Red Drop and Emerson Effect

The quantum yield falls off drastically for far-red light of wavelengths greater than 680 nm (Fig. 11.2), indicating that far-red light alone is inefficient in driving photosynthesis. This sharp drop beginning at 680 nm is called "red drop" (Emerson and Lewis 1943). On the other hand, the gross photosynthetic rate is greater when red and far-red light are given together than the sum of the rates when they are given separately (Emerson and Rabinowitch 1960). This effect is called Emerson effect. The red drop and Emersion effect need to be considered in the design and operation of LED lighting system, unlike in the case of sunlight that contains a significant amount of far-red. Because, far-red flux and the ratio of red to far-red flux of a LED strongly depend on the type of LEDs. Some LEDs emit far-red significantly, and the others emit it very little.

11.4 Action Spectrum of Plant Canopy in PFALs

11.4.1 Action Spectrum of Plant Canopy

Is the reflection or transmission of green light (photons) from the leaves of a plant canopy useless for photosynthesis? In a plant canopy with a leaf area index (LAI, ratio of leaf area to cultivation area) of 3–4, most green photons transmitted by the upper leaves are received by the middle or lower leaves. On the other hand, most

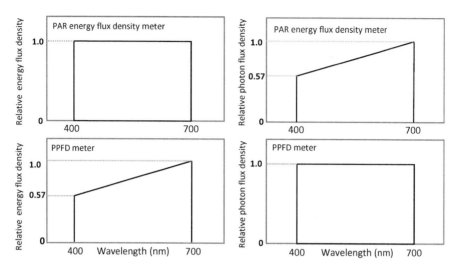

Fig. 11.3 Relative sensitivity of energy and photon flux density for PAR energy flux density meter (above) and PPFD meter (below). Relative energy flux density for PPFD meter and relative photon flux density for PAR energy flux density meter are 0.57 (=400/700) at 400 nm

blue and red photons are absorbed by the upper leaves and not transmitted, so that the lower leaves receive very few blue and red photons from the upper leaves (Yabuki and Ko 1973).

Therefore, under the same PPFD at the plant canopy surface, the net photosynthetic rate of the whole plant canopy with LAI of 3–4 should not differ significantly between receiving blue, green, and red photons. In some cases, the net photosynthetic rate of the whole plant when receiving green photons can be slightly higher than that when receiving blue and red photons (Terashima 1986), because the vertical distribution of green photons in the whole plant canopy is more uniform compared with that of blue and red photons (Kozai et al. 2015).

11.4.2 Spectral Sensitivities of PPFD and PAR Flux Density Meters

Figure 11.3 shows schematic diagrams of the spectral sensitivities of PPFD and PAR flux density meters. The vertical axis of the graph on the left is the relative energy flux density, and that on the right is the relative photon flux density.

All the sensitivity lines are straight; the line is flat for the PAR energy flux density meter with respect to relative energy flux density and for the PPFD meter with respect to relative photon flux density. In both meters, the flux density increases linearly with increasing wavelength to compensate for the associated decrease in energy per photon. Photometric unit should be taken into consideration when analyzing or controlling spectrum condition of the lighting system.

11.4.3 Action Spectrum of Plant Canopy in the PFAL

11.4.3.1 Multiple Reflection of Green Light in Cultivation Area

In most PFALs, plants are grown in a cultivation area with LED lamps on the ceiling covered with light reflective material. Reflective cultivation panels are also frequently used. In some cases, the inside surfaces of the upper side walls are covered with reflective sheets. Thus, green light reflected upward by the leaves and cultivation panels is reflected from the ceiling (and the inner surfaces of side walls) back to the plant canopy. Thus, the green light is used more efficiently in the PFAL than in a greenhouse or an open field.

11.4.3.2 PPFD at the Canopy Surface as Affected by LAI and Reflectivity of Cultivation Area

It should be noted that the light environment above and within the plant canopy in the cultivation area of the PFAL is significantly affected not only by the radiometric or photometric properties of the LED lighting system and its layout but also by the optical characteristics of the cultivation area and plant canopy (Kozai and Zhang 2016). Simulated results of the light environment as affected by the photometric properties of LED lighting systems, cultivation area, and plant canopy are given in Akiyama and Kozai (2016).

PPFD at the plant canopy surface generally decreases as the plant grows (as LAI increases) under constant PPF (photosynthetic photon flux) of the lamps when white-colored cultivation panels are used (Fig. 11.4). The reflectivity of white-colored panels is around 0.85, and that of the plant canopy (green) surface with LAI of 3 is around 0.15.

In Fig. 11.4, PPFD at day 1 (small seedlings on white-colored cultivation panels with reflectivity of 0.85) is approximately 2 times higher than that at day 9, when the cultivation panel is mostly covered with plants with canopy surface reflectivity of about 0.15. This is the reason why PPFD at the plant canopy surface decreases with increasing LAI.

It is noted that the plant canopy surface receives photosynthetic photons from two sources: one is photosynthetic photons directly from the lamps, and the other is photosynthetic photons from the multiple reflections between the plant canopy/cultivation panel surface and the ceiling of the cultivation area and/or side walls.

The PPFD by the latter increases with increasing the multiply accumulate of $(r^1 \times r^2)^n$ where r^1 and r^2 are, respectively, the reflectivity of the former (r^1: 0.9) and latter (r^2: 0.80–0.15) and n is the n-th reflection. In a case where a constant PPFD (not PPF) is required, PPF must be increased as LAI increases (see Table 11.1 for the meanings of PPFD and PPF).

From a practical point of view, PPFD at the canopy surface needs to be increased as the plants grow (as LAI increases). To keep PPFD at the same level,

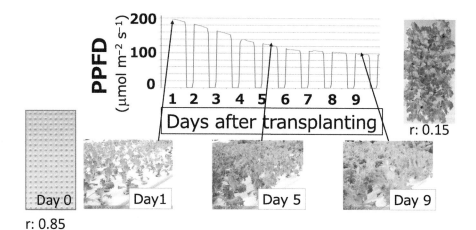

Fig. 11.4 Decrease in PPFD with passage of days at the canopy surface under constant PPF (photosynthetic radiation flux of LED lamps) after transplanting (Akiyama T, unpublished) due to the decrease in reflectivity of plant canopy surface. Reflectivity (r) is 0.85 for white-colored culture panel and is 0.15 for the panels covered fully with green leaves

photosynthetic photon flux (PPF) of the lamps must be increased by 70% during the period from day 1 to day 9. To increase the PPFD, for example, by 1.5 times, PPF must be increased by 2.55 ($= 1.7 \times 1.5$).

11.5 Light-Emitting Diodes (LEDs)

11.5.1 Fundamental Properties of LEDs

The fundamental properties of LED lamps and lighting systems (consisting of LED lamps and luminaire) to be released by LED manufacturing companies as consumer product label are listed in Table 11.1. The data on these properties represents the minimum requirements for lighting system design for PFALs. Note the difference in units between efficiency and efficacy. A product label similar to the one shown in Table 11.1 has been proposed for supplemental and sole horticultural lamps including high-pressure sodium, fluorescent, incandescent, and LED lamps (Both et al. 2017).

Both efficacy and efficiency decrease with increasing LED tip temperature, so it is essential to keep the temperature as low as possible using a heat sink plate and/or a forced stream of air or other fluid medium. Direct electric current is used to drive the LEDs, so alternating electric current (AC) is converted to direct electric current (DC) using an AC-DC converter with energy loss of 3–5% or more. In the design and operation of LED lighting systems, the price of LEDs per PPF, lifetime and easy installation, and maintenance are important factors.

Table 11.1 Characteristics of LED lamps or lighting systems (consisting of lamps and luminaire) required for lighting design of PFAL (revised after (Goto 2016)) and to be released by LED manufacturing companies as a product label. The figures in the right-hand column are examples and do not imply representative or standard values

Item	Unit	Example
Temperature of ambient air	°C	25
Power		
Voltage	V (voltage)	200
Current	A (ampere)	0.16
Effective power consumption	W (watt)	32.0
Light characteristics		
Spectral distribution (300–800 nm) as a dataset	$\mu mol\ m^{-2}\ s^{-1}\ nm^{-1}$	In excel file
Percentage of UV (300–399 nm), blue (400–499), green (500–599), red (600–700), far-red (700–799), near-infrared (800–1500)	–	0.25, 20, 40, 35, 4.5, 0.25
Angular distribution curves perpendicular to and parallel to LED tube, as a dataset	$Mol\ s^{-1}\ rad^{-1}$	In excel file
Photosynthetic photon flux (PPF)	$\mu mol\ s^{-1}$	48.0
Photosynthetically active radiant flux	$W\ (=J\ s^{-1})$	8.0
Luminous flux	Lm	450
Correlated color temperature (CCT)	K	3000
Color rendering index (CRI or Ra)	–	87.0
Efficiency/efficacy		
Photosynthetic radiation energy efficiency (also referred to as PAR energy efficiency)	$J\ J^{-1}$	0.25 (= 8/32)
Photosynthetic photon number efficacy	$\mu mol\ J^{-1}$	2.0
Luminous efficacy	$Lm\ W^{-1}$	150

Maintainability: Product age (lifetime) (h) at 10% decrease in PPF, product age at malfunction (h), and waterproof and dustproof characteristics
Thermal characteristics: Lamp or package surface temperature – PPF curve
Luminaire: The entire electrical light fitting, including all the components needed for mounting, operation, and glare prevention (Fujiwara 2016)
Size, shape, and weight: Drawing and weight of package and drawing of LED lamp

Note: Radian: SI unit of plane angle. 1 rad (1 radian) = 57.3° = 180/p = 180/3.14. lm: Lumen

11.5.2 Maximum Photosynthetic Photon Number Efficacy

When one J (joule) of electric energy is 100% converted into photons, the μmol of photons per J, P, can be calculated as a function of the wavelength (λ: nm) using the equation P = λ/119.6 (Fujiwara 2016). Then, P would be 3.8 μmol J^{-1} at 455 nm (blue) and 5.5 μmol J^{-1} at 660 nm (red) (Table 11.2). The ratio of the two values is 1.45 (5.5/3.8 = 660/455).

The actual value of the photosynthetic photon efficacy for commercially available LEDs generally ranges between 2.0 and 3.0 as of 2017, although the photosynthetic photon number efficacy varies with different factors such as electric current, tip

Table 11.2 Maximum photosynthetic photon number efficacy in theory and range of actual photosynthetic photon number efficacy of commercially available LEDs for various wavelengths. The maximum efficacy, P, was calculated by $P = \lambda/119.6$, where λ is the wavelength (nm). The actual efficacy varies with the electric current, tip temperature, PPF (photosynthetic photon flux), and other factors of the LED

Wavelength (color)	Maximum efficacy	Range of actual efficacy as of 2017
400 nm	3.34 µmol J^{-1}	2.0–3.2 µmol J^{-1}
455 nm (blue)	3.80 µmol J^{-1}	
555 nm (green)	4.64 µmol J^{-1}	
660 (red)	5.52 µmol J^{-1}	
700 nm	5.85 µmol J^{-1}	

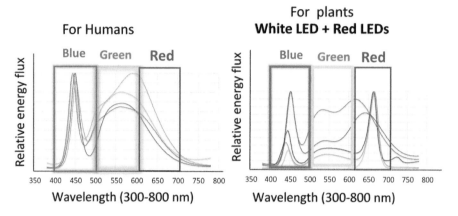

Fig. 11.5 Spectral distributions of white LEDs for humans and those with for plants

temperature, and PPF and will be steadily improved during the next 5 years or so. The photosynthetic photon number efficacy is one of the most important indices for showing the electric energy-saving characteristics of LEDs.

11.5.3 White LEDs for Plant Lighting

As a result of the improved cost performance of white LEDs (blue LED coated with a yellow phosphor), they have become increasingly popular for use in offices, residential buildings, and streets. As shown in Fig. 11.5 (left), the white LED for humans has a higher percentage of green light (500–599 nm) than red light (700–799 nm). The percentages of blue, green, and red light energy of a typical white LED are, respectively, 25, 45, and 30 (e.g., Fig. 11.5, left).

White LEDs have recently been used in many PFALs. In the PFAL, however, red LEDs are often added to white LEDs to increase the percentage of red area. This is because the percentage of red light in white LEDs for office use is too low for use in plant lighting. On the other hand, the latest development of phosphor-conversion technique from blue chips made possible to create wide-spectrum white LEDs that cover high level of red to far-red area. The merit of such wide-spectrum LEDs on plant growth and energy saving is reported (Nozue et al. 2017). It is noted that all white LEDs in Fig. 11.5 for plant lighting contain a significant portion of far-red light (700–800 nm).

Even so, the light emitted from white LEDs for plant lighting contains a significant portion of green light (Fig. 11.5, right) with different percentages. Thus, the contribution of green light to photosynthesis and the advantages and disadvantages of white LEDs over the combination of red and blue LEDs (without green light) are topics of discussion (Snowden et al. 2016). Other topics are the effect of green light on disease resistance and morphology of plants (Kudo and Yamamoto 2015), flowering control (Meng and Runkle 2016), plant growth (Kim et al. 2004), and secondary metabolite production (Ohashi-Kaneko 2015).

11.5.4 White LEDs Do Not Emit White Light

Blue, green, and red LEDs, respectively, emit blue, green, and red light with one peak wavelength and a narrow wavelength band (less than 100 nm) (Fig. 11.5). In this case, the color of the light is basically determined by the peak wavelength of the LED.

On the other hand, white LEDs do not emit white light. Instead, they emit blue, green red light, and (far-red) with multiple peak wavelengths in the range between 350 and 800 nm. In this case, "white" means the color that humans see at the surface of the LED. Human eyes are more sensitive to green light than blue and red light. Thus, white LEDs for use in offices and homes emit more green light than blue and red light to improve the color rendering index (CRI), correlated color temperature (CCT), and luminous efficacy (lm W^{-1}).

CRI is a quantitative measure of the ability of a light source to reveal the colors of various objects in comparison with an ideal or natural light source. If the light source is ideally natural, $CRI = 100$. The CCT of a light source is the temperature of an ideal black-body radiator that radiates light of a color comparable to that of the light source. Colors having a CCT over 5000 K (Kelvin, absolute temperature) are called "cool (white) colors," while those with a CCT in the range of 2700–3000 K are "warm (white) colors." White LEDs with different spectral distributions may have the same CRI and CCT.

11.6 Conclusion

Photosynthesis research has a long history during which an enormous amount of information and experience has been accumulated. On the other hand, research on LED applications for PFAL has only recently started. For smart LED lighting for plants in PFALs, we need to reconsider the fundamental characteristics of photosynthesis and LEDs from the viewpoints described in this chapter.

Acknowledgment We thank all the members of the Joint Committee on LED lighting for plant growth (chairperson: Eiji Goto) in Japan Plant Factory Association and Japan Plant Factory Industries Association. A few ideas described in this chapter were obtained during the discussion in the Committee.

References

Akiyama T, Kozai T (2016) In: Fujiwara K, Runkle E (eds) LED lighting for urban agriculture). Springer Nature, pp 91–112
Both AJ, Bugbee B, Kubota C, Lopez RG, Mitchell C, Runkle ES, Wallace C (2017) Proposed product label for electric lumps used in the plant sciences. HorTechnology 27(4):1–10
Emerson R, Lewis CM (1943) The dependence of the quantum yield of chlorella photosynthesis on wave length of light. Am J Bot 30:165–178
Emerson R, Rabinowitch E (1960) Red drop and role of auxiliary pigments in photosynthesis. Plant Physiol 35:477–485
Fujiwara K (2016) Radiometric, photometric and photometric quantities and their units (Chapter 26). In: Kozai T, Fujiwara K, Runkle E (eds) LED lighting for urban agriculture). Springer Nature, pp 367–376
Gabrielsen EK (1948) Influence of light of different wave-lengths on photosynthesis in foliage leaves. Physiol Plant 1:113–123
Goto E (2016) Measurement of photometric and radiometric characteristics of LEDs for plant cultivation. Chapter 28 of Kozai T, K. Fujiwara and E Runkle (eds.), pp 395–402
Horton P, Ruban AV, Walters RG (1996) Regulation of light harvesting in green plants. Annu Rev Plant Phys 47:655–684
Inada K (1976) Action spectra for photosynthesis in higher plants. Plant Cell Physiol 17:355–365
Kim H-H, Goins GD, Wheeler RM, Sager JC (2004) Green light supplementation for enhanced lettuce growth under red- and blue-light-emitting diodes. Hortscience 39(7):1617–1622
Kozai T, Zhang G (2016) Some aspects of the light environment (Chapter 4). In: Kozai T, Fujiwara K, Runkle E (eds) LED lighting for urban agriculture, Springer, Singapore, pp 49–55
Kozai T, Niu G, Takagaki M (eds) (2015) Plant factory: an indoor vertical farming system for efficient quality food production. Academic Press, Amsterdam, p 405
Kozai T, Fujiwara K, Runkle E (eds) (2016) LED lighting for urban agriculture, Springer, Singapore, 454 pp
Kudo R, Yamamoto K (2015) Induction of plant disease resistance and other physiological responses by green light illumination. In Kozai T, Fujiwara K, Runkle E (eds) LED lighting for urban agriculture. Springer Nature, Singapore, pp 261–273
Lichtenthaler HK, Buschmann C (2001) Chlorophylls and carotenoids: measurement and characterization by UV-VIS spectroscopy. Curr Protocol Food Anal Chem F4.3.1–F4.3.8
McCree KJ (1972) The action spectrum, absorptance and quantum yield of photosynthesis in crop plants. Agric For Meteorol 9:191–216

Meng Q, Runkle SE (2016) Control of flowering using night-interruption and day-extension LED lighting. In Kozai T, Fujiwara K, Runkle E (eds) LED lighting for urban agriculture. Springer Nature, Singapore, pp 203–217

Mitchell CA, Dzakovich MP, Gomez C, Burr JF, Hernandez R, Kubota C, Curry C, Meng Q, Runkle ES, Bourget CM, Morrow RC, Both AJ (2015) Light-emitting diodes in horticulture. Horticult Rev 42:1–87

Nozue H, Shirai K, Kajikawa K, Gomi M, Nozue M (2017) White LED light with wide wavelength spectrum promotes high-yielding and energysaving indoor vegetable production. Acta Horticuturae (GreenSys 2017 in press)

Ohashi-Kaneko K (2015) Functional components in leafy vegetables (Chapter 13 of Kozai et al. (2015)). pp 177–183

Snowden MC, Cope KR, Bugbee B (2016) Sensitivity of seven diverse species to blue and green light: interactions with photon flux. PLoS One 11(10):e0163121

Terashima I (1986) Dorsiventrality in photosynthetic light response curves of a leaf. J Exp Bot 37:399–405

Yabuki K, Ko B (1973) The dependence of photosynthesis in several vegetables on light quality. Agric Meteorol 29(1):17–23

Chapter 12
Reconsidering the Terminology and Units for Light and Nutrient Solution

Toyoki Kozai, Satoru Tsukagoshi, and Shunsuke Sakaguchi

Abstract Technical terms and units used for light and nutrient solution that are sometimes misunderstood or confusing are discussed, and more unified terminology and units are proposed for better communication and understanding among people with different academic and business backgrounds.

Keywords Equivalent · Photometry · Photonmetry · Radiometry · Valence

12.1 Introduction

The research and development of plant factories with artificial lighting (PFAL) and greenhouses with and without supplemental lighting is a multidisciplinary field, and so similar technical terms are often used with somewhat different definitions and units. To make them easier to understand and remember by people with different academic backgrounds (biology, physics, chemistry, engineering, business, etc.), the technical terms and units as well as their relationship need to be simple but logically and scientifically reasonable.

This chapter discusses the technical terms and units used for light and nutrient solution, which are sometimes misunderstood or confusing, and proposes more unified terminology and units for better communication and understanding among people with different academic and business backgrounds.

T. Kozai (✉)
Japan Plant Factory Association (NPO), Kashiwa, Chiba, Japan
e-mail: kozai@faculty.chiba-u.jp

S. Tsukagoshi
Center for Environment, Health and Field Sciences, Chiba University, Kashiwa, Chiba, Japan
e-mail: tsukag@faculty.chiba-u.jp

S. Sakaguchi
PlantX Corp, Kashiwa, Chiba, Japan
e-mail: sakaguchi@plantx.co.jp

The basic rules for terminology and units are (1) each technical term has only one clear meaning or definition, (2) each technical term has only one unit and its derivatives, and (3) mutually related technical terms must be logically understandable. For example, "action spectrum" in plant physiology means "gross photosynthetic rate" over a wavelength of 400–700 nm; however, "action" by itself does not imply "gross photosynthetic rate." Thus, "gross photosynthetic rate spectrum" may be more clearly understandable. As another example, the unit used for "daily light integral" is either mol m^{-2} or W m^{-2}, so it is not a suitable term. The word order of the term "photosynthetically active radiation (PAR)" is not a logical match to that of "photosynthetic photon," so "photosynthetic radiation" may be better than PAR when used with the term "photosynthetic photon."

12.2 Light

12.2.1 Metrics of Light

Basically, the terminology and units used for light need to match with those defined by the International Commission on Illumination (CIE) and the International Electrotechnical Commission (IEC). However, the technical terms for photosynthetic photons and photosynthetic radiation for higher plants are not fully discussed by the CIE and IEC.

12.2.1.1 Radiometry, Photometry, and Photonmetry

Table 12.1 shows fundamental quantities in radiometry, photometry, and photonmetry and their respective units according to the International System of Units (SI). *Photo*metry is a method of measuring light for human eyes, while *photon*metry in this book is a method of measuring photon in relation to plants. Note the difference in units between photometry and photonmetry and between flux and flux density. Photonmetry is used when discussing the properties of lamps for plant lighting and the effect of the light environment on plants. The meaning of each term in Table 12.1 is more fully explained in Fujiwara (2016).

Table 12.2 is a modified version of Table 12.1 with special attention given to radiometry and photonmetry of photosynthetic radiation and photosynthetic photon (wavelength range for both: 400–700 nanometers (nm)). Note the difference in meaning and unit between photosynthetic photon flux (PPF) and photosynthetic photon flux density (PPFD). PPF is a term used for the light source, and PPFD is a term used for the light environment.

The discussion above is a summary of the "Standards for terminology and units of radiometry, photometry, photonmetry and properties of light emitting diodes (LEDs) for plant lighting," proposed in 2016 by the Committee on LED Lighting for PFAL established by the Japan Plant Factory Association (JPFA) (Goto et al. 2016).

Table 12.1 Fundamental quantities in radiometry, photometry, and photonmetry and their SI units (Fujiwara 2016)

Radiometry	Radiant intensity	Radiant flux	Radiant energy	Irradiance
(Energy basis)		[(W sr^{-1}) sr]	[W s]	
	[W sr^{-1}]	= [W]	= [J]	[W m^{-2}]
Photometry	Luminous intensity	Luminous flux	Quantity of light	Illuminance
(luminosity basis)		[cd sr]		[lm m^{-2}]
	[cd]	= [lm]	[lm s]	= [lx]
Photonmetry	Photon intensity	Photon flux	Photon number	Photon flux density
(Photon basis)				(Photon irradiance)
		[(mol s^{-1} sr^{-1}) sr]	[(mol s^{-1}) s]	[(mol s^{-1}) m^{-2}]
	[mol s^{-1} sr^{-1}]	= [mol s^{-1}]	= [mol]	= [mol m^{-2} s^{-1}]
Relationship	A	A · sr	A · sr · s	A · sr · m^{-2}
		= B	= B · s	= B · m^{-2}

Note: sr, J, W, cd, lm, and lx denote, respectively, steradian (solid angle), joules, watts (=J s^{-1}), candela, lumen, and lux

Table 12.2 Radiometry and photometry of photosynthetic radiation (or photosynthetically active radiation, PAR) and photosynthetic photon (wavelength: 400–700 nm) (Fujiwara 2016)

	Intensity	Flux	Quantity	Flux density
Radiometry	Photosynthetic radiant intensity [W sr^{-1}]	Photosynthetic radiant flux [W]	Photosynthetic radiant energy [J]	Photosynthetic irradiance [W m^{-2}]
Photonmetry	Photosynthetic photon intensity [mol s^{-1} sr^{-1}]	Photosynthetic photon flux (PPF) [mol s^{-1}]	Photosynthetic photon number [mol]	Photosynthetic photon flux density (PPFD) [mol m^{-2} s^{-1}]
Relationship	A	A · sr = B	A · sr · s = B · s	A · sr · m^{-2} = B m^{-2}

Note: sr, J, and W denote, respectively, steradian (solid angle), joules, and watts (= J s^{-1})

12.2.1.2 Photosynthetic Radiation Energy Efficacy and Photosynthetic Photon Number Efficiency

The photosynthetic radiation energy efficiency of a lamp is defined as the ratio of the photosynthetic radiation energy flux (unit: W) of the lamp to the effective power consumption of the lamp (unit: W). The effective power consumption measured by a power meter should include the electric power consumed by the lighting unit, power supply, and instruments for controlling the quantity or quality of light, such as a timer clock, dimmer, or computer-programmed control system.

Note the difference between efficiency and efficacy. The term efficiency is used when the numerator and denominator units are the same, so that the unit of efficiency is dimensionless. Efficacy is used when the numerator and denominator units are different. Details about the measurement methods are given in Goto (2016).

12.2.2 Technical Terms to Be Reconsidered

This section discusses the meanings or definitions of technical terms and units used for light and nutrient solution that are often misunderstood or confused, and some new technical terms are proposed for reconsideration. Confusion or misunderstanding occurs mainly because these technical terms and units have been unsystematically introduced from different scientific fields (meteorology, plant physiology, horticultural science, agricultural engineering, illumination engineering, etc.) into the field of PFAL. Table 12.3 shows a summary of the conventional and proposed technical terms and units to be reconsidered.

12.2.2.1 PPFD vs PPF

As described above, PPF in units of mol s^{-1} is used to express the flux of photosynthetic photons emitted from a lamp, while PPFD in units of μmol m^{-2} s^{-1} is used to express the flux density of photosynthetic photons received by a virtual or real surface. However, PPF is sometimes used to mean PPFD, which is confusing and should be avoided.

12.2.2.2 Light Intensity

As shown in Tables 12.1 and 12.2, the terms radiant intensity (W sr^{-1}) and photon intensity (mol sr^{-1}) are, respectively, used to express the amount of radiant energy or photons emitted per solid angle of sr (steradian) of a light source. On the other hand, the term light intensity is still often used to mean either photosynthetic radiation flux density (W m^{-2}), PPFD (μmol m^{-2} s^{-1}), or both. This misuse of light intensity should be stopped as soon as possible, given the increasing number of people interested in the PPF and radiant/photon intensity of LEDs, in addition to PPFD as a light environment for plants.

12.2.2.3 PAR and PPF and PRF and PRFD

Although the term PAR has been widely used for many years, it may need to be reconsidered for the following reasons. Firstly, it is necessary to add "flux" as PAR flux when used with the unit W, while "flux" is included in PPF. In other words, "PAR flux" corresponds to PPF. Secondly, if the term "photosynthetically active

Table 12.3 Technical terms and units of light and nutrient solution to be reconsidered

No.	Conventional technical terms	Proposed terms for reconsideration
	Light, radiation, and photon	
1	Action spectrum (see Chap. 11 for the definition)	Spectral gross photosynthetic rate or gross photosynthetic rate spectrum
2	Daily light integral (DLI) (see Chap. 11 for the definition and units)	Daily (photosynthetic) photon integral (DPI or DLI-P) or daily (photosynthetic) radiation integral (DRI or DLI-R)
3	Photosynthetically active radiation (PAR, 400–700 nm)	Photosynthetic radiation (350–750 nm)
4	(Photosynthetic radiation energy) efficiency and (photosynthetic photon) efficiency	(Photosynthetic radiation energy) efficiency ($J\,J^{-1}$) or (photosynthetic photon) efficacy ($\mu mol\,J^{-1}$)
5	Light intensity (Group 1)	Photosynthetic radiant flux or radiant flux ($J\,s^{-1} = W$), photosynthetic photon flux (PPF) or photon flux ($\mu mol\,s^{-1}$)
6	Light intensity (Group 2) and PPF (to mean photosynthetic photon flux density or photosynthetic radiant flux density)	PPFD (photosynthetic photon flux density (PPFD), $\mu mol\,m^{-2}\,s^{-1}$) or PRFD (photosynthetic radiant flux density, $J\,m^{-2}\,s^{-1} = W\,m^{-2}$)
7	Photosynthetically active radiation (PAR)	Photosynthetic radiation, photosynthetic radiant energy (J), photosynthetic radiant flux ($J\,s^{-1} = W$), or photosynthetic radiant flux density ($W\,m^{-2}$)
8	Photosynthetic photon (400–700 nm)	Photosynthetic photon (350–750 nm)
9	Quantum yield	(Photosynthetic) photon yield
10	Quantum yield spectrum	Spectral photosynthetic photon yield, photon yield spectrum
11	–	Spectral photosynthetic radiation yield (on energy basis)
	Nutrient solution	
12	EC (electric conductivity) ($dS\,m^{-1}$)	Concentration of mEq per kg (see Chap. 12)
13	ppm ($mg\,L^{-1}$)	Ppm ($mg\,kg^{-1}$)
14	Molar concentration ($mol\,L^{-1}$ or M) (mol of solute divided by 1 L of solution)	Concentration of solute ($mol\,mol^{-1}$, $mol\,kg^{-1}$ or $kg\,kg^{-1}$): "Amount of solute" divided by the amount of solution
15	Concentration of Eq (valence × Mol of solute) per L of water	Concentration of Eq per kg of solution

radiation (PAR)" is used, it may also be necessary to use the term "photosynthetically active photon" so that both terms have the same word structure.

Photosynthetic radiation energy flux and photosynthetic radiation energy flux density are, respectively, often shortened to photosynthetic radiation flux and photosynthetic radiation flux density by omitting the word "energy," which may not cause any confusion.

The term "photosynthetic radiation flux" or "PRF" can be newly introduced as an alternative to "PAR flux" in the near future. Then, the word order of PRF would be the same as that of PPF, and the relationship between PRF and PPF would be more logical and understandable than that between PAR flux and PPF. The same idea applies to PRFD (photosynthetic radiation flux density) (W m^{-2}) and PPFD (μmol m^{-2} s^{-1}).

Photosynthetic radiation energy flux is sometimes used in units of μmol s^{-1}, which is confusing with PPF, also used in units of μmol s^{-1}. Thus, photosynthetic radiation energy flux should be used in units of W ($=$ J s^{-1}), not μmol s^{-1}, because the term "radiation" is used in units of energy (joule), not photons (mol).

12.2.2.4 DLI (Daily Light Integral)

The term DLI (daily light integral) is widely accepted and is used in either units of mol m^{-2} d^{-1} or units of W m^{-2} d^{-1}. Since the word "daily" is part of the term, "d^{-1}" can be or should be removed from the unit.

Another issue is that DLI does not indicate whether it is about photosynthetic photons in units of mol m^{-2} or photosynthetic radiation in units of W m^{-2}. To make the meaning of DLI clearer, the terms "daily photosynthetic radiation flux density integral" (daily PRFD integral) or DLI-R in units of W m^{-2} and "daily photosynthetic photon flux density integral" (daily PPFD integral) or DLI-P in units of mol m^{-2} may be newly introduced in the future.

12.2.2.5 Lumen (lm) and Color Rendering Index (CRI or Ra)

White LEDs (blue LEDs covered with plastic containing phosphors for wavelength conversion) are increasingly popular as a light source in the PFAL. Strong blue light poses a risk of eye damage due to the strong energy per photon. Meanwhile, the light emitted from red LEDs may affect human psychology when the percentage of red light energy exceeds 70% of the total light energy. This means that the light spectrum of a lamp must be selected by considering not only the growth of plants but also the health of people working in the PFAL and the effect of the color of the plants on humans.

In this sense, in addition to photosynthetic radiant energy efficiency and photosynthetic photon number efficacy, the color rendering index (CRI or Ra) and spectral light distribution in the range of 320–780 nm are important indices for the lamps used in the PFAL. CRI is a quantitative measure of the ability of a light source to faithfully reveal the colors of various objects in comparison with an ideal or natural light source.

12.2.2.6 Wavelength and Photon Number

Radiation is often characterized by its spectral distribution over a wavelength range, while photon is characterized by the photon number (or wave number), not by the wavelength. However, since the photon number is inversely proportional to the wavelength, there is no confusion even when the wavelength is used to understand the spectral characteristic of photons. It is noted, however, that the shape of the curve for a certain radiant spectrum drawn with the wavelength as the horizontal axis is largely different from that drawn with the photon number as the horizontal axis, because the photon number is not proportional but is inversely proportional to the wavelength.

12.2.2.7 Wavelength Range of Photosynthetic Radiation and Photon

It is widely accepted that the photosynthetic radiation (or photosynthetic photon) ranges between 400 and 700 nm. It is also well known that the action spectrum curve of a typical green leaf, like the one shown in Fig. 11.2 in Chap. 11, ranges between about 350 and 750 nm. This means that photosynthesis activated by the wavelengths between 350–400 and 700–750 nm has been neglected in the definition of photosynthetic radiation flux (or PPF), probably because the gross photosynthetic rate over these wavelengths was considered to be negligibly small. In reality, however, net photosynthesis in the region between 350–400 and 700–750 nm accounts for over 5% of that in the region between 350 and 750 nm (Fig. 11.2 in Chap. 11).

Recent inexpensive equipment for measuring the net photosynthetic rate has become accurate enough to measure the net photosynthetic rate activated by radiation (or photons) ranging from 350–400 to 700–750 nm. This means that the wavelength range of photosynthetic radiation needs to be reconsidered as 350–750 nm.

12.2.2.8 Quantum Yield and Photon Yield

Quantum yield is defined as the yield of photochemical products (mol m^{-2}) divided by the total amount of quanta (mol m^{-2}) or photosynthetic radiation energy (J m^{-2}) absorbed during a certain time period. The quantum yield can be renamed as the photosynthetic photon yield or photon yield, since quanta is now also called photons.

12.2.2.9 Absorptance and Absorption Coefficient

Absorptance of a leaf is defined as the fraction of incident light energy that is absorbed by the leaf, in contrast to the absorption coefficient which is the ratio of the absorbed to incident light energy. Absorption coefficient should not be confused with absorptance. Absorptance spectrum and absorption spectrum mean, respectively, a curve of absorptance or absorption coefficient for a particular wavelength over a certain waveband.

The absorption coefficient, α, is expressed by $F(x) = F(x_0) e^{-\alpha (x - x_0)}$ where $F(x)$ is the photon flux density at point x below the surface of a layer (or a solution) and $F(x_0)$ is the photon flux density at a surface point x_0.

12.3 Nutrient Solution

In this section, several fundamental technical terms and units of nutrient solution are reconsidered to make them more logically understandable by people with different academic and business backgrounds. Basically, plants absorb inorganic nutrient ions (not molecules) and water from the roots and absorb CO_2 and light energy from leaves to grow by photosynthesis. Plants can grow without any organic nutrients, whereas animals and microorganisms need organic nutrients as essential elements for growth.

Note that the roots may absorb a small amount of organic substances such as amino acids, vitamins, and mono-/disaccharides in the nutrient solution, and they may affect the plant growth in some cases (Ge et al., 2009) (this topic is not discussed in this chapter).

12.3.1 Fertilizer, Nutrient Elements (Nutrients), and (Nutrient) Ions

A fertilizer is any natural or synthetic substance containing one or more plant nutrient elements useful for plant growth. It can be a mixture of inorganic and/or organic substances. The nutrient elements are classified into macro- (major) and micro- (minor) nutrient elements. The macroelements are nitrogen (N), phosphorus (P), potassium (K), calcium (Ca), sulfur (S), and magnesium (Mg), and the microelements are boron (B), chlorine (Cl), manganese (Mn), iron (Fe), zinc (Z), copper (Cu), molybdenum (Mo), and nickel (Ni) (Mohr and Schopfer 1995). The nutrient elements contained in the fertilizer are dissolved in water (rainwater, groundwater, city water, etc.) and ionized before being absorbed by the plant roots.

When organic fertilizer exists in the nutrient solution, a portion of it gradually decomposes to nutrient ions in the presence of specific microorganisms. Since the plant roots do not distinguish the ions originating from inorganic fertilizer from those originating from organic fertilizer, they absorb the ions equally regardless of the origin (Fig. 12.1). On the other hand, in the presence of beneficial bacteria in a hydroponic nutrient solution, the bacteria may live in the plant root system and form a symbiosis, which often affects the root morphology, root function, and nutrient uptake. Then, the root system promotes the growth of aerial parts (Fang 2017). This topic is a challenging area for the next generation of hydroponics.

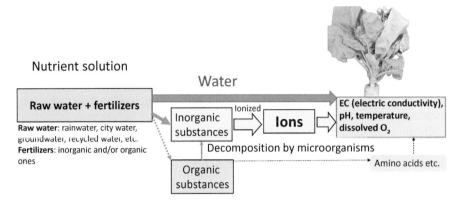

Fig. 12.1 Conversion of inorganic and organic fertilizers to nutrient ions dissolved in the nutrient solution

12.3.2 Ion, Valence, Equivalent, and Equivalent Concentration

Ions are positively or negatively charged by electrons. Examples of positively charged ions are K^+, NH_4^+, Ca_2^{2+}, and Mg^{2+}, and those of negatively charged ions are NO_3^- and $H_2PO_4^-$. KNO_3 (potassium nitrate) is ionized to K^+ and NO_3^- when dissolved in water. Plant roots absorb K^+ separately from NO_3^-.

The valence of an ion (or an atom) represents the number of electrons to achieve stability by having a full or empty outermost orbital, which is a measure of its capacity to combine with other ions to form chemical compounds or molecules. The valence of K^+ and NO_3^- is 1, and that of Ca^{2+} and Mg^{2+} is 2.

Eq (equivalent or mol equivalent) of ions in solution is calculated by multiplying the number of molecules of each ion (measured in moles) by the charge it carries. If 1 mol of KCl and 1 mol of $CaCl_2$ are dissolved in a solution, there is 1 Eq of K^+, 2 Eq of Ca^{2+}, and 3 Eq of Cl^- in that solution (the valence of calcium is 2, so that 1 mole of calcium ion has 2 Eq).

Eq concentration is expressed by Eq per kg (or L). mEq (milliequivalent) and mEq kg^{-1}, one thousandth of Eq and Eq kg^{-1}, respectively, are frequently used in hydroponics. Controlling the total sum of mEq kg^{-1} for all the nutrient ions is a major control variable and is much more important than controlling the total sum of mmol kg^{-1} or mg kg^{-1} for all the nutrient ions. Another major control variable is pH (potential of hydrogen: a scale of acidity from 0 to 14). It is to be noted that the sum of each ion concentration is not proportional to the sum of Eq kg^{-1} for each ion.

12.3.3 EC Meter

EC (electrical conductivity) quantifies how strongly a given material promotes the flow of electric current. Basically, mEq kg^{-1} is proportional to the EC of the solution when all the ions are dissociated (separated). The EC meter is robust, relatively inexpensive and can be used for its continuous measurement and control only with occasional calibration.

On the other hand, it is practically impossible to separately control the mEq kg^{-1} of each ion in nutrient solution, because the ion sensors are generally expensive and also not suited to continuous measurement and control. This is the reason why the EC meter is widely used.

12.3.4 Why Not Call the EC Meter an Eq kg^{-1} Meter?

Since the EC meter is widely used in hydroponics to estimate the total Eq kg^{-1} of the nutrient solution, why not call the EC meter an "Eq meter" simply by changing its scale for the readings? The value we want to know is mEq kg^{-1}, not EC. At least the EC meter should have a reading for mEq kg^{-1} in addition to the reading for dS m^{-1}. The development of an Eq kg^{-1} meter for measuring the mEq kg^{-1} of each ion and/or an Eq kg^{-1} meter for measuring the mEq kg^{-1} of nutrient and non-nutrient ions separately is expected. "Section 3.2 Ion balance" describes one such trial.

12.3.5 Definition of ppm

When substance "A" (e.g., solute) with a mass (or weight) or volume of W_1 is contained in "B" (e.g., solvent) with a mass or volume of W_2, the concentration of A is expressed by ($W_1/(W_1 + W_2)$) where the units of the numerator and denominator are the same, and thus the unit is dimensionless.

The unit "ppm" (parts per million, $1/10^6$ or 1.0×10^{-6}) is widely used for the dilute concentration of a substance such as CO_2 in the atmospheric air (about 400 ppm), although ppm is not an SI unit (International System of Units). The SI unit for ppm is μmol mol^{-1} or μL L^{-1}. Note that the SI unit for CO_2 and the air containing CO_2 are the same, as mentioned above.

The unit "ppm" is also widely used with the unit mg L^{-1} for expressing the concentration of a nutrient element in the nutrient solution for hydroponics. In this case, the units of the numerator and denominator are different. However, the numerical value expressed with mol L^{-1} is almost equal to the value expressed with mol kg^{-1}, since the mass of water per liter is 0.9982 kg at 20 °C and 0.9957 kg at 30 °C. Nowadays, the mass (weight) of a liquid can be measured by an inexpensive electronic balance as accurately as its volume measured with a graduated cylinder. Also, the unit mg kg^{-1} is more easily understandable and measurable than mg L^{-1} or mol L^{-1}. Thus, the use of mg kg^{-1} and mol kg^{-1} is recommended in an interdisciplinary field such as PFAL.

12.4 Conclusion

We hope that this chapter contributes the first small step in reconsidering the terminology and units of light and nutrient solution. By using unified terminology and units in the area of PFALs, easier understanding and communication among people with different academic backgrounds can be expected. The same applies for many areas other than light and nutrient solution.

Acknowledgment We thank all the members of the Committee for LED Lighting for plants (chairperson: Eiji Goto) and those on the Committee for Hydroponic solution controller (chairperson: Yutaka Shinohara). Some important ideas presented in this chapter were obtained during discussions at the committee meetings.

References

Fang W (2017) From hydroponics to bioponics. In: Proceedings of international forum for advanced protected horticulture (2017 IFAPH) Organized by National Taiwan University 1–17

Fujiwara K (2016) Radiometric, photometric and photonmetric quantities and their units (Chapter 26). In: Kozai T, Fujiwara K, Runkle E (eds) LED lighting for urban agriculture). Springer Nature, pp 367–376

Ge T, Song S, Roberts P, Jones DL, Huang D, Iwasaki K (2009) Amino acids as a nitrogen source for tomato seedlings: the use of dual-labeled (13C, 15N) glycine to test for direct uptake by tomato seedlings. Environ Exp Bot 66:357–361

Goto, E. 2016. Guideline for presenting LED grow lights properties. Committee on LED lighting for plant factories with artificial light. Abstract book for annual meeting of the Japanese Society of Agricultural, Biological and Environmental Engineers and Scientists. pp 48–49 (in Japanese)

Goto E, Fujiwara K, Kozai T (2016) Proposed standards developed for LED lighting, vol 16. Urban Ag News online Magazine, pp 73–75

Mohr H and Schopfer H. 1995. Plant nutrition (Chapter 2). In: Plant physiology. (translated into English by G. Lawlor and D.W. Lawlor). Springer, Berlin, pp 9–30

Part IV
Advances in Research on LED Lighting

Chapter 13
Usefulness of Broad-Spectrum White LEDs to Envision Future Plant Factory

Hatsumi Nozue and Masao Gomi

Abstract The effect of the spectrum of a light source on plant growth and the importance of the control of color quality of light-emitting diodes (LEDs) are described. Plant growth is greatly facilitated by the addition of far-red light to conventional blue-red monochromatic LEDs, and a good yield of whole plants is promoted by the addition of green light. This knowledge suggests the usefulness of broad-spectrum white LEDs. Tremendous progress in blue light source phosphor-conversion technology has made the color control of white LEDs possible. Spectral features favorable for plant growth are characterized by a systematic survey of commercially available white LEDs. It is suggested that the balance among blue, red, and far-red colors that constitute white-emitting LEDs is a key factor in determining a suitable LED spectrum. The color temperature (CCT) and the color rendering index (CRI) can serve as a rough and limited indicator for the choice of a white LED. In addition to the benefits of using white LEDs in a plant factory, the accompanying complications and potential strategies are discussed.

Keywords White LED · Spectral PFD · Far-red light · Green light · Correlation color temperature · Color rendering index

13.1 Introduction

Plants on earth have been evolutionarily adapted to sunlight. The complexity of plant photosystems is regarded as a tuning system to maintain photosynthetic efficiency under fluctuating sunlight. The plants protect photosystems by the loss of excess energy through heat dissipation (Horton et al. 1996). In plant factories with artificial

---

H. Nozue (✉)
Research Center for Advanced Plant Factory, Shinshu University, Ueda, Japan
e-mail: htmi-kj@shinshu-u.ac.jp

M. Gomi
Suwa Technology Center, Nichia Corporation, Shimosuwa, Japan
e-mail: masao.gomi@nichia.co.jp

lighting (PFAL), it is expected that the absolute maximum input energy will be put to use for photosynthetic reactions. To achieve progress with efficient lighting, the two primary factors that must be considered are the quality of the illumination system itself and the energy-use efficiency of plants. The desirable LEDs for general use in residences and offices are not always good for plant growth because plant growth is greatly dependent on the spectral features of the LEDs (Chap. 11 in this book). It is generally understood that one of the merits of LEDs is their high electricity-to-light energy conversion, which results in less heat released, and long lifetime. An additional merit of LEDs is their technical capability to create various kinds of spectra (Kozai et al. 2016). Closely related concepts have been recently proposed for mainstream agricultural LED applications, including the importance of the choice of the LED spectrum (Cocetta et al. 2017) and the technological possibility of spectrum control using LEDs (Pattison et al. 2016; Ahlman et al. 2016).

Plant responses to various light spectra have been of great interest to researchers as signaling systems; however, activity by agricultural interests regarding the practical benefits has only just begun. LED applications in horticulture began with monochromatic red and blue. The control of the red-to-blue ratio to improve photosynthetic reactions and plant growth has been one of the considerations for LED use (Matsuda et al. 2004; Hogewoning et al. 2010; Naznin et al. 2016a; Naznin et al. 2016b). Additionally, the visual merits of white light for lettuce and the enhancement of growth were noted (Kim et al. 2004). The usefulness of far-red light for photosynthesis and production has also been tested using an LED light (Park and Runkle 2016; Zhen and Iersel 2017). In addition to the red/blue (R/B) and red/far-red (R/FR) ratios, green light is also an important factor for the lighting of PFALs (Chap. 11 in this book). A possible reason that horticultural LEDs were not transitioned to white may be the low luminous efficiency and high cost of green LEDs. On the other hand, progress in luminescent phosphor techniques has led to the commercial availability of LED lighting equipment with high-color rendering properties that cover the PAR (400–700 nm) region and beyond (Xavier et al. 2017). This chapter describes how LED spectra contribute to plant growth and the usefulness of white LEDs with broadband spectra for horticulture. In addition, the difficulties arising during the research and potential strategies for resolving them are described.

13.2 A Combination of Monochromatic LEDs Facilitates Plant Growth

13.2.1 *Characteristic Features of Growth Under Monochromatic LEDs*

Plants utilize visible light (380–780 nm) as an energy source for growth. The irradiation in this range can be fully used for photosynthesis, although its efficiency is dependent on the wavelength (Chap. 11 in this book). Recent experimental data

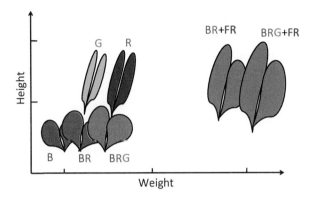

Fig. 13.1 Growth features of lettuce under monochromatic and combination LEDs. The leaf color indicates color vision under each LED lighting. B, 445 nm; G, 540 nm; R, 660 nm; FR, 730 nm

with monochromatic LEDs have more clearly revealed the differences in the effectiveness on plant growth due to wavelength (Naznin et al. 2016a; Naznin et al. 2016b; Park and Runkle 2016). The characteristic aspect of the lettuce grown under monochromatic LEDs and their combination is summarized in Fig. 13.1. Under blue light, the leaves open facing the light. The plants appear healthy but are of modest size. Under red and green light, leaves are little holding with longer shape. The plants are fragile-looking but relatively heavier. Plants grow better under a blue and red combination with normally shaped leaves. However, the morphology of the whole plant is not the same as in sunlight. The plant height is apparently limited. Regarding the effect of the ratio of red to blue (R/B) on the rapidity of growth and plant morphology, a higher R/B promotes growth but leads to a long, spindly form. Under lights that do not contain green wavelengths, the leaves are not green in color, but rather dark. Therefore, it is difficult to notice abnormal changes in the leaves.

13.2.2 Effect of Far-Red Light

The additive far-red effects of R and B LED radiation on leaf and root growth associated with stem elongation and leaf expansion is clearly shown in the early growth stages of snapdragon seedlings (Park and Runkle 2016). The supplemental effect of far-red to RB and RGB LEDs on lettuce production is obvious as well. The phenotype under far-red-supplemented LED is very similar; the stems and leaves are elongated, and leaves are apparently enlarged, which increase the reception of light energy and accelerate the rate of growth. R/FR becomes important for the effective application of far-red light, because R/FR is responsible for the most effective light signaling related to phytochromes.

13.2.3 Effect of Green Light

In contrast to the distinct effect of warm color lights, the effect of blue and green on the growth of an individual plant varies among species and the irradiance photon flux density (PFD) level (Bugbee 2016). Still, the effect of green light on plants is generally understood as positive. Lettuce growth was significantly enhanced by the addition of 24% green fluorescence to RB LEDs (Kim et al. 2004). However, under our cultivation conditions, the effect of green light on lettuce growth is found to be much less than the far-red effect. Optimization of the green level to B or R will be a future subject of research. Such positive effects of green on yields are clearer when the plants are cultivated under higher irradiation doses.

13.3 Usefulness of Broad-Spectrum White LEDs

13.3.1 Technical Principles for Visually Creating a White Color with an LED

Two types of white LED lights that use technically different methods exist in principle. For the creation of lights that seem white to the eye, a mixture of RGB primary colors with monochromatic LEDs (Fig. 13.2a) or a mixture of monochromatic blue and phosphor emitting complement yellow colors (Fig. 13.2b) is generally used. The cost of production and electricity consumption of the former is ordinarily higher than for the latter type. The conventional phosphor white LEDs contain a yellow peak (at approximately 580 nm) with a broad base. As a result, their spectra are continuous from blue to the beginning of red and include enough green but little red. It is noteworthy that recent progress in phosphor technology makes possible the emission of longer wavelengths of red to far-red lights (Fig. 13.3). The white LEDs with a wide variety of spectra are created by such a wavelength conversion technique. Some of those are supplied for the special needs of high-color rendering to display clearly. Intelligent control of the chromaticity of solid-state lighting has been proposed for horticultural lighting (Pattison et al. 2016). As a next step, the theoretical and experimental elucidation of the relationship between the LED spectrum and its potency is expected shortly. Special techniques for designing spectra are helpful in providing ideal LEDs for horticultural needs.

13.3.2 Influence of Spectral Differences on Vegetable Production

In terms of the effects of lighting systems on plant development, many factors are involved other than the properties of the light source itself. To verify how spectral

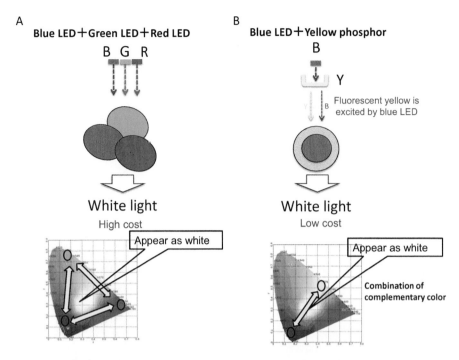

Fig. 13.2 Principles of the color emission of white-appearing LEDs. Combination of monochromatic three light primary color LEDs (**a**) and a blue basal color LED supplemented with complementary phosphor-conversion yellow (**b**)

differences of LEDs affect plant production, uniformity of the culture equipment and the structure and lighting control systems is required. A systematic survey on the production efficiency of 18 phosphor white LEDs has been used to collect useful information about the choice of the spectrum (Nozue et al. 2017a).

The spectral dependency of the yield is clearly shown among white LEDs with a variety of fluorescence emission spectra as well as LED lighting that consists of monochromatic colors. Examples of LED spectra that produce lower and higher fresh weight of lettuce are shown in Fig. 13.4. Conventional white LEDs that have lower-color rendering properties (Fig. 13.4a, LP-1 and Lp-2) and LEDs drastically unbalanced between blue and longer wavelength fluorescent colors (Fig. 13.4a, LP-3 and LP-4) had almost the same or less productivity than the FL. White LEDs that have more balanced spectra had an average 1.3-fold higher than the FL. In terms of the spectral continuity of white LEDs, it is difficult to distinguish the difference between good and poor yields. The ratio of the integrated PFD values of the R/B and the R/FR help to distinguish LEDs for both good and poor yields. When the R/B (x-axis) and R/FR (y-axis) values of each LED spectrum are plotted, the good-yield LEDs are clustered in a range of approximately 1.5–4.5 for R/B and 2.5–7.5 for R/FR (Fig. 13.5). Neither higher nor lower values beyond these ranges resulted in a good yield.

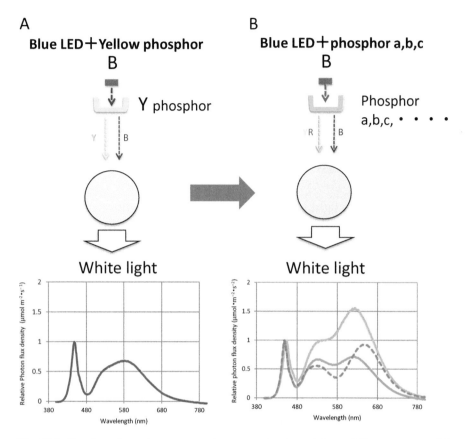

Fig. 13.3 Creation of a variety spectra using new technology. Advanced fluorescent material technology has made possible the creation of white LEDs with a variety of spectra. The recent red conversion approach extended the spectral area to longer wavelengths. Wavelength spectra show the altered examples from an original low-color rendering white LED (**a**) to broad-spectrum white LEDs (**b**). Spectral arrangement to suit a desired form is possible with this technique

13.3.3 Relationship of R/B Ratio to Plant Height and Yield and R/FR Ratio to Plant Height and Yield

Linear relationships are obtained between the R/B ratio, the height, and the yield in a limited range of R/B from 1 to 5 (Table 13.1). The positive correlations are modest, and a large variation among experiments is found in the correlations between R/B and height. The R/FR ratio is inversely associated with the height and the yield in a limited range of R/FR from 1 to 7. Correlation to the height is strong and to the yield is modest (Table 13.1). A statistical analysis suggests that contribution of tall morphology to productivity is limited, and the enhancement of plant production due to R and RF occurs within defined spectral conditions. It is indicated that there is a good balance among B, R, and FR of white LEDs. The spectral choice for better

A. Lower production (≤FL)

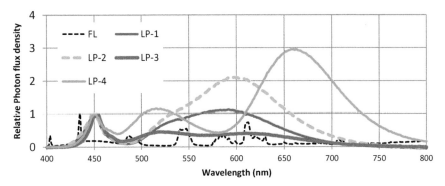

B. Higher production (> FL)

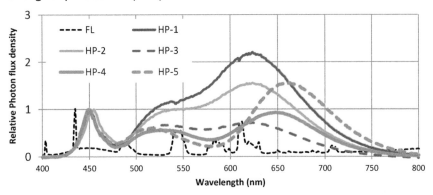

Fig. 13.4 Examples of a variety of white LED spectra distinguished by their capacity for plant production. LEDs whose spectra contain apparently less or excessive 620–780 nm light (R-FR) stimulated less production (**a**). LEDs whose spectra contain R and FR in balance stimulate higher production (**b**). FL, fluorescent light; LP, lower production; HP, higher production

plant production could be guided by optimizing the R/B and R/FR values. However, it should be noted that these values vary depending on the measurement and calculation methods. Still, they are useful as a tool for comparative analysis. The unit of light intensity should be a photon-based PFD instead of the energy-based $Wm^{-2} s^{-1}$.

13.3.4 CCT and CRI

To characterize a light source, the color temperature expressed in Kelvin (K) is often used. CRI shows the effect of an illuminant on the color appearance of objects in comparison with a natural light source. The parameter Ra (i.e., the average of the

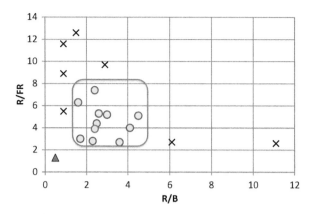

Fig. 13.5 Effect of balancing the R/B and R/FR values of white LEDs on plant growth. △, FL; ○, better production than FL; ×, less production than FL. A square shows approximate area for good production. R, integrated PFD values of 620–700 nm; B, integrated PFD values of 400–490 nm; FR, integrated PFD values of 700–780 nm

Table 13.1 Correlation coefficients of the R/B and the R/FR ratios with the plant height and yield

	R/B	R/FR
Height	(+) 0.54 ± 0.25	(−) 0.81 ± 0.13
Fresh weight	(+) 0.48 ± 0.09	(−) 0.53 ± 0.17

The data are expressed as the average of three independent experiments and the variation (SD) between them. (+) indicates a positive correlation, and (−) indicates a negative correlation

Table 13.2 Relative lettuce fresh weight among white LEDs that have different K or Ra

Ra \ K	3000	3500	4000	5000	6500
80	–	–	–	1.00	–
90	1.45	1.31	1.23	1.20	1.11
95	–	–	–	1.28	–

rendering index) is often used as the index value. The K and CRI (Ra) are indispensable light characteristic traits generally in use for every lighting application. Along with the extension of LED applications to general illumination, horticultural lighting has emerged as a next-generation LED application (Pattison et al. 2016; Han et al. 2017; Carney et al. 2016). Still, there is little information about whether the CCT and CRI of a white LED can be useful indexes for horticultural use. An example of systematic assays with white LEDs that have the same K or Ra value indicate that a relatively higher yield is achieved with a lower K and a higher Ra (Table 13.2).

The percentage of fluorescence emission with a longer wavelength than blue (400–490 nm) in total PFD decreases in an inverse proportion to K. For example, the rate of the integrated PFD value to that of B of 5000 K is less than half of 3000 K (Table 13.3, Fig. 13.6a). The percentage of red area light designated R plus FR

Table 13.3 Integrated PFD of fluorescent emission light relative to blue light

	3000 K	3500 K	4000 K	5000 K	6500 K
F/B	13.9	9.3	8.9	5.6	4.0

F, fluorescence emission with a wavelength of 490–780 nm; B, blue light with a wavelength of 400–490 nm

Table 13.4 Integrated PFD of R + FR light relative to blue light

	Ra80	Ra90	Ra95
(R + FR)/B	0.24	1.08	1.61

(R + FR), light with a wavelength of 620–780 nm; B, blue light with a wavelength of 400–490 nm

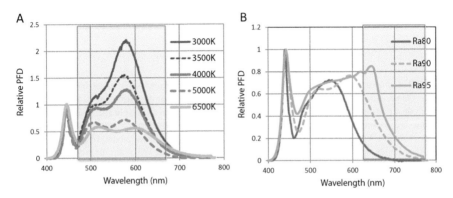

Fig. 13.6 Examples of spectral differences among white LEDs arising from the difference in K or Ra. LED spectra with Ra90 but different K (**a**) and with 5000 K but different Ra (**b**). The squares show comparison ranges of fluorescence emission and red to far-red area light

(620–780 nm) increases in proportion to Ra. For 5000 K LEDs, calculation indicates that the rate of R plus FR to B of Ra90 is 4.5 folds of that of Ra80 (Table 13.4, Fig. 13.6b).

Taken together, the difference in the productivity observed with these white LEDs that are arranged by K and Ra values is thought to reflect the differences in the color distributions. A few white LEDs that are not included in this line are commercially available. Basically, the spectral features of LEDs must be assessed.

13.3.5 Power Consumption of White LEDs

The beneficial aspects of the lower electrical energy of phosphor white LEDs are generally well known. Our results indicate a potential benefit in plant growth, too. Significant difference in their light-use efficiency between conventional lighting systems is found. When the irradiance PFD level is low, the difference is clear. The electrical power consumptions for producing 1 g of marketable aerial part of

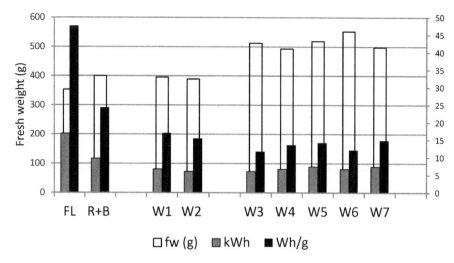

Fig. 13.7 Comparison of electrical power consumption for production of marketable aerial part of lettuce plants. Six plants were cultivated under each lightings at 130 µmol m^{-2} s^{-1}. Integrated power consumption during cultivation (kWh) is divided by the fresh weight (g). Wattages of each lightings are obtained from measured values. Total productions of edible parts of lettuce are used as the fresh weight. FL and R + B indicate fluorescent light and combined red and blue (2.5:1) monochromatic LEDs, respectively. W1-2 are low-color rendering, and W3-7 are high-color rendering types of white LEDs

lettuce plants were less than those of RB LEDs and nearly one-third those of fluorescent light (Fig. 13.7). The energy consumption values are affected by high irradiance. The values obtained under 180–200 µmol m^{-2} s^{-1} were 1.3–1.5-folds of those under 110–130 µmol m^{-2} s^{-1}. Disappearing of wavelength dependency to photosynthetic characteristics can in principle be understood. If a grower choses higher PFD more than 200 µmol m^{-2} s^{-1}, the difference energy would be reduced. It is still apparent that the use of white LEDs can promote energy savings.

13.4 Notes on LED Use for Plant Cultivation

13.4.1 Adequate PFD Dose for a Light Source

The effective radiation dose varied due to spectrum-related factors. The growth rate is accelerated with an increase of PFD until photosynthesis reactions become saturated. However, the light curve of each LED is different because of the differences of the capacity of each light source to contribute to photosynthesis. In some cases, light saturation occurs early, and the supply of higher PFD is merely wasted. In other cases, greater irradiation promotes growth (Fig. 13.8).

Under close-set conditions, plants are sometimes extremely sensitive to LED irradiation. The risk of injury is dependent on the quantity and quality of the light and

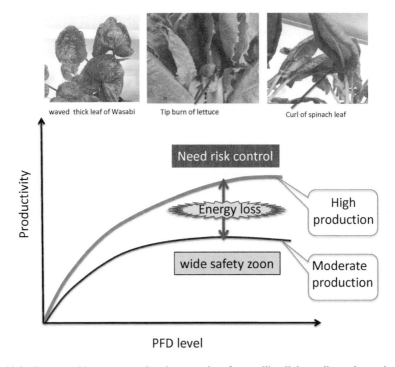

Fig. 13.8 Conceptual image supporting the necessity of controlling light quality and quantity in a plant factory. Excess energy that does not support photosynthesis causes an increase in the risk of product damage

the botanical variety and additionally influenced by other culture conditions such as temperature, humidity, ventilation, and possibly the CO_2 concentration and nutrition. The situations that are unfavorable to plants seem extremely complicated. PFD dose control of plants is an effective means to avoid damage. What little is known about actual methods for PFD control from experiment has resulted from work in indoor plant factories in Japan. The color temperature (K) seems to be ordinarily associated with the frequency of damage during production. As previously described, a lower K and higher PFD engender rapid growth but increase the risk of quality deterioration and result in considerable damage in the productivity. It should be noted that the response of cultured plants is variable among cultivars and growth age.

13.4.2 Damages Controllable by Choice of LED Spectrum

Red light is well known to facilitate plant growth via the promotion of photosynthetic activity. However, if plants are grown under a single monochromatic red LED that has an emission peak at 660 nm, the plants cannot sense excess light that causes photodamage to the leaves because of the mechanistic specificity of plant photosystems. The plants sense excess energy by means of blue light absorbance, which stimulates damage avoidance reactions in the photosystems and cellular functions (Gruszecki et al. 2010). The question "Is a combination of red and blue enough?" might arise. To answer this question, we must understand photosynthetic function in three dimensions. Blue and red lights are thought to be absorbed on the surface mainly by the palisade tissue of the leaves; in contrast, green and far-red lights penetrate beneath the leaf surface deeply into the foliage (Sun et al. 1998; Terashima et al. 2009; Broadersen and Volgelmann 2010). Therefore, the application of white LEDs with a broad wavelength spectrum covering the visible light area (380–780 nm) is a reasonable approach to achieve highly efficient conditions in a plant factory. In other words, broad-spectrum LEDs are safer and more easily employed tools for agricultural use.

13.4.3 Growth Phase and Troubles

During the early growth stages, horticultural plants are easy to cultivate under LEDs in a disease- and pest-free plant factory. Most nursery plants grow under any type of LED, possibly because young leaves respond to light conditions in a more flexible fashion than mature leaves (Nozue et al. 2017b). Trouble in the presence of LED lights usually occurs at later growth stages, or sometimes just before harvest. A few signs that seem directly or indirectly related to unfavorable light conditions are tip burn of lettuce leaves, curing of spinach leaves, and the thickness and shrinking of wasabi leaves (Fig. 13.8).

13.5 Steps on the Path to the Future of PFALs

Many studies on the influence of the spectral quality of LEDs on plants have been carried out. However, less attention has been paid to white LEDs. With the appearance of commercially available broad-spectrum white LEDs at a reasonable cost, PFALs have become more realistic systems for horticulture. The white LED has already spread throughout the world; therefore, it is expected that LEDs with good usability in agricultural systems can soon be provided. The benefits of phosphor white LEDs, other than overall good usability, include familiar natural color, low cost, and ease of replacement, all of which will help to establish a realistic expectation for PFALs. In addition, the optimization of LED lighting as an instrument for horticulture, in other words, a structural adjustment for improving the usage efficiency of plants, will be needed. It is hoped that PFALs will play a broad and important role as a tool for solving real-world problems such as food insecurity, natural disasters, demands for medical treatment, and nutrient fortification. To cope with such difficult situations, low-cost facilities and running costs are necessary. A better selection of LEDs and the appropriate application of lighting systems could be a critical requirement of future plant factories.

Acknowledgment The authors are grateful to all our colleagues for their help and assistance in the research for this study. We especially express our thanks to Professor Kozai, who provided us the opportunity to write this chapter, to Mr. Kajikawa and Dr. Shirai, who supported the basic part of the research, and Professor Nozue for all-around advice.

References

Ahlman L, Bankestad D, Wik T (2016). LED spectrum optimization using steady-state fluorescence gain. Acta Hortic. 1134. ISHS 2016. DOII 10.17660

Brodersen GR, Volgelmann TC (2010) Do changes in light direction affect absorption profiles in leaves? Funct Plant Biol 37(5):403–412

Bugbee B (2016) Toward an optimal spectral quality for plant growth and development: the importance of radiation capture. Acta Hortic 1134. ISHS 2016. DOII 10.17660

Carney MJ, Venetucci P, Gesick E (2016) LED lighting in controlled environment agriculture. Conservation applied research & development (CARD) final report COMM-20130501-73630. Outsourced Innovation

Cocetta G, Casciani D, Bulgari R, Musante F, Kolton A, Rossi M, Ferrante A (2017) Light use efficiency for vegetables production in protected and indoor environments. Eur Phys J Plus 132:43

Gruszecki WI, Luchowski R, Zubik M, Grudzinski W, Janik E, Gospodarek M, Goc J, Gryczynski Z, Gryczynski I (2010) Blue light-controlled photoprotection in plants at the level of the photosynthetic antenna complex LHCII. J Plant Physiol 167:69–73

Han T, Vaganov V, Cao S, Li Q, Ling L, Cheng X, Peng L, Zhang C, Yakovlev AN, Zhong Y, Tu M (2017) Improving "color rendering" of LED lighting for the growth of lettuce. Sci Rep 7:45944. https://doi.org/10.1038/srep45944

Hogewoning SW, Trouwborst G, Maljaars H, Poorter H, van Ieperen W, Harbinson J (2010) Blue light dose–responses of leaf photosynthesis, morphology, and chemical composition of Cucumis sativus grown under different combinations of red and blue light. J Exp Bot 61 (11):3107–3117

Horton P, Ruban AV, Walters RG Regulation of light harvesting in green plants. Annu Rev Plant Physiol Plant Mol Biol 1996, 1996, 47:655–684

Kim H-H, Goins GD, Wheeler RM, Sager JC (2004) Green light supplementation for enhanced lettuce growth under red- and blue-light-emitting diodes. Hortscience 39(7):1617–1622

Kozai T, Fujiwara K, Runkle E (eds) (2016) LED lighting for urban agriculture. Springer, pp 454 pp

Matsuda R, Ohashi-Kaneko K, Fujiwara K, Goto E, Kurata K (2004) Photosynthetic characteristics of rice leaves grown under red light with or without supplemental blue light. Plant Cell Physiol 45(12):1870–1874

Maznin MT, Lefsrud M, Gravel V Hao X (2016a) Using different ratio of red and blue LEDs to improve the growth of strawberry plants. Acta Hortic 1134. ISHS 2016. DOII 10.17660

Maznin MT, Lefsrud M, Gravel V Hao X (2016b) Different ratios of red and blue LED light effects on coriander productivity and antioxidant property. Acta Hortic 1134. ISHS 2016. DOII 10.17660

Nozue H, Shirai K, Kajikawa K, Gomi M, Nozue M (2017a) White LED light with wide wavelength spectrum promotes high-yielding and energysaving indoor vegetable production. Acta Hortic. (GreenSys 2017, In press)

Nozue H, Oono K, Ichikawa Y, Tanimurab S, Shirai K, Sonoike K, Nozue M, Hayashida N (2017b) Significance of structural variation in thylakoid membranes in maintaining functional photosystems during reproductive growth. Physiol Plant 160:111–123

Park Y, Runkle ES (2016) Investigating the merit of including far-red radiation in the production of ornamental seedlings grown under sole-source lighting. Acta Hortic 1134. ISHS 2016. DOII 10.17660

Pattison PM, Tsao JT, Krames MR (2016) Light emitting diode technology status and directions: opportunities for horticultural lighting. Acta Hortic 1134. ISHS 2016. DOII 10.17660

Sun J, Nishino JN, Volgelmann TC (1998) Green light drives CO_2 fixation deep within leaves. Plant Cell Physiol 39(10):1020–1026

Terashima I, T Fujita T Inoue, W S and O Oguchi. 2009. Green light drives leaf photosynthesis more efficiently than red light in strong white light: revisiting the enigmatic question of why leaved are green. Plant Cell Physiol 50 (4), 684–697

Xavier D, Wakui S, Takuya N (2017) Future performances in CRI for indoor and CCT for outdoor LED lightings. In LED lighting technologies – smart technologies for lighting innovations. Luger research e.U – institute for innovation & Technology – Dornbirn 2017, pp 224–235. ISBN 978-3-9503209-8-5

Zhen S, van Iersel MW (2017) Far-red light is needed for efficient photochemistry and photosynthesis. J Plant Physiol 209:115–122. https://doi.org/10.1016/j.jplph.2016.12.004

Chapter 14
LED Lighting Technique to Control Plant Growth and Morphology

Tomohiro Jishi

Abstract Light with diverse spectral distribution can be designed with narrowband lights from light-emitting diodes (LED) in plant factories with artificial lighting. The spectral distribution of the light affects plant growth mainly via photosynthesis, and it affects plant morphology mainly via other light receptor reactions. Photosynthetic reactions and the reactions of light receptors against spectral distribution are described. Plant net photosynthetic rate (P_n) is affected by the amount of light received per plant; hence, plant morphology is important to increase the amount of light received by plants and their growth rates. Plant morphology is affected by spectral distribution via multiple photoreceptors, and it is also affected by photon flux density. Therefore, there may be an interaction between these effects. Understanding these effects and these interactions is required to control plant growth and morphology using LED.

Keywords Plant factory with artificial lighting (PFAL) · Photoreceptor · Spectral photon flux density

14.1 Introduction

In plant factories with lighting via light-emitting diodes (LED), the spectral photon flux density (SPFD) is determined by the selection of LED. SPFD of the light affects plant growth and morphology. The size and shape of agricultural crops are important factors in determining their value; hence, lighting methods affect the value of products. Adoption of suitable LED lighting techniques, which control plant growth and morphology, is essential to maximize profit in the plant factories with LED.

The effects of SPFD on plants have been well investigated. Although not mentioned in this chapter, it is known to affect pest control, flower bud formation, and useful component content in plants. In this chapter, the existing knowledge on

---

T. Jishi (✉)
Central Research Institute of Electric Power Industry, Chiba, Japan
e-mail: jishi@criepi.denken.or.jp

controlling the growth rate and morphology of plants by the LED lighting method has been summarized from a physiological point of view with reference to previous studies.

14.2 Photosynthesis

Carbohydrates, which make up the bulk of the stored food in plants, are obtained by photosynthesis, and the energy used for protein synthesis is also derived from the light energy harnessed during photosynthesis. Plant growth depends on photosynthesis. Therefore, it is essential to increase the plant net photosynthetic rate (P_n) in order to promote plant growth. It is important to be aware that the P_n per plant is affected by both the P_n per leaf area and the amount of light received per plant.

14.2.1 Light Absorption Rates

Light with wavelength in the range of 400–700 nm is mainly absorbed by chlorophyll and used for photosynthesis and is called photosynthetically active radiation (PAR). We call the photon flux density of 400–700-nm light as the photosynthetic photon flux density (PPFD) [unit: mol m^{-2} s^{-1}]. However, light of 400–700 nm is not used for photosynthesis with equal efficiency. In the photosynthetic action spectrum (McCree 1972), the relative quantum yield is high at red light and low at green light. Far-red light with wavelength of more than 700 nm is also used for photosynthesis. Therefore, considering this fact, the yield photon flux density (YPFD) is sometimes used as an index of light intensity (Sager et al. 1982). YPFD is corrected with the relative quantum yield of each wavelength.

Because green light can penetrate further into the leaf than blue and red light, in strong white light, any additional green light absorbed by the lower chloroplasts would increase leaf photosynthesis to a greater extent than would absorb additional red or blue light (Terashima et al. 2009). Following the same logic, because of its high transmittance, green light may increase the plant P_n more than that by blue and red light. However, such P_n-promoting effect is observed only under light saturation. As long as we grow plants with ordinary PPFD, green light would not significantly increase plant P_n compared to blue and red light because of its high reflectance.

14.2.2 Angle of Leaves

When lighted with blue light, leaves become horizontal and direct the leaf surfaces in the direction of the light (Fig. 14.1; Inoue et al. 2008). This leaf positioning increases the light-receiving area and promotes growth, when leaves are not covering each

Fig. 14.1 Cos lettuce plants under blue LED light and red LED light. The picture at the top was taken 8 h after the picture at the bottom

other. However, when leaves are covering each other, plant P_n may be lowered with horizontal leaves (Long et al. 2006). The efficiency of utilizing the absorbed light for photosynthesis generally decreases at high PPFD. A part of the light energy intercepted by the upper horizontal leaves is wasted, although lower leaves can use the light energy efficiently for photosynthesis. When leaves are more vertical and the direction of light and the leaf surfaces are almost parallel, PPFD on the leaf surfaces decrease (Lambert's cosine law), and more light penetrates to the lower layers of the canopy. Plant P_n is expected to increase when each leaf of the plant uniformly receives light.

14.2.3 Opening of the Stomata

Blue light promotes stomatal opening (Kinoshita et al. 2001), so that more CO_2 can enter the leaves. However, P_n is saturated at high intercellular CO_2 concentrations. The P_n-promoting effect of stomatal opening can be expected under conditions where CO_2 concentration is the limiting factor for photosynthesis (Shibuya et al. 2018). The effects of stomatal opening on P_n would be small in plant factories with artificial lighting (PFAL), where cultivation occurs at relatively low PPFD and high CO_2 concentrations.

14.2.4 PPFD

The response curve of P_n to PPFD usually follows a concave function (Fig. 14.2), and P_n per PPFD (P_n/PPFD) reaches the maximum value at a certain PPFD. This

Fig. 14.2 Scheme showing the effect of photosynthetic photon flux density (PPFD) on net photosynthetic rate (P_n)

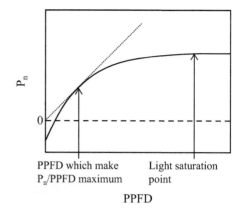

value is the contact point of a tangent drawn to the curve from the point of (PPFD, P_n) = (0,0). This PPFD value maximizes the efficacy of P_n/PPFD, but does not maximize the benefit of PFAL. During cultivation, in addition to electricity costs for LED lighting, the running costs such as air-conditioning fee, building use fee, and labor cost are involved. Hence, the longer it takes to cultivate, the greater the running costs for production. Profit will be maximized at certain growth rate with PPFD values between the "maximum P_n/PPFD" point and the light saturation point.

14.3 Morphology and Other Reactions

A plant with a large light-receiving area has a high P_n and a high growth rate. It is important to control the plant's morphology, not only to ensure the appearance commands a high commercial value but also to increase the plant growth rate. Besides photosynthesis, plants are affected by the light environment that influences their morphology and other reactions via stimulation of multiple photoreceptors. The SPFD of light affects plants via multiple reactions. In this section, I first briefly describe the photoreceptors and then summarize the effects of light of each color on the photoreceptors. We often call 400–500-nm light as blue light, 500–600-nm light as green light, 600–700-nm light as red light, and 700–800-nm light as far-red light. This convention of light classification has been followed in this chapter also.

14.3.1 Light Receptors

Phototropin, cryptochrome, and phytochrome are known as photoreceptors which affect plant morphology. Phytochrome b (hereinafter referred to as phytochrome) is a photoreceptor related to the shade avoidance effect. Phytochrome reaction is strongly affected by red and far-red light, but it is also weakly affected by blue and green light. Phototropin and cryptochrome are blue light receptors, but their absorption ranges include ultraviolet and green light as well.

14.3.1.1 Phototropin

Phototropin, a blue light receptor, is involved in phototropism that affects morphology in which the stem extends toward the light. It is also involved in stomatal opening. Furthermore, it relates to leaf flattening and leaf angling.

14.3.1.2 Cryptochrome

Another blue light receptor, cryptochrome, is involved in elongation suppression (Kang and Ni 2006). Lettuce (Hoenecke et al. 1992) and cucumber (Hernández and Kubota 2016) grown under artificial light were elongated without blue light. Lettuce is suggested to be a species whose morphology is strongly affected without blue light (Dougher and Bugbee 2001).

14.3.1.3 Phytochrome

Phytochrome is affected by light of a wide wavelength range, but the effects of red and far-red light are remarkably large, so the effects of light of other wavelengths are usually ignored. Therefore, the ratio of the photon flux densities of 600–700-nm red light and 700–800-nm far-red light, i.e., the R/FR ratio, is used as an index of the effects of SPFD on phytochrome. There is another index to more closely describe the effects of SPFD on phytochrome, the phytochrome photostationary state (PSS), also called the phytochrome photoequilibrium. The PSS is an index that considers the influence of the entire 350–800-nm spectrum of light on the phytochromes. It takes a value from zero to one, and the lower value indicates a light environment suitable for the promotion of elongation.

14.3.2 Effects of Light Quality

14.3.2.1 Blue Light

Blue light has an elongation suppressing effect via cryptochrome reaction (Table 14.1). Lettuce elongation was suppressed by adding blue light to red light (Hoenecke et al. 1992), which is supposed to have been the action of cryptochromes. On the contrary, sole blue light decreases PSS and has an elongation promotion effect via phytochrome reaction. However, the effect of blue light on phytochromes is weak and is canceled by red light of comparable PPFD (Table 14.1). Cucumber

Table 14.1 Spectral characteristics of blue, green, red, far-red, and blue and red LED lights

		LED light				
		Blue (400–500 nm)	Green (500–600 nm)	Red (600–700 nm)	Far red (700–800 nm)	Blue and red (PPFD; 1:1)
Light receptor reaction						
Phototropin	Stomatal opening					
	Leaf flattening					
Cryptochrome	Elongation suppression					Elongation suppression
Phytochrome	Elongation promotion (sole blue)	Elongation suppression (sole green)	Elongation suppression	Elongation promotion	Elongation suppression	
Parameter value						
YPFD (relative value when PFD is 100)	71	76	90	13	81	
PSS	0.48	0.86	0.90	0.10	0.86	

The yield photon flux density (YPFD) and phytochrome photostationary state (PSS) were calculated according to Sager et al. (1988) with spectral photon flux density of LED light units (ISL-305X302 series, CCS Inc.)

stems were elongated under sole blue light (Hernández and Kubota 2016), which was possibly by the action of phytochromes.

Flattening of the leaves via phototropin increases in the light-receiving areas and is expected to promote growth.

14.3.2.2 Green Light

Green light does not strongly affect the photoreceptors. PSS has a high value under sole green light, but the effect of green light on phytochromes is negligibly small compared with those under red and far-red light of comparable PPFD.

It was reported that leaf lettuce growth was promoted by adding green light to blue and red light (Kim et al. 2004). The mechanism of growth promotion by green light is unclear. The other study reported that the addition of green light to blue and red light induced effects similar to shade avoidance effects (Zhang et al. 2011). In addition, green light has an advantage of increasing the visibility of plants during cultivation.

14.3.2.3 Red Light

Red light has a strong effect on phytochromes and suppresses plant elongation. Lighting with blue and red light simultaneously would suppress plant elongation more than that caused by lighting with sole blue light and sole red light (Craver and Lopez 2016), because elongation is suppressed by the actions of both cryptochrome and phytochrome reaction (Table 14.1).

14.3.2.4 Far-Red Light

Far-red light has a strong effect on phytochromes and promotes plant elongation. Far-red light has a higher transmittance rate to leaves compared to that of red light. Therefore, plants hidden behind other plants receive light with a high proportion of far-red light. Such plants, which experience disadvantaged photosynthesis, elongate to escape the environment, and this reaction is called the shade avoidance syndrome.

From the above discussion, it is clear that if there is far-red light, plants shaded by taller plants would elongate and "catch up" with the taller plants, whereas shaded plants do not elongate and "catch up" with taller plants without far-red light. Therefore, we can reduce the variations in the size of seedlings and produce uniformly sized seedlings by adding far-red light (Shibuya et al. 2016).

14.3.3 PPFD

Generally, plants grown under weak light elongate, and plants grown under strong light show compact shapes. A reaction with sugar concentration as a signal should partly affect this morphology (Kozuka et al. 2005). A relative SPFD promoting elongation would make plants too elongated under weak light conditions, whereas a relative SPFD suppressing elongation would make plants too compact and dwarf under strong light conditions. We can balance the plant's form elongation and contraction by selecting a specific PPFD and relative SPFD. For example, even under a strong light with a PPFD of 300 mol m^{-2} s^{-1}, which naturally makes cos lettuce too compact, cos lettuce grew into a normal shape with sole blue light that promotes plant elongation (Fig. 14.3). Plants develop normal shapes when

Fig. 14.3 Cos lettuce plants grown under sole blue light with a PPFD of 300 μmol m^{-2} s^{-1}

elongation and contraction are balanced with PPFD and relative SPFD. In other words, the optimal relative SPFD depends on PPFD value.

14.4 New Lighting Technique

Various LED lighting methods have been proposed. I will introduce some of them that change SPFD over time.

14.4.1 Pulsed Light (Intermittent Light)

Pulse width modulation (PWM) is used for dimming LED lights. PWM can adjust light intensity linearly over a wide range, and the energy loss in the circuit is small. PWM should also be considered for LED light control in PFAL. Most studies reported that pulsed light with low frequency is disadvantageous for photosynthesis (Tennessen et al. 1995; Jishi et al. 2015). On the contrary, these studies also reported that the P_n under pulsed light with frequencies more than approximately 100 Hz is comparable with the P_n under continuous light. PWM for general home lighting is usually designed with a frequency of much higher than 100 Hz to prevent humans from blinking. Therefore, when the general PWM system for home lighting is used in PFAL, P_n would neither decrease nor increase. There are reports of algal growth being promoted by pulsed light having a specific frequency (Nedbal et al. 1996). Pulsed light may promote plant growth through factors other than photosynthesis.

14.4.2 SPFD Setting Depends on Time of Day

When we use multiple LED, we can change SPFD over time by independently adjusting the PPFD of each LED light. Cos lettuce growth was promoted by varying the lighting times of blue and red light compared to that under simultaneous lighting with blue and red light (Jishi et al. 2016) (Figs. 14.4–14.5, and 14.6). Plant growth may be promoted by a simple method that only controls lighting time independently using timers, which calls for further research in future.

On the contrary, the temporal changes in the P_n and growth rate of *Arabidopsis* are synchronized with the circadian rhythm of about 24 h, and the growth rate decreases under conditions where the plants cannot synchronize with environmental changes (Dodd et al. 2005). We should design the LED lighting methods in a way that does not disturb the plant circadian rhythms so as not to decrease plant growth rates.

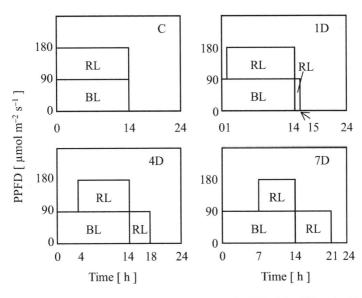

Fig. 14.4 Lighting patterns adopted by Jishi et al. (2016). Blue LED light (BL) and red LED light (RL) were used, with a PPFD of 90 µmol m^{-2} s^{-1} for 14 h per day for each light. Lighting with RL and BL started simultaneously in (C), or the starting time for lighting with RL was delayed by 1 h (in 1D), 4 h (in 4D), and 7 h (in 7D) from that of lighting with BL

Fig. 14.5 Cos lettuce plants grown under lighting patterns portrayed in Fig. 14.3. (Jishi et al. 2016)

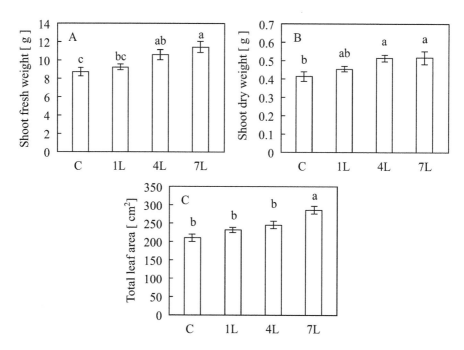

Fig. 14.6 Aerial part fresh weight (**a**), aerial part dry weight (**b**), and total leaf area (**c**) of cos lettuce plants under lighting patterns portrayed in Fig. 14.3. Bars represent standard errors of the means ($n = 7$–8). Means labeled with different small letters in each panel differ significantly from each other at the 5% level, as determined using the Tukey-Kramer HSD test

14.4.3 SPFD Settings Depend on the Duration of Cultivation

There is a study that changed SPFD depending on the duration of cultivation (Chang and Chang 2014). This study focused on the effects of UV, but SPFD settings, depending on the duration of cultivation, may have promoted growth apart from controlling morphology. For example, the leaf morphology in which the leaves are thinly expanded outward is expected to promote growth at the beginning of cultivation. At the late stage of cultivation, promoting P_n with a high PPFD and controlling shape to increase the commercial value are more important. There are few studies focusing on the setting environment dependent on the duration of cultivation, and future research in this area is necessary.

References

Chang CL, Chang KP (2014) The growth response of leaf lettuce at different stages to multiple wavelength-band light-emitting diode lighting. Scientia Horticulturae 179:78–84. https://doi.org/10.1016/j.scienta.2014.09.013

Craver JK, Lopez RG (2016) Control of morphology by manipulating light quality and daily light integral using LEDs, Chapter 15. In: Kozai T, Fujiwara K, Runkle E (eds) LED lighting for urban agriculture. Springer Nature, Singapore, pp 203–217

Dodd AN, Salathia N, Hall A, Kevei E, Toth R, Nagy F, Hibberd JM, Millar AJ, Webb AAR (2005) Plant circadian clocks increase photosynthesis, growth, survival, and competitive advantage. Science 309(5734):630–633. https://doi.org/10.1126/science.1115581

Dougher TAO, Bugbee B (2001) Differences in the response of wheat, soybean and lettuce to reduced blue radiation. Photochem Photobiol 73(2):199–207 https://doi.org/10.1562/0031-8655(2001)073<0199:DITROW>2.0.CO;2

Hernández R, Kubota C (2016) Physiological responses of cucumber seedlings under different blue and red photon flux ratios using LEDs. Environ Exp Bot 121:66–74. https://doi.org/10.1016/j.envexpbot.2015.04.001

Hoenecke ME, Bula RJ, Tibbitts TW (1992) Importance of "blue" photon levels for lettuce seedlings grown under red-light-emitting diodes. Hortscience 27(5):427–430

Inoue SI, Kinoshita T, Takemiya A, Doi M, Shimazaki K (2008) Leaf positioning of Arabidopsis in response to blue light. Mol Plant 1(1):15–26. https://doi.org/10.1093/mp/ssm001

Jishi T, Matsuda R, Fujiwara K (2015) A kinetic model for estimating net photosynthetic rates of cos lettuce leaves under pulsed light. Photosynth Res 124(1):107. https://doi.org/10.1007/s11120-015-0107-z

Jishi T, Kimura K, Matsuda R, Fujiwara K (2016) Effects of temporally shifted irradiation of blue and red LED light on cos lettuce growth and morphology. Sci Hortic 198:227–232. https://doi.org/10.1016/j.scienta.2015.12.005

Kang X, Ni M (2006) Arabidopsis SHORT HYPOCOTYL UNDER BLUE1 contains SPX and EXS domains and acts in cryptochrome signaling. Plant Cell 18(April):921–934. https://doi.org/10.1105/tpc.105.037879

Kim, H. H., Goins, G. D., Wheeler, R., M., Sager, J. C., (2004). Green-light supplementation for enhanced lettuce growth under red-and blue-light-emitting diodes. Hortscience 39(7), 1617–1622

Kinoshita T, Doi M, Suetsugu N (2001) Regulation of stomatal opening. Nature 414 (December):0–4

Kozuka T, Horiguchi G, Kim GT, Ohgishi M, Sakai T, Tsukaya H (2005) The different growth responses of the Arabidopsis Thaliana leaf blade and the petiole during shade avoidance are regulated by photoreceptors and sugar. Plant Cell Physiol 46(1):213–223

Long SP, Zhu XG, Naidu SL, Ort DR (2006) Can improvement in photosynthesis increase crop yields? Plant Cell and Environ 29(3):315–330. https://doi.org/10.1111/j.1365-3040.2005.01493.x

McCree KJ (1972) The action spectrum, absorptance and quantum yield of photosynthesis in crop plants. Agric Meteorol 9(C):191–216. https://doi.org/10.1016/0002-1571(71)90022-7

Nedbal L, Tichý V, Xiong F, Grobbelaar JU (1996) Microscopic green algae and cyanobacteria in high-frequency intermittent light. J Appl Phycol 8(4–5):325–333. https://doi.org/10.1007/BF02178575

Sager JC, Edwards JL, Klein WH (1982) Light energy utilization efficiency for photosynthesis. Trans ASAE 25:1737–1746

Sager JC, Smith WO, Edwards JL, Cyr KL (1988) Photosynthetic efficiency and Phytochrome Photoequilibria determination using spectral data. Trans ASAE 31:1882–1889

Shibuya T, Kishigami S, Takahashi S, Endo R, Kitaya Y (2016) Light competition within dense plant stands and their subsequent growth under illumination with different red:far-red ratios. Sci Hortic 213:49–54. https://doi.org/10.1016/j.scienta.2016.10.013

Shibuya T, Kano K, Endo R, Kitaya Y (2018) Effects of the interaction between vapor-pressure deficit and salinity on growth and photosynthesis of Cucumis sativus seedlings under different CO_2 concentrations. Photosynthetica 56:893

Tennessen DJ, Bula RJ, Sharkey TD (1995) Efficiency of photosynthesis in continuous and pulsed light emitting diode irradiation. Photosynth Res 44(3):261–269. https://doi.org/10.1007/BF00048599

Terashima I, Fujita T, Inoue T, Chow WS, Oguchi R (2009) Green light drives leaf photosynthesis more efficiently than red light in strong white light:revisiting the enigmatic question of why leaves are green. Plant Cell Physiol 50(4):684–697. https://doi.org/10.1093/pcp/pcp034

Zhang T, Maruhnich SA, Folta KM (2011) Green light induces shade avoidance symptoms. Plant Physiol 157(3):1528–1536. https://doi.org/10.1104/pp.111.180661

Chapter 15
The Quality and Quality Shifting of the Night Interruption Light Affect the Morphogenesis and Flowering in Floricultural Plants

Yoo Gyeong Park and Byoung Ryong Jeong

Abstract The effects of the quality of night interruption (NI) on the morphogenesis and flowering were investigated in *Petunia hybrida* Hort. "Easy Wave Pink" (qualitative long-day plant), *Pelargonium* × *hortorum* L.H. Bailey "Ringo 2000 Violet" (day-neutral plant), and *Dendranthema grandiflorum* "gaya yellow" (qualitative short-day plant). Plants were grown in a closed-type plant factory under a light intensity of 180 µmol·m^{-2}·s^{-1} PPFD (photosynthetic photon flux density) provided by white (W) light-emitting diodes (LEDs) under a condition of either long day (LD, 16 h light/8 h dark), short day (SD, 10 h light/14 h dark), or SD with 4 h NI. NI was provided by 10 ± 3 µmol·m^{-2}·s^{-1} PPFD.

In the first experiment, NI was provided by blue (NI-B), green (NI-G), red (NI-R), far-red (NI-Fr), or white (NI-W) LEDs. The shoot length and plant height of the petunia and the geranium were the greatest in NI-Fr, while those of the chrysanthemum were the greatest in LD. In the petunia, flowering was observed in LD, NI-G, NI-Fr, and NI-W. Flowering of the geranium was not affected by the night interruption light (NIL) quality, and all plants flowered in all treatments. Flowering of the chrysanthemum was observed in SD, NI-B, and NI-Fr. These results suggest that the morphogenesis, flowering, and transcriptional factors of these plants were highly affected by the quality of the NIL, especially in the chrysanthemum. Furthermore, NI-R or NI-W was the most suitable in the NI strategy of controlling the morphogenesis and flowering of the long-day plants during SD seasons.

Y. G. Park
Institute of Agriculture and Life Science, Gyeongsang National University, Jinju, Republic of Korea

B. R. Jeong (✉)
Institute of Agriculture and Life Science, Gyeongsang National University, Jinju, Republic of Korea

Division of Applied Life Science (BK21 Plus Program), Graduate School, Gyeongsang National University, Jinju, Republic of Korea

Research Institute of Life Science, Gyeongsang National University, Jinju, Republic of Korea

© Springer Nature Singapore Pte Ltd. 2018
T. Kozai (ed.), *Smart Plant Factory*, https://doi.org/10.1007/978-981-13-1065-2_15

In the second experiment, the quality of the NIL was shifted from one to another after the first 2 h in sets of two qualities among B, R, Fr, and W. LD and SD were used as the control. Twelve SD treatments with shifting of the NIL quality by LEDs were as follows: from blue to red (NI-BR), from red to blue (NI-RB), from red to far-red (NI-RFr), from far-red to red (NI-FrR), from blue to far-red (NI-BFr), from far-red to blue (NI-FrB), from white to blue (NI-WB), from blue to white (NI-BW), from far-red to white (NI-FrW), from white to far-red (NI-WFr), from red to white (NI-RW), and from white to red (NI-WR). The plant height of the chrysanthemum was greater in NI treatments that consisted of Fr than that in other NI treatments, and it was the least in NI-WB among the NI treatments. Flowering of the chrysanthemum was observed in NI-RB, NI-FrR, NI-BFr, NI-FrB, NI-WB, NI-FrW, NI-WFr, NI-WR, and SD and was especially pronounced in NI-BFr and NI-FrB. The photoperiod affected both the morphogenesis and flowering in the chrysanthemum. While the first NIL did not affect either the morphogenesis or the flowering, the second NIL significantly affected both.

Keywords Day-neutral plant · Flowering control · Light quality · Long-day plant · Morphogenesis · Short-day plant

15.1 Introduction

The production of most floricultural crops requires that plants are delivered to markets as flowers on predetermined dates. In retail flower shops or garden centers, consumers prefer to purchase ornamental plants that are available as flowers than in the vegetative stage. The morphogenesis and flowering of ornamental plants are influenced by the intensity, quality, direction, and duration of light. Control of the photoperiod (light duration) during the vegetative stage is an important method for controlling the flowering time, thereby achieving a more efficient production. For an economical production, growers also need a cultural technique that affects the flowering and crop characteristics.

Night interruption (NI) breaks up the long dark period to deliver photoperiodic lighting, resulting in modified long-day (LD) conditions for plants (Vince-Prue and Canham 1983). In general, NI more effectively induces flowering than adding the same quantity or duration of light to the end of the photoperiod (Oh et al. 2008). The full moon, which can be considered as a natural NI and has a 0.3 lux photometric intensity, may have at most only a slight delaying effect on the flowering of short-day plants (Bunning and Moser 1969; Evans 1971; Kadman-Zahavi and Peiper 1987). According to literature, *Cymbidium aloifolium* photosynthesized during a 4-h NI with a light intensity as low as 3–5 $\mu mol \cdot m^{-2} \cdot s^{-1}$ PPFD (Kim et al. 2011), which is slightly higher than that of moonlight, suggesting that the increased growth and accelerated flowering under NI can be attributed to the increased net photosynthesis during this NI period (Park et al. 2013). Therefore, a NI with a low light intensity was effective in economically controlling the flowering of the species.

The flowering response of long-day plants and short-day plants to the light quality supplied during the daily photoperiod has been extensively studied. The response of flowering plants to the light quality supplied during NI has also been studied for *Antirrhinum majus* (Craig and Runkle 2012), *Arabidopsis thaliana* (Bagnall et al. 1996), *Dendranthema grandiflorum* (Higuchi et al. 2013; Park et al. 2015), *Eustoma grandiflorum* (Yamada et al. 2009), *Oryza sativa* (Tong et al. 1990), *Pelargonium × hortorum* (Park et al. 2017), and *Petunia hybrida* (Park et al. 2016). The effects of the light quality on the regulation of flowering depend on the plant species. However, there is not much literature on the effects of the quality and quality shifting of the night interruption light (NIL) using light-emitting diodes (LEDs) on the morphogenesis, flowering, and especially transcription of photoreceptor genes in long-day plants, day-neutral plants, and short-day plants. LEDs enable an easy control of the light duration, intensity, and quality. New practical applications of LEDs as the NIL for floricultural crop production are needed.

This study focused on the effects of the NIL quality on the flowering of photoperiodic plants and those processes mediated by photoreceptors for potential applications in floricultural crop production. The effects of NI with different light qualities and quality shifting on the morphogenesis, flowering, and transcription of photoreceptor genes were investigated in three representative photoperiodic model plants, each with a distinct photoperiodic response, i.e., petunia [*Petunia hybrida* Hort. "Easy Wave Pink" (qualitative long-day plant)], ger*anium* [*Pelargonium × hortorum* L.H. Bailey "Ringo 2000 Violet" (day-neutral plant)], and ch*rysanthemum* [*Dendranthema grandiflorum* "gaya yellow" (qualitative short-day plant)].

15.2 The Effects of the Light Quality of Night Interruption on the Morphogenesis and Flowering in Floricultural Plants

15.2.1 Materials and Methods

15.2.1.1 Plant Materials and Growth Conditions

The plants used were petunia [*Petunia hybrida* Hort. "Easy Wave Pink" (qualitative long-day plant)], geran*ium* [*Pelargonium × hortorum* L.H. Bailey "Ringo 2000 Violet" (day-neutral plant)], and chrysan*themum* [*Dendranthema grandiflorum* "gaya yellow" (qualitative short-day plant)]. The selected plants have distinct photoperiodic responses. The critical day length is 14 h per day in the long-day plant and 12 h per day in the short-day plant. The plants were propagated in a glasshouse and were transferred to a closed-type plant factory, first for acclimatization under an environmental condition of 20 ± 1 °C, 60 ± 10% RH, and 140 ± 20 µmol·m^{-2}·s^{-1} PPFD provided by fluorescent lamps (F48 T12-CW-VHO, Philips Co Ltd., Eindhoven, the Netherlands) and then for photoperiodic treatments with LED lighting systems installed at a 25 cm distance above the plant canopy.

Seeds of petunia and geranium (Ball Horticultural Co., Ltd., West Chicago, IL, USA) were sown in 288-cell plug trays containing a commercial medium. Seedlings of petunia and geranium were grown on a glasshouse bench and transplanted into 50-cell plug trays in a closed-type plant factory 28 and 20 days after sowing, respectively. Chrysanthemum cuttings were stuck in 50-cell plug trays containing the same commercial medium, placed on a glasshouse bench, and rooted cuttings were transferred 12 days after sticking. During the acclimatization period in the plant factory, petunia and geranium plants were grown with a 10-h photoperiod (SD condition), and chrysanthemum was grown with a 16-h photoperiod (LD condition) to suppress flower initiation. After 14, 19, and 12 days, the plants (approximately 6.0, 7.7, and 7.9 cm in shoot length and plant height for petunia, geranium, and chrysanthemum, respectively) were transferred to the photoperiodic light treatments. The plants were fertigated once a day with a greenhouse multipurpose nutrient solution [in $mg \cdot L^{-1}$ $Ca(NO_3)_2 \cdot 4H_2O$ 737.0, KNO_3 343.4, KH_2PO_4 163.2, K_2SO_4 43.5, $MgSO_4 \cdot H_2O$ 246.0, NH_4NO_3 80.0, Fe-EDTA 15.0, H_3BO_3 1.40, $NaMoO_4 \cdot 2H_2O$ 0.12, $MnSO_4 \cdot 4H_2O$ 2.10, and $ZnSO_4 \cdot 7H_2O$ 0.44] throughout the experiment.

15.2.1.2 Photoperiodic Light Treatments

During the photoperiodic treatments, the plants were grown under a light intensity of 180 ± 10 $\mu mol \cdot m^{-2} \cdot s^{-1}$ PPFD provided by W LEDs (MEF50120, More Electronics Co. Ltd., Changwon, Korea) during the light period under either LD (16 h light/8 h dark), SD (10 h light/14 h dark), or SD with a 4-h NI. The critical day length required for flowering in the short-day plant used throughout this study is about 12 h, and therefore 14 h of darkness was sufficient to initiate flowering. The average PPFD, measured with a photo-radiometer (HD2102.1, Delta OHM, Padova, Italy) at a 25 cm distance above the NIL benchtop, was adjusted to provide the same 10 ± 3 $\mu mol \cdot m^{-2} \cdot s^{-1}$ PPFD with either B (450 nm, NI-B), G (green, 530 nm, NI-G), R (660 nm, NI-R), Fr (730 nm, NI-Fr), or W (400–700 nm, NI-W) LEDs from 23:00 to 03:00 (Fig. 15.1). Light spectral distributions of all treatments were scanned using a spectroradiometer (USB 2000 Fiber Optic Spectrometer, Ocean Optics Inc., Dunedin, FL, USA) at the same 25 cm distance above the benchtop at an interval of 1 nm.

15.2.1.3 Data Collection and Analysis

After 33, 37, and 46 days of initiating the respective photoperiodic treatments in petunia, geranium, and chrysanthemum, the shoot length or plant height, number of leaves per plant, chlorophyll content, percent flowering, days after treatment

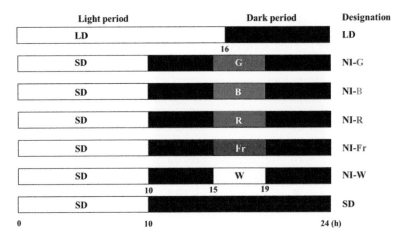

Fig. 15.1 The night interruption light (NIL) treatments by light-emitting diodes (LEDs) used in this study and the NIL qualities used for 4 h: NI-G, green; NI-B, blue; NI-R, red; NI-Fr, far-red; and NI-W, white. LD and SD, respectively, indicate a 16-h long day and a 10-h short day

initiation to visible flower bud or days to visible buds, and number of flowers per plant were measured.

15.2.2 Long-Day Plant (Petunia)

The greatest shoot length (7.4 cm) was found in NI-G and NI-W and the smallest (4.7 cm) in NI-R (Table 15.1) (Park et al. 2016). The number of leaves per plant was the fewest when plants were grown under NI-G (90.6) (Table 15.1). Leaf area was the greatest when plants were grown under NI-R (610.3) and the smallest under SD (441.3) (Table 15.1). Red (R) light is important for the development of the photosynthetic apparatus and accumulation of starch (Saebo et al. 1995). R light was observed to promote growth of *Cymbidium* leaves (Tanaka et al. 1998). Chlorophyll contents for plants grown under NI treatments were lower than those grown in LD (Table 15.1). Chlorophyll contents were lower for plants under all NI treatments except NI-W, in which the chlorophyll contents were greater than those for plants grown in LD (Table 15.1), suggesting that W light is effectively used for photosynthesis at night due to its broad spectrum (400–700 nm). The current results indicate that the light quality used during NI, despite its low intensity, induced photosynthesis in petunia.

Flowering was observed for plants in LD, NI-G, NI-R, NI-Fr, and NI-W treatments (Fig. 15.2). A 100% flowering rate was observed for plants in LD. NI-R, NI-Fr, NI-G, and NI-Fr showed a 33.3, 33.3, 16.6, and 16.6% of plants flowering, respectively. Days after treatment initiation to visible flower bud or days to visible

Table 15.1 The effects of the quality of the night interruption light (NIL) provided at 10 μmol·m^{-2}·s^{-1} PPFD on the morphogenesis and flowering of petunia, geranium, and chrysanthemum measured at 30, 37, and 46 days after treatment, respectively

Night interruption (NI)	Shoot length or plant height (cm)	No. of leaves	Leaf area (cm^2)	Chlorophyll (μg · mg^{-1}FW)
Petunia				
LD	5.8 ab[a]	109.3 a	448.0 d	2.412 a
NI-G	7.4 a	90.6 a	447.1 d	1.542 bc
NI-B	7.1 a	102.6 a	546.7 b	1.767 bc
NI-R	4.7 b	104.0 a	610.3 a	1.631 bc
NI-Fr	7.0 a	111.6 a	491.3 c	1.425 c
NI-W	7.4 a	107.6 a	595.7 a	1.755 b
SD	5.9 ab	114.3 a	441.3 d	1.044 d
Geranium				
LD	29.5 ab	28.3 a	522.3 a	2.565 ab
NI-G	26.5 bc	19.6 bc	445.0 b-c	1.466 b
NI-B	32.0 a	18.6 c	499.9 ab	2.037 b
NI-R	28.6 ab	19.6 bc	457.7 a-c	2.185 b
NI-Fr	32.1 a	15.6 c	376.1 d	3.449 a
NI-W	25.0 bc	19.6 bc	411.5 cd	1.688 b
SD	21.5 c	25.0 ab	423.7 b-c	1.897 b
Chrysanthemum				
LD	19.4 a	155.3 a	387.9 a	2.828 b
NI-G	16.2 cd	127.6 b	355.6 a	3.120 ab
NI-B	15.6 d	109.0 b	245.0 b	3.015 b
NI-R	18.2 ab	120.3 b	353.5 a	3.562 a
NI-Fr	17.7 bc	82.6 c	173.6 c	1.411 c
NI-W	17.0 b-d	121.3 b	370.0 a	2.949 ab
SD	11.4 d	80.3 c	174.7 c	3.536 ab

[a]Mean separation within columns for each spieces by Duncan's multiple range test at 5% level

buds were higher for all flowered plants in NI treatments compared to those for plants in LD.

In summary, the effects of light treatment on the morphogenesis were as follows: NI-G increased shoot length, NI-B, NI-R, and NI-W increased leaf expansion, and NI-Fr increased plant height and decreased chlorophyll content (Park et al. 2016). The morphogenesis and flowering were significantly affected by light quality during NI. Despite the reduced number of flowers per plant, flowering was promoted by G, R, Fr, and W lights, whereas it was inhibited by B light, implying that the light quality has different effects on NI-induced day extension. To obtain high-quality plants, however, treatment with NI using a high light intensity should be considered.

Fig. 15.2 The effects of the quality of the night interruption light (NIL) provided at 10 μmol·m^{-2}·s^{-1} PPFD on flowering of petunia (**a**, **b**), geranium (**c**), and chrysanthemum (**d**) measured at 30, 37, and 46 days after treatment, respectively. See Fig. 15.1 for details on the quality of the NIL

15.2.3 Day-Neutral Plant (Geranium)

The greatest plant height was observed in NI-Fr (32.1 cm) and NI-B (32.0 cm) (Park et al. 2017) (Table 15.1). When supplied during the photoperiod, R and B lights suppress growth, while Fr light promotes it. The Fr inhibition of hypocotyl elongation during the photoperiod was mediated by phytochrome A encoded by the *phyA* (Quail et al. 1995), and *phyB* was the major sensor contributing to the R:Fr responses; a loss of the *phyB* function results in a plant that displays a phenotype similar to that of a plant in a constitutive shade avoidance (Reddy and Finlayson 2014). The so-called shade avoidance responses have been well characterized by Vandenbussche et al. (2005) and Franklin (2008): increased hypocotyl, stem and petiole elongation, a more erect leaf position, pronounced apical dominance, and early flowering. When plants are grown in close proximity, many species develop elongated stems and smaller leaves; this behavior is known as the shade avoidance response (Hersch et al. 2014). This response increases plants' reach to sunlight above other plants and thus constitutes a competitive advantage (Hersch et al. 2014). This implies that the increased plant height observed in NI-Fr in this study may be mediated by *phyA*, giving a similar response as the shade avoidance. Park et al. (2016) reported that light sources with low R:Fr ratios promote flowering and stem elongation in petunia but reduce its ornamental value due to overgrowth and poor branching. In addition, B light-mediated inhibition of stem elongation during the photoperiod was reported to be caused by a reduction of cell division (Dougher and Bugbee 2004; Muneer et al. 2014), but increased plant height of geranium observed in this study suggests that the effect of B light night interruption is species-specific. The number of leaves per plant was generally the lowest in NI than in LD or SD and was the greatest in LD and the least in NI-Fr (Table 15.1). Among the NI treatments, the increase in the leaf area was more pronounced than the increase in the number of leaves per plant in NI-B and NI-R (Table 15.1). The result implies that leaf expansion increased in NI-B and NI-R and decreased in NI-Fr. It was thought that photosynthetic pigments mainly absorb photons from the R and B light ranges of the visible spectrum during the photoperiod (Possart et al. 2014). In shade-intolerant plants such as *Arabidopsis thaliana*, the reduction in the R:Fr ratio has a number of striking effects on plant growth and development (Franklin 2008). Reduced leaf expansion in NI-Fr observed in this study is similar to a shade avoidance response. The chlorophyll content was particularly high and the greatest in NI-Fr and the least in NI-G (Table 15.1). Chlorophyll synthesis is associated with phytochrome activity (Huq et al. 2004; Monte et al. 2004; Tepperman et al. 2004), which is primarily affected by the ratio of the R and Fr lights (Smith 2000). It was often assumed that G light was not important in driving photosynthesis mainly due to the low absorptivity coefficient in the absorption spectra of purified chlorophylls (Sun et al. 1998).

Flowering was induced by all treatments (Fig. 15.1). The percent flowering was not significantly affected by the NIL quality (Table 15.1). Days after treatment initiation to visible flower bud or days to visible buds decreased in NI-G, NI-B, NI-R, and NI-W treatments compared to those in the SD. NI with photosynthetically active radiation (green, B, R, and W) and its photon fluxes lower than the compensation point (50 $\mu mol \cdot m^{-2} \cdot s^{-1}$ PPF, Yue et al. 1993) may not help vegetative growth other than the leaf area (Table 15.1) as expected but can hasten flowering (Fig. 15.1) even in day-neutral plants such as the geranium used in this study.

15.2.4 Short-Day Plant (Chrysanthemum)

The height of plants grown under LD and all NI treatments was greater than that of plants grown in SD (Table 15.1). In NI treatments, plant height was the greatest in NI-R (18.2 cm), followed by NI-Fr (17.7 cm). These results are in agreement with results reported by Kim et al. (2004) for chrysanthemum in which the stem and the length of the bottom 3rd internode were the greatest under R and R + Fr lights during the photoperiod. However, the effect of R light on stem elongation was inconsistent. Meanwhile, Heo et al. (2002) found that the stem length in marigold decreased under R light during the photoperiod, and this difference may be due to different synergistic interactions of the phytochromes acting on the inhibition of stem elongation. The level of phytochrome photo-equilibrium (Φ = PFr/P total ratio: ratio of PFr to total phytochrome) during the photoperiod decreased chrysanthemum height (Heins and Wilkins 1979). The results of the current study suggest that plant growth in response to the light quality given during the photoperiod and during NI are different.

The number of leaves per plant increased by 93% when the plant was grown in LD as compared to in SD (Table 15.1), and was the lowest in NI-Fr (82.6) and SD (80.3). The leaf area showed a similar tendency as the number of leaves per plant (Table 15.1). Plants grown under NI-W (370.0 cm^2) showed the greatest leaf area, followed by plants grown under NI-G (355.6 cm^2), and then plants grown under NI-R (353.5 cm^2).

The chlorophyll content was the greatest for plants grown in SD (Table 15.1) and was lower for plants in all NI treatments, especially in NI-Fr; NI-R was an exception where the plants showed a similar chlorophyll content compared to that in plants grown in SD. The chlorophyll content was 60% lower when grown in NI-Fr when compared to being grown in SD. Similar results were reported by Li and Kubota (2009) who described that the decrease in these phytochemicals with supplemental Fr during the photoperiod was attributed to 'dilution', since the enhancement of dry weight was shown under W + Fr light.

Flowering was completely induced by NI-B, NI-Fr, and SD (Fig. 15.1). Similar results were reported for chrysanthemums in previous studies (Higuchi et al. 2012; Jeong et al. 2012; Stack et al. 1998). It was reported that the *hy4* mutant alleles in the Columbia ecotype background flowered late in both SD and LD with either day extensions or NI, and that NI with B light had a stronger effect than NI with W or R lights (Bagnall et al. 1996). Days after treatment initiation to visible flower bud or days to visible buds observed in this study was prolonged by 1 day and 4 days in NI-Fr and NI-B, respectively (Fig. 15.1). The number of flowers per plant increased by 33% in NI-Fr compared to that in NI-B and SD. The early flowering in SD for the *phyB* mutants presumably reflects the constitutive shade avoidance phenotype of these plants (Franklin 2008; Franklin and Quail 2010).

15.3 The Effects of Shift in the Light Quality of Night Interruption on the Morphogenesis and Flowering in Chrysanthemum

15.3.1 Materials and Methods

15.3.1.1 Plant Materials and Growth Conditions

Chrysanthemum cuttings were stuck in 50-cell plug trays and were placed on a glasshouse bench. The rooted cuttings were transferred to a closed-type plant factory 12 days after transplanting and sticking, respectively. After 12 days of acclimatization in the plant factory, the plants (approximately 13.3 cm in plant height) were subjected to the photoperiodic light treatments. Growth conditions, such as the light

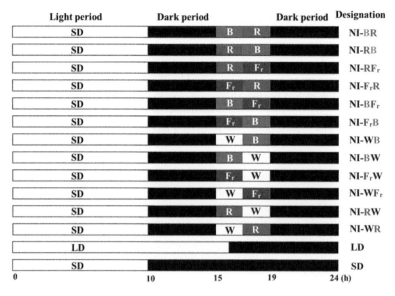

Fig. 15.3 Shifting of the night interruption light (NIL) quality by light-emitting diodes (LEDs) during a 4-h night interruption (NI) in 10-h short-day (SD) treatments: NI-BR, blue to red; NI-RB, red to blue; NI-RFr, red to far-red; NI-FrR, far-red to red; NI-BFr, blue to far-red; NI-FrB, far-red to blue; NI-WB, white to blue; NI-BW, blue to white; NI-FrW, far-red to white; NI-WFr, white to far-red; NI-RW, red to white; and NI-WR, white to red. LD indicates a 16-h long-day treatment

intensity, medium, and nutrient solution used were the same as those described in Sect. 15.2.1.1.

15.3.1.2 Photoperiodic Light Treatments

Plants were grown under 180 μmol·m^{-2}·s^{-1} PPFD provided by W LEDs under either LD (16 h light/8 h dark), SD (10 h light/14 h dark), or SD with NI for 4 h total (from 23:00 to 03:00) using LEDs at an intensity of 10 μmol·m^{-2}·s^{-1} PPFD. The quality of the NIL was shifted from one to another after the first 2 h in sets of two qualities among blue (B, 450 nm), red (R, 660 nm), far-red (Fr, 730 nm), and white (W, 400–700 nm). LD and SD were used as the control, and 12 SD treatments with shifting of the NIL quality by LEDs were as follows: from blue to red (NI-BR), from red to blue (NI-RB), from red to far-red (NI-RFr), from far-red to red (NI-FrR), from blue to far-red (NI-BFr), from far-red to blue (NI-FrB), from white to blue (NI-WB), from blue to white (NI-BW), from far-red to white (NI-FrW), from white to far-red (NI-WFr), from red to white (NI-RW), and from white to red (NI-WR) (Fig. 15.3). The methods of assessing the intensity and the quality of the NIL were the same as those described in Sect. 15.2.1.2. After 49 days of initiating the photoperiodic treatments, the plant height, number of leaves per plant, chlorophyll content, percent

Table 15.2 The effects of quality shifting of the night interruption light (NIL) provided at 10 μmol·m^{-2}·s^{-1} PPFD on the morphogenesis and flowering of chrysanthemum measured at 49 days after treatment, respectively

Night interruption (NI)	Plant height (cm)	No. of leaves	Leaf area (cm^2)	Chlorophyll (μg · mg^{-1} FW)
LD	21.8 cd[a]	389 a	461 de	3.172 ab
NI-BR	20.8 d	399 a	719 ab	3.058 ab
NI-RB	21.9 cd	352 ab	762 a	3.237 ab
NI-RFr	24.5 a	354 ab	614 bc	2.762 ab
NI-FrR	24.2 ab	324 ab	707 ab	2.958 ab
NI-BFr	24.7 a	171 c	464 de	2.565 ab
NI-FrB	21.9 cd	102 d	262 f	3.213 ab
NI-WB	18.0 e	218 c	440 de	3.014 ab
NI-BW	22.5 b–d	331 ab	676 a–c	3.181 ab
NI-FrW	22.2 cd	336 ab	402 e	2.305 b
NI-WFr	23.4 a–c	199 c	407 e	2.916 ab
NI-RW	21.9 cd	294 b	556 cd	3.484 a
NI-WR	22.3 cd	295 b	562 cd	3.485 a
SD	17.9 e	219 c	441 de	3.667 a

[a]Mean separation within columns by Duncan's multiple range test at 5% level

flowering, days after treatment initiation to visible flower bud or days to visible buds, and number of flowers per plant were measured.

15.3.2 Morphogenesis

The plant height was higher under all NI treatments except NI-WB, when compared to that under SD (Table 15.2). It was especially greater in NI containing Fr light, regardless of the order of light quality used, than in other treatments. These results are in agreement with those reported by Oyaert et al. (1999) and Kim et al. (2004) in chrysanthemum. The use of B light during the photoperiod to suppress plant elongation has been previously documented in chrysanthemum. Folta and Spalding (2001) found that the inhibition response was associated with the involvement of cryptochromes and phototropins and demonstrated that the inhibition process was a de-etiolated response. In this study, the suppression of plant height observed in NI-WB might have involved the effect of the B light in the second NIL.

The number of leaves per plant was the lowest in NI-FrB, probably due to the termination of vegetative growth caused by the early flowering (Table 15.2). Constitutively, the number of leaves was dependent on the changes in days to flowering, especially early flowering (Lin 2000). The leaf area was significantly larger for plants in NI-BR and NI-RB, and lower for plants in NI-FrB (Table 15.2), suggesting that leaf expansion is the result of the combined effect of the first and second NILs.

Fig. 15.4 The effects of quality shifting of the night interruption light (NIL) provided at 10 μmol·m^{-2}·s^{-1} PPFD on flowering of chrysanthemum measured at 49 days after treatment, respectively. See Fig. 15.1 for details on the quality of the NIL

For example, the combination of B and R light promotes leaf expansion, while the combination of Fr and B suppresses it. Matsuda et al. (2007) found a higher photosynthetic capacity in spinach leaves when grown under 300 μmol·m^{-2}·s^{-1} mixed R/B irradiance containing 30 μmol·m^{-2}·s^{-1} B than in leaves grown under R alone. Therefore, in this study, the increase of relative growth rate in NI-RB could be attributed to the synergistic effect of R and B lights.

The overall chlorophyll content was lower for plants in NI provided by Fr light, regardless of the NIL order used compared to that for plants in other treatments (Table 15.2). The chlorophyll content was 37% lower for plants in NI-FrW compared to plants in SD. This response suggests that Fr fluorescence consists of a low light intensity compared to the higher levels of incident light reflected from leaves in this spectral region (Smorenburg et al. 2002).

15.3.3 Flowering

Flowering was induced by NI-RB, NI-FrR, NI-BFr, NI-FrB, NI-WB, NI-FrW, NI-WFr, NI-WR, and SD, and the days after treatment initiation to visible flower bud or days to visible buds were shortened in NI-BFr and NI-FrB (Fig. 15.4). This suggests that flowering of the short-day plant (chrysanthemum) was promoted by the synergistic effect of B and Fr lights. In a combined shifting treatment of B and R lights (NI-RB) or B and W lights (NI-WB), NI terminated by B light-induced flowering. Thus, it is probable that the B light receptor can strengthen the floral-inducer activity, resulting in a higher energy requirement to inhibit flowering. However, in the combined treatment of the R and Fr lights (NI-RFr and NI-FrR) or Fr and W lights (NI-FrW and NI-WFr), flowering was not affected by quality shifting of the NIL. These findings suggest that the P_r/P_{fr} ratio may be more important than the effects of quality shifting of the NIL. In many short day plants, NI becomes effective only when the supplied dose of light is sufficient to saturate the photoconversion of P_r (phytochrome that absorbs R light) (Purohit and Ranjan 2002). A subsequent exposure to the Fr light, which photoconverts the pigment back to the physiologically inactive P_r form, restores the flowering response. The days after treatment initiation to visible flower bud or days to visible buds increased in NI-RB, NI-FrR, NI-FrW, and NI-WR. This finding suggests an antagonistic effect of the B, R, W, and Fr light receptors in the inhibition and promotion of chrysanthemum flowering. The results also showed that the number of flowers per plant was 32% higher in NI-BFr compared to that in SD.

15.4 Conclusions

To summarize, NI-G increased the shoot length in the long-day plant, increased the petiole length and decreased the chlorophyll content in the day-neutral plant, and increased the plant height and leaf expansion in the short-day plant. NI-B increased leaf expansion in the long-day plant, day-neutral plant, and short-day plant. NI-R increased leaf expansion in the long-day plant, day-neutral plant, and short-day plant and increased plant height in the short-day plant. NI-Fr increased plant height in the long-day plant, day-neutral plant, and short-day plant; decreased the chlorophyll content in the long-day plant and short-day plant, but not in the day-neutral plant; and decreased leaf expansion in the day-neutral plant and short-day plant. NI-W increased leaf expansion in the long-day plant, day-neutral plant, and short-day plant. In conclusion, the morphogenesis and flowering of the three photoperiodic model plants were significantly affected by the NIL, especially in the short-day plant. The quality of the NIL in the day-neutral plant had a more pronounced effect on the morphogenesis than on the flowering. These findings suggested that NI-R or NI-W was the most suitable in the NI strategy of controlling the morphogenesis and flowering of the long-day plants during SD seasons. Moreover, B light may have applicability for flowering control in short-day plants.

The plant height was the smallest in NI-WB among all NI treatments. The impact of changing the light quality at low-intensity NI treatments on leaf expansion was greater in treatments using a combination of B and R or R and W lights, regardless of their order of use. Flowering was observed in NI-RB, NI-FrR, NI-BFr, NI-FrB, NI-WB, NI-FrW, NI-WFr, NI-WR, and SD treatments and was especially promoted in NI-BFr and NI-FrB treatments. In a combined shifting treatment of either B and R lights or B and W lights, NI concluding with B light (NI-RB and NI-WB) exposure induced flowering. Thus, it is likely that the B light receptor enhances the activity of floral inducers, resulting in a higher energy requirement for the promotion of flowering. Statistically, the light quality of the first 2 h of NI exposure affected neither the morphogenesis nor flowering, while the light quality of the last 2 h of NI exposure significantly affected both the morphogenesis and flowering, as we hypothesized. Further studies are required to investigate the effects of B light supplementation on the flowering promotion, photoreceptor gene expression, and protein production in other cultivars of chrysanthemum.

References

Bagnall DJ, King RW, Hangarter RP (1996) Blue-light promotion of flowering is absent in hy4 mutants of *Arabidopsis*. Planta 200:278–280
Bunning E, Moser I (1969) Interference of moonlight with the photoperiodic measurement of time by plants, and their adaptive reaction. Proc Natl Acad Sci U S A 62:1018–1022
Craig DS, Runkle ES (2012) Using LEDs to quantify the effect of the red to far-red ratio of night-interruption lighting on flowering of photoperiodic crops. Acta Hort (956):179–185
Dougher TAO, Bugbee B (2004) Long-term blue light effects on the histology of lettuce and soybean leaves and stems. J Amer Soc Hort Sci 129:497–472
Evans LT (1971) Flower induction and the florigen concept. Ann Rev Plant Physiol 22:365–394
Folta KM, Spalding EP (2001) Unexpected roles for cryptochrome 2 and phototropin revealed by high-resolution analysis of blue light-mediated hypocotyl growth inhibition. Plant J 26:471–478
Franklin KA (2008) Shade avoidance. New Phytol 179:930–944
Franklin KA, Quail PH (2010) Phytochrome functions in *Arabidopsis* development. J Expt Bot 61:11–24
Heins RD, Wilkins HF (1979) The influence of node number, light source, and time of irradiation during darkness on lateral branching and cutting production in 'bright golden Anne' chrysanthemum. J Amer Soc Hort Sci 104:265–270
Heo JW, Lee CW, Chakrabarty D, Paek KY (2002) Growth responses of marigold and salvia bedding plants as affected by monochromic or mixture radiation provided by a light emitting diode (LED). Plant Growth Regul 38:225–230
Hersch M, Lorrain S, Wit M, Trevisan M, Ljung K, Bergmann S (2014) Light intensity modulates the regulatory network of the shade avoidance responses in *Arabidopsis*. Proc Natl Acad Sci U S A 111:6515–6520
Higuchi Y, Sumitomo K, Oda A, Shimizu H, Hisamatsu T (2012) Days light quality affects the night-break response in the short-day plant chrysanthemum, suggesting differential phytochrome-mediated regulation of flowering. J Plant Physiol 169:1789–1796
Higuchi Y, Narumi T, Oda A, Nakano Y, Sumitomo K, Fukai S, Hisamatsu T (2013) The gated induction system of a systemic floral inhibitor, antiflorigen, determines obligate short-day flowering in chrysanthemums. Proc Natl Acad Sci U S A 110:17137–17142
Huq E, Al-Sady B, Hudson M, Kim C, Apel K, Quail PH (2004) Phytochrome-interacting factor 1 is a critical bHLH regulator of chlorophyll biosynthesis. Sci Signaling 305:1937–1941
Jeong SW, Park S, Jin SJS, Seo O, Kim GS, Kim YH, Bae H, Lee G, Kim ST, Lee WS, Shin SC (2012) Influences of four different light-emitting diode lights on flowering and polyphenol variations in the leaves of chrysanthemum. J Agric Food Chem 60:9793–9800
Kadman-Zahavi AVISHAG, Peiper D (1987) Effects of moonlight on flower induction in *Pharbitis nil*, using a single dark period. Ann Bot 60:621–623

Kim SJ, Hahn EJ, Heo JW, Paek KY (2004) Effects of LEDs on net photosynthetic rate, growth and leaf stomata of chrysanthemum plantlets in vitro. Sci Hort 101:143–151

Kim YJ, Lee HJ, Kim KS (2011) Night interruption promotes vegetative growth and flowering of *Cymbidium*. Sci Hort 130:887–893

Li Q, Kubota C (2009) Effects of supplemental light quality on growth and phytochemicals of baby leaf lettuce. Environ Expt Bot 67:59–64

Lin CT (2000) Plant blue-light receptors. Trends Plant Sci 5:337–342

Matsuda R, Ohashi-Kaneko K, Fujiwara K, Kurata K (2007) Analysis of the relationship between blue-light photon flux density and the photosynthetic properties of spinach (*Spinacia oleracea* L.) leaves with regard to the acclimation of photosynthesis to growth irradiance. Soil Sci Plant Nutr 53:459–465

Monte E, Tepperman JM, Al-Sady B, Kaczorowski KA, Alonso JM, Ecker JR, Li X, Zhang Y, Quail PH (2004) The phytochrome-interacting transcription factor, PIF3, acts early, selectively, and positively in light-induced chloroplast development. Proc Natl Acad Sci U S A 101:16091–16098

Muneer S, Kim EJ, Park JS, Lee JH (2014) Influence of green, red and blue light emitting diodes on multi protein complex proteins and photosynthetic activity under different light intensities in lettuce leaves (*Lactuca sativa* L.). Intl J Mol Sci 15:4657–4670

Oh W, Rhie YH, Park JH, Runkle ES, Kim KS (2008) Flowering of cyclamen is accelerated by an increase in temperature, photoperiod and daily light integral. J Hort Sci Biotechnol 83:559–562

Oyaert E, Volckaert E, Debergh PC (1999) Growth of chrysanthemum under coloured plastic films with different light qualities and quantities. Sci Hort 79:195–205

Park YJ, Kim YJ, Kim KS (2013) Vegetative growth and flowering of *Dianthus*, *Zinnia*, and *Pelargonium* as affected by night interruption at different timings. Hort Environ Biotechnol 54:236–242

Park YG, Muneer S, Jeong BR (2015) Morphogenesis, flowering, and gene expression of *Dendranthema grandiflorum* in response to shift in light quality of night interruption. Intl J Mol Sci 16:16497–16513

Park YG, Muneer S, Soundararajan P, Manivnnan A, Jeong BR (2016) Light quality during night interruption affects morphogenesis and flowering in *Petunia hybrida*, a qualitative long-day plant. Hort Environ Biotechnol 57:371–377

Park YG, Muneer S, Soundararajan P, Manivnnan A, Jeong B (2017) Light quality during night interruption affects morphogenesis and flowering in geranium. Hort Environ Biotechnol 58:212–217

Possart A, Fleck C, Hiltbrunner A (2014) Shedding (far-red) light on phytochrome mechanisms and responses in land plants. Plant Sci 217-218:34–46

Purohit SS, Ranjan R (2002) Flowering. In: Purohit SS, Ranjan R (eds) Phytochrome and flowering. Agrobios, Jodhpur, pp 52–61

Quail PH, Boylan MT, Parks BM, Short TW, Xu Y, Wagner D (1995) Phytochromes: Photosensory perception and signal transduction. Science 268:675–680

Reddy SK, Finlayson SA (2014) Phytochrome B promotes branching in *Arabidopsis* by suppressing auxin signaling. Plant Physiol 164:1542–1550

Saebo A, Krekling T, Appelgren M (1995) Light quality affects photosynthesis and leaf anatomy of birch plantlets in vitro. Plant Cell Tissue Organ Cult 41:177–185

Smith H (2000) Phytochromes and light signal perception by plants an emerging synthesis. Nature 407:585–591

Smorenburg K, Bazalgette CLG, Berger M, Buschmann C, Court A, Bello UD, Langsdorf G, Lichtenthaler HK, Sioris C, Stoll MP, Visser H (2002) Remote sensing of solar induced fluorescence of vegetation. Proc SPIE 4542:178–190

Stack PA, Drummond FA, Stack LB (1998) Chrysanthemum flowering in a blue light-supplemented long day maintained for biocontrol of thrips. Hortscience 33:710–715

Sun J, Nishio JN, Vogelmann TC (1998) Green light drives CO_2 fixation deep within leaves. Plant Cell Physiol 39:1020–1026

Tanaka M, Takamura T, Watanabe H, Endo M, Yanagi T, Okamoto K (1998) In vitro growth of *Cymbidium* plantlets cultured under superbright red and blue light-emitting diodes (LEDs). J Hort Sci Biotech 73:39–44

Tepperman JM, Hudson ME, Khanna R, Zhu T, Chang SH, Wang X, Quail PH (2004) Expression profiling of *phyB* mutant demonstrates substantial contribution of other phytochromes to red-light-regulated gene expression during seedling de-etiolation. Plant J 38:725–739

Tong Z, Wang T, Xu Y (1990) Evidence for involvement of phytochrome regulation in male sterility of a mutant of *Oryza sativa* L. Photochem Photobiol 52:161–164

Vandenbussche F, Pierik R, Millenaar FF, Voesenek LA, Van Der Straeten D (2005) Reaching out of the shade. Curr Opin Plant Biol 8:462–468

Vince-Prue D, Canham AE (1983) Horticultural significance of photomorphogenesis. p. 518-544. In: Shropshire W, Mohr H (eds) Encyclopedia of plant physiology (NS). Springer-Verlag, Berlin

Yamada AT, Tanigawa T, Suyama T, Matsuno T, Kunitake T (2009) Red:far-red light ratio and far-red light integral promote or retard growth and flowering in *Eustoma grandiflorum* (Raf.) Shinn. Sci Hort 120:101–106

Yue D, Gosselin A, Desjardins Y (1993) Effects of forced ventilation at different relative humidities on growth, photosynthesis and transpiration of geranium plantlets in vitro. Can J Plant Sci 73:249–256

Part V
Advanced Technologies to Be Implemented in the Smart Plant Factory

Part V

Chapter 16
Mechanization of Agriculture Considering Its Business Model

Tamio Tanikawa

Abstract In recent years, it is expected that practical application of technologies may significantly change business models such as IoT, big data, AI, and robotics. Combining these technologies is expected to make agriculture more efficient and strong industries, but it is difficult to realize with discussions only in the agriculture field. In this chapter, we discuss how to realize the strong agricultural business, centering on the discussion of backcasting from the business model.

Keywords Mechanization agriculture · Robotics · Data analysis · Business model

16.1 Introduction

In Japan and many developed countries, there are major challenges to respond to the aging society due to the rapid aging of the population. In particular, the declining labor force due to the declining birthrate and aging population will cause domestic industrial competitiveness to deteriorate. Automation technology has been actively introduced to secondary industries such as cars and electric appliances. One of the things that supported industries in Japan is automation technology represented by industrial robots. In order to expand the robot industry cultivated in the industrial field, technology development as a service robot to the living field, nursing care, and welfare fields has been advanced as support for an aging society. It is mainly the utilization of robot technology for tertiary industry. Particularly in the tertiary industry, it is also a labor-intensive-type industry, and robot technology as support of workers is expected. Meanwhile, also in the primary industry, labor shortage due to aging becomes serious, and many of the tasks that are easy to mechanize are shifting to automation. Many of agriculture are also labor intensive, so in the aging society of the future, the productivity will inevitably decrease. Even in the primary industry, mechanization like a robot must be further promoted.

T. Tanikawa (✉)
National Institute of Advanced Industrial Science and Technology (AIST), Tokyo, Japan
e-mail: tamio.tanikawa@aist.go.jp

16.2 Mechanization of Agriculture

In the mechanization of agricultural work, there is a long history at the research level. It is technically not inferior, and it is not missing technology development. On the other hand, even if the technology is sufficient, practical application will not always succeed.

Figure 16.1 shows an obstacle from basic research to business. It is well known that there is a "valley of death" from basic research to technology development toward practical use. Technological development here will proceed with the "addition" approach to develop necessary technologies. On the other hand, there is "the Darwin sea" imitating the cost competition of commercialization as an obstacle to commercialization. This is the idea of "subtraction" which reduces the cost of developed technology according to the business model. In other words, if you get too negative by requiring too much subtraction, it means that it is not a business. We have to make business models that do not subtract much. Specifically, for practical application, it is important whether or not there is a large market that can be mechanized. For example, since the secondary industry has an international market, it is expected to have ample capital investment as a facility for mass production. Therefore, automation technology typified by robot technology is a big business in the secondary industry. On the other hand, in agriculture, there is not a huge international market at the moment. Even if there is high technical capability, we cannot expect the amount of capital investment to be established as business. In other words, if the balance between benefits and costs associated with mechanization does not tilt toward the benefit side, even if there is high technical capability, it is difficult to put it into practical use. Until now, as technology development of service robots, robots are expected to be utilized in areas such as life support and tertiary care support in tertiary industry, but the same problem is faced.

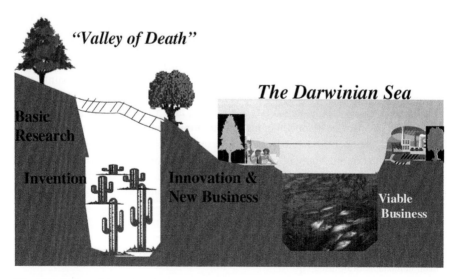

Fig. 16.1 Valley of death in technology development and the Darwin sea for practical application. C. Wessner OECD lecture material

Even if we aim to orient merit of mechanization in agriculture in the same direction as mass production in industrial field, as international market does not exist, as mentioned in the previous section, it is difficult to commercialize. Reconsidering the benefits of mechanization toward business is an important point in advancing mechanization of agriculture. Traditional mechanization was aimed at substituting human labor phrases. Therefore, the cost comparison is compared with labor costs of workers. If the business market is not wide, machine sales stations are few, and unit prices do not decrease. As a result, the employment cost of workers is lower than mechanization, and furthermore, machinery work is part of agricultural work, so the benefits of mechanization are low.

In order for agricultural producers to make capital investment, it is important that stable profits are always obtained. It is not a mechanization aiming for a large profit by mass production; there is also a notion of mechanization to maintain a stable supply every year without being influenced by the weather. By maintaining the market price of agricultural products, if we can realize a stable profit for the year, we can estimate necessary capital investment. Currently, costs may still be severe, but the plant factory is one solution. The plant factory is not influenced by the weather, a stable production amount can be secured, and the production amount is easy to predict. Considering the total cost to consumption, such as there is no need to wash pesticides because they do not use pesticides, there is also the possibility that the plant factory will be established as a business. The plant factory is not yet a production technology that can deal with various agricultural crops. However, mechanization to a steady production volume like a plant factory is an important concept in agriculture.

Stabilization of agricultural output is the first step in mechanization. Secondly, in order to obtain higher profit, another purpose of mechanization is to increase the added value of agricultural crops and its visualization. A business that sells agricultural products at a high price by raising added value, such as organic agriculture or organic pesticide cultivation, is established. This is a business model that is built on credit. If it can clearly visualize added value, it seems that international competitiveness can be obtained especially for Japanese quality agricultural crops. Here, an important element is data in the production process.

16.3 Precision Agriculture Based on Production Data

While smartphones are producing a big business, there have been equipment of the same function as smartphones like electronic notebooks in the past. Why did not electronic notepads become popular at that time? On the other hand, why are smartphones which have the same concept very popular now? One of the major reasons is the spread of the Internet and wireless technology for mass data transfer. The IoT technology infrastructure that things are connected to the Internet has been developed, and many business models that have never existed have been born. Even in agriculture, this trend is likely to cause major reforms. By adding the Internet to the mechanization in the previous section, the data obtained by mechanization can be widely used, and new added value is created. Up to now, mechanization was simply a cost comparison with workers' labor costs. However, by mechanizing, it is possible to leave the history in agricultural work as accurate data that cannot be obtained by human work. That is a new benefit of mechanization. Work content in a labor-intensive field like agriculture depends greatly on the experience of workers, and it is difficult to formalize. Agriculture is particularly difficult to formalize because it is necessary to select optimal conditions for production of products in various environmental changes. In recent years, various environmental data and work history information in agricultural work as big data are analyzed by using AI technology, and precision agriculture based on scientific evidence is advanced. Precision agriculture here is to formalize work contents that were an empirical tacit knowledge by using AI technology, etc. from the obtained data and to perform optimum agriculture based on scientific theory. Furthermore, by combining the data in the production process obtained here with the traceability of the product, it is possible to visualize the quality of the product, such as the safety and taste. Thereby, there is a possibility that it can be sold as a high value with high added value corresponding to the labor of the production process. In particular, when aiming for international markets, this traceability system guarantees the brandification of Japanese products, and it is possible to establish a business model of agricultural products of high added value which does not fall into price competition. Mechanization of agriculture means not only discussing mechanization within the scope of production but also designing a total business model including distribution up to the table. And then the necessary mechanization should be discussed in that business model.

16.4 Conclusion

While the recent IoT, big data, and AI technology are greatly expected, the original strength will be shown by cross-industry and mutual use of each data. On the other hand, it is very difficult to connect the industry sideways in the Japanese vertical integration business model. However, smart agriculture is not a debate of only conventional agriculture, but optimal design among industries such as logistics and food service (sixth sector industrialization) is required. In other words, the new business model cannot be done with discussion of forecast that creates business from conventional technology development. First of all, we need to design a business model that realizes strong agriculture. Next, we examine the technology utilization (Cloud, IoT, AI, robot, etc.) necessary for the business and the social system to utilize the technology. Discussions on such backcasting should proceed.

Reference

Wessner C (1998) Public/private partnerships for innovation: experiences and perspectives from the U.S. OECD lecture material

Chapter 17
Quantifying the Environmental and Energy Benefits of Food Growth in the Urban Environment

Rebecca Ward, Melanie Jans-Singh, and Ruchi Choudhary

Abstract The environment and energy consumption of indoor urban farms within the city's infrastructure cannot at present be adequately simulated using typical building simulation models as they do not include the ability to simulate the potentially significant heat and mass transfer between plants and the internal air. On the other hand, tools developed for the simulation of climate-controlled greenhouses do not allow complex interactions with existing buildings and infrastructure. In this chapter, we present the development of an urban-integrated greenhouse model, with the ability to simulate the response of the indoor climate to crop growth. We validate the model against data from an urban farm 50 m underground. Applying an analysis of resource needs and availability through a numerical simulation model allows us to investigate mechanisms to optimise the environment and energy benefits of growing food within the city space.

Keywords Urban-integrated agriculture; Numerical simulation; Co-benefit potential

17.1 Introduction

City life is already pressured due to increases in urban population and the impacts of climate change; the future demands on cities are expected to reshape both our urban and rural environments. Urban populations heavily draw on local and remote natural resources, with knock-on effects on biodiversity and ecosystem services. A heavily urbanised world will also face pressure from climate change manifested through changing weather patterns, demographic shifts and social changes. However, more than 60% of the urban area projected for 2030 has yet to be built, thus providing the opportunity to transform our future cities with resilience in mind (Secretariat of the Convention on Biological Diversity 2012).

R. Ward · M. Jans-Singh · R. Choudhary (✉)
Engineering Department, University of Cambridge, Cambridge, UK
e-mail: rmw61@cam.ac.uk; mkj32@cam.ac.uk; ruchi.choudhary@eng.cam.ac.uk

Globally, arable land available per person has decreased from 0.39 ha in 1965 to 0.20 ha today (World Bank 2016), and increasing urban populations mean that food miles are tending to rise. Food production closer to the point of use, i.e. urban farming, offers an elegant way to relieve some of these stresses, as argued throughout this book. The rise in attraction of locally produced food may be linked to the cost of food production which has risen by 9% over the last 10 years, with modern cities relying almost exclusively on imports for food. With increasingly globalised trade, countries are rarely self-sufficient, and complex supply and transport systems govern how food arrives on our plates. For example, the average food product travels 1650 km in the United States (Weber and Matthews 2008), and only 54% of the food consumed in the United Kingdom is produced there (Department of Agriculture Food and Rural Affairs 2016). Landlocked or highly populated areas are particularly vulnerable. Singapore, Hong Kong, the UAE, Egypt and Norway are a few examples of countries with less than 5% arable land (World Bank 2016). Hydroponic urban farms thus offer a promising new technology for the growth of food in urban environments in a manner which is efficient both in terms of land and environmental impact. Touliatos et al. (2016) showed that vertical hydroponic systems could significantly increase the yield of lettuce per unit area, despite the spatial variation in environmental conditions. Such urban farms may be sited closer to the point of use and need not always be in dedicated buildings but could be integrated seamlessly into the city's fabric. For instance, a recent study showed that a hydroponic rooftop greenhouse could halve the energy use compared with the existing supply chain for a tomato in Lisbon (Benis et al. 2017).

There are potential benefits of urban farming in addition to improved food security and reduced food miles. Urban sprawl puts stress on urban infrastructure, as cities try to manage air quality, water, energy supply and movement of goods. For instance, cities currently rely heavily on large grey infrastructure for storm water management due to their low porosity and high built density. Urban agriculture could be a productive alternative for more sustainable urban drainage solutions by providing local reuse of storm water (Zahmatkesh et al. 2015). Furthermore, energy consumption in cities is set to rise by up to 30% due to increased cooling requirements to mitigate urban heat island (UHI) effects and temperature increases due to climate change (Kolokotroni et al. 2012). Increasing the proportion of vegetation using green roofs (Xu et al. 2016) or trees (Skelhorn et al. 2016) has been shown to reduce the UHI effect and thereby to reduce the building energy demand for cooling. Kikegawa et al. (2006) used building energy simulation to demonstrate that side wall greening was an effective countermeasure against UHI as it reduced the baseline temperature of the cityscape. The impacts may not just be observed in the reduction of cooling demand; a small-scale study of a rooftop greenhouse showed that integrating hydroponics to the building's ventilation design could reduce energy demand for heating by 41% (Delor 2011).

The objective of this chapter is to present mechanisms for designers to quantify the benefit of small-scale food farms located in urban spaces, with a particular emphasis on the integration of a closed environment greenhouse into unused city space. Following this exploration of the motivations and opportunities for urban agriculture projects, unused urban spaces and the corresponding urban farm designs are discussed in Sect. 17.2. The resources required and strategies for synergising growing environments with the available urban resources are explained in Sect. 17.3. Simulation models for controlled environment agriculture (CEA) have thus far been developed for stand-alone greenhouses, especially in the Netherlands (Vanthoor 2011), and the CEA research group in Arizona (Kacira 2016), but models simulating the climate of an urban-integrated greenhouse are nascent (Benis and Ferrao 2017; Graamans et al. 2017). Section 17.4 thus reviews the development of simulation and introduces Sect. 17.5: a detailed look into the energy simulation and monitoring of an urban-integrated farm in London.

17.2 Derelict and Vacant Space in Cities

Urban farming can be found in different forms: allotments, conventional greenhouses or "building-integrated agriculture" (BIA). The focus of this work is the reuse of derelict or vacant space for urban farming, and the different typologies of spaces for BIA are listed below. These are illustrated in Fig. 17.1.

1. Flat, unused rooftops;
2. Underground: unused tunnels and basements;
3. Vacant space within buildings (abandoned or empty floors);
4. Occupied space within buildings: co-existent space such as atriums, corridors, and open plan spaces.

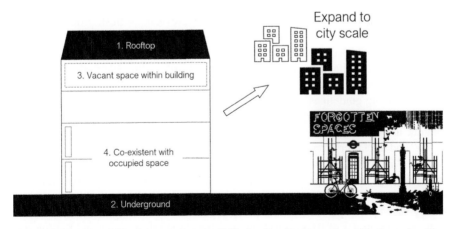

Fig. 17.1 The four types of vacant space available in cities for integration of hydroponics. The "Forgotten Spaces" image refers to the exhibition at Somerset House designed by Studio Glowacka (2013) for the RIBA (Royal Institute of British Architects), exploring how vacant spaces in London could be repurposed

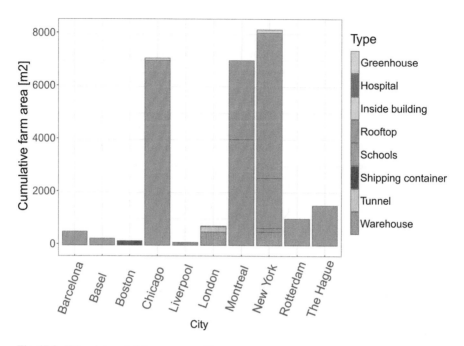

Fig. 17.2 Urban-integrated farms across cities

Derelict spaces are often forgotten by city planners and developers but offer resource availabilities that could be usefully harnessed for food production purposes. The "Forgotten Spaces" exhibition in London showcased innovative concepts to purpose this "lost" space in 2013 (Studio Glowacka 2013). In the United States, Grewal and Grewal (2012) found that reusing commercial rooftops and vacant lots for urban farming could generate up to 100% of Cleveland's fresh produce needs, improving self-sufficiency from 0.1% to 17%. In Europe, the population is already urbanised to 70–80%, and urban sprawl is decreasing the available land for agriculture. However, availability of space is not necessarily a limiting factor for urban farming, especially if reuse of space for multifunctional purposes becomes integral to city planning guidelines (Viljoen et al. 2010; Custot et al. 2012).

The growth of urban farming has varied across the world (Opitz et al. 2016), in part due to the difference between the established urban development in the "Global North" and the faster growing cities in the "Global South". In Asia, high-tech "plant factories" in dedicated warehouses have been spreading exponentially, while cities in developing countries are rife with small urban gardens (Orsini et al. 2013). Interest in hydroponics and new technologies has been growing in Western cities, typically led by start-up companies and non-profit organisations catering to niche markets. The major examples of BIA projects across cities in North America and Europe are presented by type and surface area in Fig. 17.2. As can be seen in the figure, the majority of urban farming ventures are in North America and on rooftops. Indeed, agricultural knowledge in urban areas has faded in Europe; in 1918, "Victory

Fig. 17.3 Urban farmers have re-purposed the top floor of an abandoned Philip's factory in The Hague. It is the largest aquaponic farm in Europe (UrbanFarmers AG 2016)

Gardens" sprouted everywhere in London producing up to 2 million tons of vegetables (House of Commons 1998), but city planning decisions since World War 2 have led to an expansion of grey infrastructure with a corresponding loss of green space for horticulture in European cities (Barthel et al. 2010).

Some examples of recent urban farming projects at different scales are illustrated in Figs. 17.3, 17.4, 17.5, 17.6, 17.7, and 17.8. The Netherlands are at the forefront of greenhouse engineering, and Europe's largest urban farm was opened on the top floor of an abandoned factory in The Hague in April 2016 (Fig. 17.3). Figure 17.4 is another example of a rooftop greenhouse, this time on a smaller scale in London, UK. While rooftop farming is not exclusive to western countries, no examples were found of controlled environment agriculture integrated with abandoned/unused space in other cities. However, the rural exodus in developing countries does bring agricultural knowledge to big cities. For instance, in Kathmandu, many people grow their vegetables on the roof and in any available space to hand. The different climates outside of Europe and North America sometimes render a greenhouse unnecessary and advanced forms of farming too expensive. The photographs in Figs. 17.5 and 17.6 show how the flat rooftops in Kathmandu (Nepal) are routinely used to grow greens. Derelict space is not always associated with buildings; Fig. 17.7 shows a hydroponic farm in disused underground tunnels in London, UK. And urban farms may serve more than just a food-producing purpose and also promote well-being; Fig. 17.8 shows the dual benefits of an aquaponic system enjoyed by children in a hospital in Liverpool, UK.

Fig. 17.4 The hydroponic greenhouse on the roof of the seawater greenhouse offices in London

Fig. 17.5 The rooftops in Kathmandu are reused to grow food year long

17.3 Resource Needs for a Hydroponic Urban Farm

The complexity of existing buildings in dense urban areas, including structural constraints, justifies the attraction for hydroponics. By circulating nutrient-rich drainage water at regular intervals to the base of the plants, hydroponics can produce

Fig. 17.6 Rural urban exodus has brought agricultural knowledge in growing cities

Fig. 17.7 Hydroponic farm in London operated by Growing Underground

a high density and turnover of plants in a small area (Kozai et al. 2015) while using three times less water than conventional growing methods (Grewal et al. 2011). Aquaponics, a different version of hydroponics, combines fish farming by recirculating the water from the fish tanks to feed the plants. These systems can

Fig. 17.8 Aquaponics system designed by farm urban integrated on balconies in the Alder Hey children's hospital, Liverpool

either be used with artificial lighting in closed environments or harness sunlight through glass windows.

Despite requiring energy for irrigation, lighting and ventilation, hydroponics need not necessarily be a drain on resources: in Manhattan, USA, a 120 m^2 greenhouse operated successfully without being connected to the grid by harvesting rainwater and using renewable energy through a wind turbine and solar panels (Nelkin and Caplow 2008). To link climate control of a hydroponic farm optimally with the urban environment, the physical properties of the resources and how to source them need to be understood. Figure 17.9 shows the six major resources required for the performance of an urban farm that can be found in cities: heat, light, CO_2, water, air flow and space. The availability of space was discussed in Sect. 17.2. Out of the remaining five required resources, three – heat, water and CO_2 – are available in abundance in cities as waste streams, whereas light and air flow are site dependent. Availability of these resources is discussed in more detail below. Nutrient intake is also important for plant growth and development (Kozai et al. 2015; Teitel et al. 2010; Tei et al. 1996), but as it independent of the scope of climate control, it is assumed here that the plants are given the right quantity of nutrients at all times.

17.3.1 Heat

The main sources of sensible heat in an urban-integrated farm are the incoming air, heat emitted by the light source, heat transmitted from the surrounding infrastructure

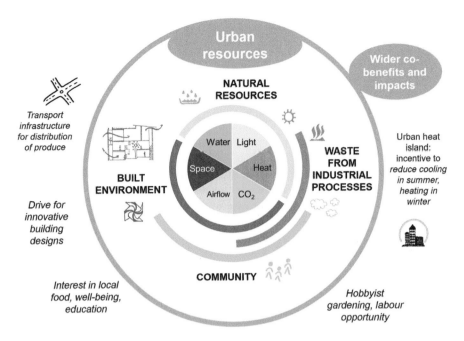

Fig. 17.9 The six major resources needed to run a hydroponics farm in a colour-coded wheel are shown with possible urban resources available within a city. Within the grey circle, the urban sources for the farm requirements are indicated by the nearest arc. The wider co-benefits and impacts of integrating urban farming are proposed in the outer circle

and any supplementary heating supplied. In the case of a semiopen farm, sunlight is the main source of radiation, whereas closed environments rely on artificial lighting. Heat exchange between a plant and its surroundings occurs through direct radiation reflection from the plant canopy, plant transpiration (sometimes referred to as evaporation), heat conduction and convection. As shown in Fig. 17.9, natural resources like sunlight and waste heat from industrial processes can be used to keep a greenhouse at adequate temperatures.

Low-grade waste heat has been reused for greenhouses since the 1970s, especially in large hydroponic greenhouses for tomatoes in North America and Northern Europe; however, these synergistic systems have remained limited due to technology and location (Parker and Kiessling 2016). There is a growing amount of recoverable heat, with 11.4 TWh in the United Kingdom, for instance, a quarter of which is from the food and drinks processing sector (Law et al. 2013), and heat exchangers can easily reuse the waste heat. It is more economical for greenhouses in cold climates to operate by making use of such waste heat streams, despite the difficulties of relying on another business or activity for the heat source.

17.3.2 Water

Water is also essential for plant growth to enable photosynthesis. In an enclosed farm environment, water circulates due to plant transpiration, evaporation from any other water surfaces, condensation of the water on surfaces and of course the incoming water vapour content of the air. The amount of water used by the plants is thus defined by the latent heat flux, which will be explained in more details in the following sections of this chapter.

Water for irrigating plants in urban farms is often sourced from the grid (e.g. Growing Underground in London, Fig. 17.7, or plant factories in Japan) but can also be sourced from the waste water streams of a city. Storm water and waste water are often combined in cities for ease of treatment but at a great cost for the services and infrastructure owners. Instances of reusing rainwater locally for greenhouses have been shown to be successful.

17.3.3 Airflow

The movement of air around the crop can enhance convective processes such as heat dissipation at the leaf surface and gas exchange (Wheeler et al. 1994; Van Iersel 2003). It influences temperature, humidity and CO_2 concentration in the air and thus impacts on crop growth and development. The ventilation design can influence optimal diurnal temperature variation by exploiting the natural buoyancy of air (Chaudhary et al. 2011). Mechanically assisted ventilation is very common, but sufficient analysis is generally required to ensure adequate air changes per hour.

17.3.4 Light

Plant growth processes, particularly photosynthesis, require light, specifically photosynthetically active radiation (PAR). Leaves preferentially absorb red and blue light, a process which is optimal at specific temperatures governed by the amount of incident radiation. In a completely closed environment, artificial lighting must replicate the components of solar radiation which allow optimal plant growth, in terms of light energy and length of the photoperiod. Martineau et al. (2012) compared the two most widely used technologies to replace natural light – light-emitting diodes (LED) and high-pressure sodium (HSD) lamps – and found that although the latter produce more mols of light, both technologies have similar yields per mol of light, and LEDs use 33% less energy than their counterpart. Consequently, LEDs are a keen subject of research to find the ideal red and blue light balance and the optimal photoperiod (Davis 2015).

In a traditional or semiopen greenhouse, artificial lighting need only supplement daylight, and the PAR can be calculated using Eq. 17.1. Sufficient knowledge of the incident sunlight angles can maximise the use of sunlight and minimise the supplementary lighting required.

$$PAR_{supp} = PAR_{req} - PAR_{Daylight} \quad (Wm^{-2}) \quad (17.1)$$

The latent heat flux in a greenhouse is mostly determined by the transpiration of the canopy. If the transpiration is assumed to be decoupled from the difference in moisture content between the inside and outside air (Boulard et al. 2000), the definition by Graamans et al. (2017) adapted from the Penman-Monteith formula (Teitel et al. 2010; Boulard et al. 2000) may be used to describe the transpiration flux with:

$$Q = LAI \, h_{fg} \frac{C_v - C_a}{r_s + r_a} \quad (Wm^{-2}) \quad (17.2)$$

where LAI is the leaf area index, h_{fg} is the latent heat of evaporation of water (J/kg) and C_v and C_a are the water vapour content at the transpiring surface and in the air respectively ($kg \, m^{-3}$). The additional resistance to vapour transfer stems from the stomatal and aerodynamic resistances r_s and r_a ($s \, m^{-1}$).

17.3.5 CO_2

Carbon dioxide concentration varies in a greenhouse due to the occupants (respiration), the plants (photosynthesis and respiration) and the CO_2 concentration in the air supplied through ventilation. Atmospheric air comprises 0.04% CO_2, while the air humans exhale is between 2% to 5.3% rich in CO_2. The most recent global average of CO_2 concentration was recorded to be 406 ppm (parts per million) in April 2017 (Earth System Research Laboratory 2017). During the photoperiod CO_2 can significantly decrease due to plant photosynthesis, while human activity in the farm can provide significant peaks (Portis 1982). Average comfort levels for humans are between 350 and 540 ppm, and the ASHRAE guidance for human comfort is under 1000 ppm, but the regulation specifies a health risk only beyond 5000 ppm (ASHRAE 2010).

Carbon dioxide enrichment for greenhouses has been shown to increase yields by 30% for CO_2 levels maintained at 700–1000 ppm (Hartz et al. 1991). Integrating a small-scale plant factory with an occupied space can reuse air rich in CO_2; indeed occupied buildings often have CO_2 levels over 1000 ppm (Cerón-Palma et al. 2012), and the occupant contribution to CO_2 exchange can be determined from the following equation for the oxygen consumption rate of an occupant (ASHRAE 2013):

$$V_O = \frac{0.00276 \cdot A_D M}{0.23 \cdot RQ + 0.77} \quad (Ls^{-1}) \quad (17.3)$$

where RQ is the respiratory coefficient, generally 0.83, M is the metabolic rate (taken as 2 for an activity level of walking) and A_D is the DuBois equivalent surface area of the occupant estimated for an occupant of height H (m) and weight W (kg) by the following:

$$A_D = 0.203 \cdot H^{0.725} W^{0.425} \quad (m^{-2}) \quad (17.4)$$

The carbon dioxide generation rate is taken as $V_0 \cdot RQ$ ($L\ s^{-1}$). On the other hand, reusing waste CO_2 from industrial processes can have a more regular and greater contribution.

17.3.6 Optimal Conditions

Building integrated agriculture can help alleviate urban problems with maximum efficiency by coupling the demands of a greenhouse (heat, water, CO_2), with the available resources in buildings (waste heat, rainwater, CO_2 from occupants) (Benis et al. 2017; Specht et al. 2013; Nadal et al. 2017). While most existing urban greenhouses do not exploit waste urban resources (Pons et al. 2015), some have showed promise in harvesting rainwater (Lufa farms in Montreal) or reusing the building's thermal mass (London, Fig. 17.4). In Barcelona, the first purpose-built integrated rooftop greenhouse opened in 2016, and the effect of the greenhouse on the building physics is being closely monitored (Nadal et al. 2017). To allocate the resources and create an optimal growing environment, it is essential to determine the optimal range of the most influential environmental factors. Table 17.1 shows the recommended values for lettuce at the early stage of growth as an example.

Table 17.1 Optimal environmental factors found in literature for the early stages of growth of lettuce

	Environmental factor	Lettuce	Source
1	Temperature	Between 20°C and 25°C, optimal water and air at 24°C	Thompson and Langhans (1998)
2	Humidity	Between 60% and 80%	Stine et al. (2005)
3	CO_2 concentration	Keep above 400 ppm, optimal 1000 ppm. If more than 3.5 air changes per hour, it was found that CO_2 enrichment over 400 ppm was not cost-effective	Ferentinos et al. (2000)
4	Air current speed	Increase speed from 0.01 to 0.2. No improvement after 0.5 m/s	Kitaya et al. (2000)
5	Light	High light use efficiency at early growth stages: photoperiods of 18 hours are optimal	Tei et al. (1996)

Integrating urban farms within existing spaces is considered challenging because two (or more) separate systems must be jointly considered for possible co-benefits and trade-offs. There is a need to extend research into controlled environment agriculture to investigate more closely the feasibility of integration with urban infrastructure. Benis et al. (2017) have developed a resource-use model to select farm locations in a city based on building height and climate. Graamans (2015) and Nadal et al. (2017) have researched how to integrate CEA into the building metabolism. A simulation model that quantifies the amount of urban resources *and* enables building integration, with full consideration of the physics involved could be pivotal for building confidence in the feasibility of urban farms.

17.4 Role of Simulation Models in Urban-Integrated Agriculture

17.4.1 A Historical Context

The traditional approach of greenhouse climate control is to tweak environmental settings under different weather conditions and fine-tune them with trial and error. Thus greenhouse cultivation becomes an art, and crop yield is maximised under the care of an expert greenhouse grower. Druma (1998) outlines progression from this approach following the use of computer algorithms over the latter part of the twentieth century. An intermediate step is the continuous monitoring of the state of the environment together with the actions of such an expert to use pattern recognition and identify optimal environmental settings. However, physics-based numerical simulation can improve upon this; an appropriate well-calibrated simulation may be used to understand the response of the environment and crop growth under a range of scenarios and thus provide a mechanism for consistent optimisation of the growing environment. Control of environmental conditions in a greenhouse using computer-based control systems is now commonplace within the horticultural industry, but full simulation of the dynamic response of the greenhouse environment to changes in control parameters and boundary conditions is still primarily limited to academic research.

The earliest work on the physics associated with the greenhouse environment explored in-depth the heat exchange processes and calculated the required energy consumption to compensate for the total thermal loss at equilibrium (Morris 1964). These so-called static models and later developments of the same approach (Jolliet and Bailey 1992) are quick to analyse and simple to use but have been shown to decrease in accuracy for shorter time periods, when the transient energy exchange terms become more significant (Fernández and Bailey 1992). By comparison, a dynamic model can simulate the time-dependent changes in the system by modelling explicitly the transient nature of the energy exchange. This enables prediction of the greenhouse environmental conditions, energy demand and crop growth, thereby facilitating optimisation of the system according to the local boundary conditions and climate. The earliest dynamic models were developed in the 1970s and 1980s (Takakura et al. 1971; Kindelan 1980; Bot 1983) and have formed the basis for development of the dynamic modelling approach over the last 30–40 years. Recently, computational fluid dynamics models of the greenhouse environment have been explored. While these can be extremely computationally intensive, they offer useful insights into local airflows within greenhouse environments, which can have a significant influence on crop yield (Boulard et al. 2017).

Amongst the central physical process governing heat and mass transfer between the plants and their environment, the simulation of plant transpiration has been a consistent topic of research over the past 40 years. A comprehensive review of transpiration models is given by Katsoulas and Kittas (2011). The earliest work in this area was that of Penman (1948) who proposed a model for open-field plants characterised by crop conductance, which was then modified by Monteith (1966) to incorporate the stomatal resistance of the crop. This leads to the Penman-Monteith formula for modelling plant transpiration. In its original form, this equation is not specific to plants grown within controlled greenhouse environments and requires input of a number of crop-specific parameters which are difficult to quantify. However, the approach used to generate this equation, in which the crop is assumed to behave as a "big leaf" with aggregate parameter values, does lend itself to simulation of greenhouse crops particularly when those crops do not exhibit significant variation in properties over the height of the plant. Consequently, much effort has been expended to simplify this equation and apply it to controlled greenhouse environments. The Penman-Monteith formula for modelling plant transpiration still forms the basis for most greenhouse simulations developed to date, e.g. Stanghellini (1987), Jolliet and Bailey (1992), and Kittas et al. (1999).

As per the Penman-Monteith formula and its more recent versions, plant transpiration is dependent amongst other parameters on the leaf stomatal conductance, i.e. the rate of gaseous exchange across the leaf stomata. The variation in stomatal conductance according to physiological and environmental factors has meant that development of mechanistic models for conductance is complex and still at a relatively early stage; a comprehensive review of current models based on leaf water potential, hydraulic models and hydromechanical models has been presented by Damour et al. (2010). However, incorporation of such a mechanistic model into a greenhouse climate model is not yet an established practice. Current simulation models use empirically derived and plant-specific parameters. Research has tended to focus on those crops which are of local commercial interest, for example, tomatoes (Stanghellini 1987; Jolliet and Bailey 1992), lettuce (van Henten 1994; Yang et al. 1990), cucumber (Yang et al. 1990) and rose (Katsoulas et al. 2001). Stomatal conductance is also dependent on the vapour pressure deficit at the leaf surface and hence on the temperature and ventilation regime of the greenhouse; models developed for one location may not be applicable for another location if the prevailing climatic conditions are significantly different (Katsoulas et al. 2001).

Simulation of crop development and yield requires models of photosynthesis and biomass development. Investigation into the mechanisms of photosynthesis and its simulation dates from the nineteenth century (El-Sharkawy 2011), and extensive research is still ongoing in this field (Von Caemmerer 2013). The model used most widely as a basis for investigation and further development is that of Farquhar et al. (1980). This model is a steady-state biochemical model for C3 plants and is based on the kinetic properties of the enzyme RuBisCo, whose importance in determining the rate of photosynthesis has been known since the 1960s (Bjorkman 1968; Wareing et al. 1968). A photosynthesis model was first incorporated into a greenhouse simulation in the 1990s (Gijzen et al. 1990) and led to the development of a steady-state simulation tool for greenhouse tomato production TOMSIM (Heuvelink 1996). This was later incorporated by Vanthoor (2011) into a dynamic simulation model.

Currently these models are being adapted, so they can be applied for optimisation of greenhouse design for a range of plants, under different climate conditions, e.g. Vanthoor (2011). A web-based application developed by Fitz-Rodríguez et al. (2010) serves as a tool for education in the physics of greenhouses and environmental control principles. Optimising production is important; Körner and Hansen (2012) developed an online decision support system that uses monitored climate data in conjunction with greenhouse design data to analyse the energy performance and identify strategies for improving crop yield.

17.4.2 Simulation Models for Urban-Integrated Agriculture

The above section described the development of simulation models for stand-alone greenhouses. As outlined in the introduction to this chapter, different typologies of urban farms exist, but a similar modelling approach may be used for all. The primary difference across the different typologies is the boundary conditions that govern the interaction between the crops and its surroundings. Three broad typologies of urban farm can be identified, namely:

Stand-alone: uncoupled from its surroundings, closest to a traditional greenhouse (Fig. 17.10a).

Urban-integrated: coupled with urban infrastructure, such as greenhouses on rooftops, vacant spaces, etc (Fig. 17.10b).

Building-integrated: fully coupled with the building such that the growing environment utilises the building's waste streams of CO_2, water and/or heat and contributes to the building comfort level by cooling the space and supplying O_2 to improve the air quality (Fig. 17.10c).

The simulation models appropriate to the three different typologies are illustrated in Fig. 17.10. For a stand-alone urban farm (Fig. 17.10a), the boundary conditions are the local microclimate, including the influence of any obstructions from neighbouring buildings and their influence on solar radiation levels, temperature and wind speed. This in itself can present a difficulty as it is often not straightforward

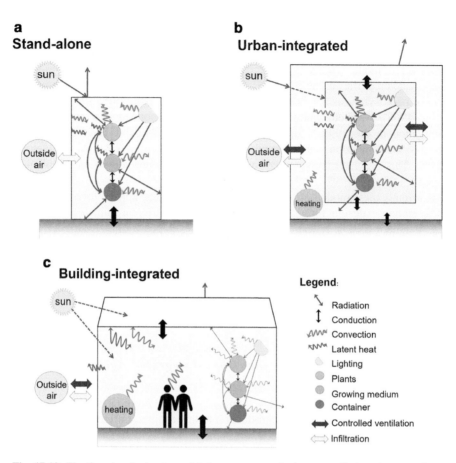

Fig. 17.10 The three typologies to model the energy transfer in a controlled environment farm integrated to the urban landscape. (**a**) Model for a "stand-alone" controlled environment agriculture (CEA) setup in a dedicated building. (**b**) Model for an "urban-integrated" CEA setup, in an empty space, reusing urban resources where available. (**c**) Model for a "building-integrated" CEA setup, the heat exchange processes are linked with those already present in the building such as heating, ventilation and conduction

to obtain local weather conditions without in situ measurements. The simulation model appropriate for an urban-integrated farm (Fig. 17.10b) is dependent on the level of integration. For a rooftop greenhouse, it may be necessary to incorporate any influence of heat exchange between the buildings through its roof, while in an underground farm, the influence of the stable ground temperatures will be important. If the urban farm is within an existing enclosed space, the boundary conditions can change significantly, and one may need to consider local heat sources/sinks. Simulation of a fully building-integrated greenhouse is complex, and the exchange mechanisms between the plants and their surroundings become an integral part of the simulation (Fig. 17.10c). Simulation of heat transfer processes and environment within a building is a well-established topic and has led to the development of a number of open-source and commercial building energy simulation software packages, e.g. EnergyPlus (Crawley et al. 2000), TRNSYS (Klein 2017) and IES-VE (McLean 2017). However, standard building energy simulation models cannot be used for a greenhouse owing to the complex interactions of the plants with their environment and the significant contribution of time-dynamic latent heat transfer processes on human thermal comfort owing to plant transpiration.

17.5 Simulation of Urban-Integrated Agriculture

A simulation model of an urban-integrated farm serves three objectives: (a) it allows optimisation of the growing environment; (b) it can be used to quantify the exchange of heat and other resources if integrated with a building or existing urban infrastructure; and (c) it can be used to optimise yield as a function of resources such as CO_2 and light. The first two objectives require modelling the exchange of energy as heat and the exchange of mass as water vapour in the simulation. If the third objective (plant yield) is the desired quantity of interest, it is also important to consider the exchange of CO_2 between a plant and its surroundings and to include an appropriate photosynthesis and growth model.

17.5.1 The Heat and Mass Exchange Model

Similar to buildings, the heat and mass exchange between plants and their environment requires mathematical representation of the following processes: radiation, convection, conduction and latent heat transfer (illustrated in Fig. 17.11).

17.5.1.1 Radiation

All bodies exchange heat as radiation by virtue of the temperature difference between the body and surrounding surfaces. For a traditional greenhouse, solar radiation is a source of heat (infra-red radiation) and photosynthetically active radiation (PAR) essential for plant growth. The net radiation exchange between two surfaces is dependent on their temperature, emissivity and the extent to which radiation from one surface will strike the second surface according to their relative geometry and area. It can be simulated using a modified form of the Stefan-Boltzmann law for black body radiative exchange as:

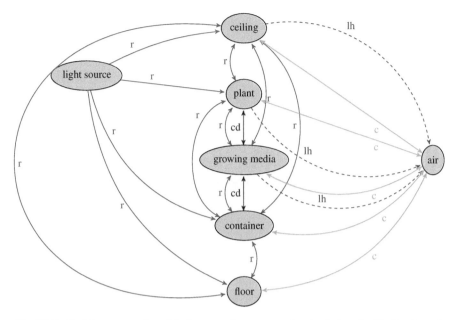

Fig. 17.11 Nodal representation of the heat transfer processes for simulation of a 1D slice through the greenhouse. Radiation (red) is r, conduction (black) is cd, convection (green) is c and latent heat transfer (blue) is lh. Each element floor, container, growing media, plant, ceiling, light source and ceiling constitutes a layer j of the model

$$q_{1 \to 2} = F_{1 \to 2}\, \sigma(T_1^4 - T_2^4) \quad (Wm^{-2}) \tag{17.5}$$

where T_1 and T_2 are the temperatures of surface 1 and surface 2 in Kelvin, σ is the Stefan-Boltzmann constant ($Wm^{-2}K^{-4}$) and $F_{1 \to 2}$ is a view factor, calculated based on the emissivities and geometries of the two surfaces (Moss 2007).

17.5.1.2 Conduction

Where surfaces at different temperatures are in contact with each other, heat transfer will occur as a result of conduction from the hotter surface to the cooler surface. In a greenhouse, this can be the heat transfer between the floor and the growing medium. Conduction of heat through a slab of homogeneous material with thermal conductivity λ ($Wm^{-1}K^{-1}$) can be represented as per Fourier's law:

$$q_{1 \to 2} = \frac{\lambda}{L}(T_1 - T_2) \quad (Wm^{-2}) \tag{17.6}$$

where T_1 and T_2 are the temperatures of the two surfaces of the material (K), separated by a thickness L (m).

17.5.1.3 Convection

Heat is transferred as a result of movement of air, a process known as convection. In a greenhouse, convection may be free or forced: free convection occurs as a result of the natural buoyancy of heated air, and the ventilation system (extraction and circulation fans) drives air flow which gives rise to forced convection. For a ventilated space, convection exchange across the boundary contributes significantly to temperature changes.

Convective heat transfer occurs both within the growth unit (the plants) due to air movement and across the boundaries of the unit as a result of ventilation. If the ventilation rate R_a (s^{-1}) is known, then the heat exchange resulting from the replacement of internal air can be calculated from the heat content of the quantity of air replaced, i.e.:

$$q_{i \to e} = R_a \rho_i (V/A_g) c_i (T_i - T_e) \quad (Wm^{-2}) \qquad (17.7)$$

where V is the volume of the plants, ρ_i is the density of the internal air (kgm^{-3}) and c_i is the heat capacity of the internal air ($Jkg^{-1}K^{-1}$). Convective heat transfer between a surface and the internal air is also dependent on the relative temperatures, this time of the surface and the air. The corresponding convective heat transfer equation may be written as:

$$q_{1 \to a} = h(T_1 - T_a) \quad (Wm^{-2}) \qquad (17.8)$$

where h ($Wm^{-2}K^{-1}$) is a heat transfer coefficient to be determined. In reality this heat transfer coefficient is a complex function of temperature, flow characteristics, properties of the air and surface geometry, but following the approach used in the development of the Gembloux Dynamic Greenhouse Climate Model (GDGCM) (Pieters and Deltour 1999), it may be approximated by

$$h = \frac{Nu \, \lambda}{d} \quad (Wm^{-2}K^{-1}) \qquad (17.9)$$

in which λ is thermal conductivity of the air ($Wm^{-1}K^{-1}$) and d is a characteristic length (m) of the surface. Nu, the non-dimensional Nusselt number, may be calculated from the general expressions for free convection $Nu = A(GrPr)^m$, or for forced convection $Nu = BRe^n Pr^m$, where A, B, n and m are constants dependent on the geometry and flow type. In these expressions, Pr is the non-dimensional Prandtl number which is the ratio of kinematic viscosity to thermal diffusivity for air, and similarly Gr is the non-dimensional Grashof number which approximates the ratio of buoyancy to viscous forces as:

$$Gr = \frac{\beta g d^3 (T_1 - T_a)}{v^2} \quad (17.10)$$

where β is the thermal expansion coefficient for air (K^{-1}); acceleration due to gravity, g (ms^{-2}); and the coefficient of kinematic viscosity of air, v ($m^2 s^{-1}$). The constants A, B, n and m for a range of geometries may be found in the literature. Pieters and Deltour (1999) suggest the use of the values proposed by Monteith (1966) for flow along a flat plate for Prandtl numbers of at least 0.7 as typical for moist air, namely:

	Laminar flow	Turbulent flow
Free convection	$0.54(GrPr)^{1/4}$	$0.15(GrPr)^{1/3}$
Forced convection	$0.66Re^{1/2}Pr^{1/3}$	$0.036Re^{4/5}Pr^{1/3}$

17.5.1.4 Latent Heat Transfer

Latent heat transfer is associated with changes in the moisture content of the internal air of the greenhouse. This can occur in three ways:

- as a result of ventilation, in which the moisture content of the incoming air differs from the air being replaced – for a ventilated space, removal of moist air also removes heat from the space.
- as a result of condensation or evaporation from surfaces within the unit if they are at a significantly different temperature from that of the surrounding air,
- as a result of plant transpiration – the primary cooling mechanism associated with plants in enclosed spaces

The ventilation latent heat transfer may be calculated from a knowledge of the air exchange rate and the difference between the moisture contents of the air leaving the unit and the replacement air. Similarly, the potential for condensation and evaporation from surfaces within the unit, and hence the latent heat transfer, is a function of the difference between the saturation vapour pressure at the temperature of the surface of interest and the moisture content of the air.

Transpiration is the most complex contributor to the latent heat transfer process. As described previously in Sect. 17.4.1, the Penman-Monteith (P-M) formula takes as its basis the assumption that the vegetative layer can be treated as a homogeneous and extensive "big leaf", which permits the equations derived for the transpiration of a single leaf to be extended to the entire plant canopy, albeit with modified parameters. The equation for latent heat transfer for a hyperstomatal leaf (stomata on only one side of the leaf) according to the P-M approach is

$$q_{v \to a} = LAI \, h_{fg} \left(\frac{1}{r_a + r_s}\right)(C_v - C_a) \quad (Wm^{-2}) \qquad (17.11)$$

where LAI is the leaf area index, i.e. the ratio of leaf area to the area of the growing medium, h_{fg} is the latent heat of evaporation of water (Jkg^{-1}), r_a and r_s are the aerodynamic and stomatal resistances to vapour transfer, respectively (sm^{-1}), and C_v is the moisture content of the air at the temperature of the vegetation (kgm^{-3}). The equation in this form requires estimation of the resistance values. There are a number of different ways to estimate the aerodynamic resistance, e.g. as a function of the air velocity (Graamans et al. 2017), as a function of the crop sensible energy balance and temperature differential (Seginer 1984) or using the classical theory of heat transfer in a boundary layer as described by Stanghellini (1987) and implemented in the GDGCM (Pieters and Deltour 1999). In a similar manner to the convection equations described above, this latter approach involves the calculation of the non-dimensional Sherwood number, Sh, which represents the ratio of convective to diffusive mass transport, from the Nusselt number and another non-dimensional parameter, the Lewis number, Le, which is equal to the ratio of thermal diffusivity to mass diffusivity. Using these parameters, $Sh = Nu \, Le^m$. Again, m is a parameter which is affected by the flow regime and which can be estimated from experimental data reported in the literature. Using the Sherwood and Lewis numbers, the aerodynamic resistance can be expressed as follows:

$$r_a = \rho_a \, c_a \, \frac{d \, Le}{\lambda \, Sh} \quad (sm^{-1}) \qquad (17.12)$$

There is evidence to suggest that the aerodynamic resistance has a less significant impact on the transpiration of greenhouse plants than the stomatal resistance (Stanghellini 1987). Stomatal resistance is crop-specific, but for greenhouse crops, it appears that the resistance is dependent primarily on irradiance and in a large part on the vapour pressure deficit at the leaf surface, together with some dependency on the temperature and CO_2 concentration in the air. The most common relationships used for simulation of stomatal resistance, incorporating local environmental effects, are based on the multiplicative model of Jarvis (1976). As an example, a multiplicative model for the stomatal resistance of a tomato crop that has been shown to be representative under a range of conditions is that of Stanghellini (1987). In this model, stomatal resistance is expressed as:

$$r_s = r_{min} \, f(I_{sol}) \, f(T_v) \, f(CO_2) \, f(vpd) \quad (sm^{-1}) \qquad (17.13)$$

where r_{min} is the minimum resistance and the functions of mean irradiation $f(I_{sol})$, vegetation temperature $f(T_v)$, CO_2 levels, $f(CO_2)$, and vapour pressure deficit, $f(vpd)$ describe the relative increase of the resistance if any of the relevant variables are limiting the vapour transfer rate. For the tomato model of Stanghellini (1987), the parameters are specified as follows:

$$r_{min} = 82.0 \quad (sm^{-1}) \tag{17.14}$$

$$f(I_{sol}) = \frac{I_{sol} + 4.30}{I_{sol} + 0.54} \tag{17.15}$$

$$f(T_v) = 1 + 2.3.10^{-2}(T_v - 24.5)^2 \tag{17.16}$$

$$\begin{aligned} f(CO_2) &= 1 & I_{sol} &= 0 \; Wm^{-2} \\ &= 1 + 6.1.10^{-7}(CO_2 - 200)^2 & CO_2 &< 1100 ppm \\ &= 1.5 & CO_2 &\geq 1100 ppm \end{aligned} \tag{17.17}$$

$$\begin{aligned} f(vpd) &= 1 + 4.3(vpd)^2 & vpd &< 800 Pa \\ &= 3.8 & vpd &\geq 800 Pa \end{aligned} \tag{17.18}$$

In addition to the above, a number of models have been proposed for different plants based on experimental data. Yang et al. (1990) present an empirical model for stomatal resistance of a cucumber crop, and several models have been studied for lettuce, including Pollet et al. (2000), and Graamans et al. (2017). As increasingly plants are being grown under artificial light, the response of plants to different ratios of red and blue light has been studied (Wang et al. 2016), indicating that stomatal resistance decreases with increasing irradiance with a rate of decrease depending on the red/blue light ratio.

17.5.1.5 The Overall Model

The environmental conditions of a greenhouse are a result of highly complex 3D interactions between the plants and their surroundings due to the processes described above. However, assuming that a typical greenhouse be highly regular in layout and orientation, it is useful to simplify the problem to consider a 1D slice and to calculate the time-dependent temperature variation through the different layers, for example, the growing medium, vegetation, air and cover, subject to external fluctuating boundary conditions, as illustrated in Fig. 17.11.

Using this approach, the heat balance for each layer j in time t may be calculated from

$$\frac{dT_j}{dt} = \frac{A_g}{m_j c_j} \sum_i q_{i,j} \qquad (17.19)$$

where A_g is the surface area (m^2), and for each layer T is the temperature (K), m is the mass (kg), c is the heat capacity ($Jkg^{-1}K^{-1}$) and $q_{i,j}$ are the heat flows from the i different heat transfer processes – radiation, conduction, etc. – between the layers, j (Wm^{-2}).

The heat balance is linked directly to the mass balance for the system, as changes in the moisture content of the air are associated with the transfer of latent heat. A similar equation may be used for the moisture content of the internal air, namely:

$$\frac{dC_a}{dt} = \frac{A_g}{h_{fg}V} \sum_k q_k \qquad (17.20)$$

where C_a is the moisture content of the air (kgm^{-3}), h_{fg} is the latent heat of condensation of water (Jkg^{-1}), V is the volume of the unit (m^3) and q_k are the heat flows from the k possible different latent heat transfer processes as illustrated in Fig. 17.11.

Using these two balance equations, the time-varying temperature and moisture levels within a greenhouse can be calculated. Or, given a set of control set points for temperature and humidity, one can compute the required total energy needed in the form of heat, moisture or ventilation.

17.5.2 The Plant Growth Model

As aforementioned, CO_2 exchange between the plants and the environment may also be of interest, as it helps improve indoor air quality and at the same time, promotes plant growth. An estimation of plant yield may also be useful, especially for commercial greenhouses.

17.5.2.1 CO_2 Exchange

Photosynthesis involves the conversion of CO_2 and water in the presence of PAR into carbohydrates and oxygen, leading to the generation of biomass. Conversely, respiration supplies the plants with energy, consuming part of the biomass generated during photosynthesis and releasing CO_2. Although CO_2 levels in the air may have a small impact on transpiration rates, the primary reason for simulating the CO_2 exchange of the crop is in order to simulate crop growth and potential yields; indeed, simulation is potentially a useful tool for optimising growth conditions to maximise the potential profit.

Crop yield is dependent on plant photosynthesis and respiration rates, and optimising yield is a case of maximising the net photosynthetic conversion of photosynthetically active radiation into biomass or carbohydrate (Castilla 2013). The photosynthesis process is dependent on the following:

- Light quality, intensity and duration – specifically the incidence of photosynthetically active radiation (PAR).
- Temperature – photosynthesis is at a minimum at about 5°C and maximum at 25°C to 35°C, decreasing thereafter. Optimum temperature increases with radiation and CO_2 levels, i.e. heating is redundant if PAR levels are low.
- CO_2 – if radiation and temperature are not limiting, the rate of photosynthesis is almost proportional to CO_2 concentration at low CO_2 levels.

Photosynthesis is not directly dependent on humidity (or vapour pressure deficit), provided there is sufficient water supply to the roots, and also does not appear to be significantly dependent on the crop (Castilla 2013).

The Farquhar-von Caemmerer-Berry biochemical leaf photosynthesis model (Farquhar et al. 1980) has been regularly used for simulation of photosynthesis at the canopy scale (Wang et al. 2016) and has been adapted for greenhouse application and simulation of crop growth by Vanthoor (2011). In this model, photosynthesis is characterised by the movement of electrons excited by the incoming PAR intercepted by the canopy. The net photosynthesis rate, which can be equated to the rate at which the CO_2 concentration of the air changes, is proportional to the gross photosynthesis rate, P minus photo-respiration, R, with P limited by the slowest of three biochemical processes dependent upon the environmental conditions of light, temperature, CO_2 and O_2 levels. The three biochemical processes are the following:

- the maximum rate of carboxylation under RuBisCo catalysis,
- the rate of ribulose 1,5-biphosphate (RuBP) regeneration via electron transport, and
- the rate of RuBP regeneration via triose phosphate utilisation.

The rate of photosynthesis is dependent on CO_2 levels, and it is common practice for a commercial grower to provide supplemental CO_2. At the extremes of CO_2 concentration, limiting equations are required, but under typical commercial growing conditions, the gross photosynthesis rate of the canopy, P, can be adequately described by the second of these biochemical processes, in which the electron transport limited rate of assimilation is calculated as follows:

$$P = \frac{J(CO2_{stom} - \Gamma)}{4(CO2_{stom} + 2\Gamma)} \quad (\mu mol(CO_2)m^{-2}s^{-1}) \tag{17.21}$$

where J is the canopy electron transport rate, 4 is the number of electrons per fixed CO_2 molecule, $CO2_{stom}$ is the CO_2 concentration in the stomata, which may be assumed to be a fixed fraction of the CO_2 concentration in the air (Vanthoor et al. 2011) and Γ is the CO_2 compensation point, i.e. the point at which $P - R = 0$. The photorespiration term, R, is given by the following:

$$R = P \frac{\Gamma}{CO2_{stom}} \quad (\mu mol(CO_2)m^{-2}s^{-1}) \tag{17.22}$$

The canopy electron transport rate, J, is a function of the maximum electron transport rate for a leaf and the PAR intercepted by the plants and can be estimated from an empirical equation:

$$J = \frac{J^{POT} + \alpha PAR_{can} - \sqrt{(J^{POT} + \alpha PAR_{can})^2 - 4\Theta J^{POT} \alpha PAR_{can}}}{2\Theta} \tag{17.23}$$

$$(\mu mol(e^{-1})m^{-2}s^{-1})$$

Here α and Θ are empirical parameters (Ogren 1993); α represents the fraction of incident PAR absorbed by the leaves and is a function of the leaf absorptance, typically about 0.85, and an interaction factor representing the distribution of the absorbed light to the photosystem processes. Θ is an empirical curvature factor with an average value of around 0.7 (Von Caemmerer 2013). J^{POT} is the potential rate of electron transport for the canopy which depends on temperature (Farquhar et al. 1980). There have been numerous equations proposed for this parameter typically of a similar Arrhenius function type, as exemplified by the equation used by Vanthoor (2011) such as:

$$J^{POT} = J^{MAX}_{25,can} e^{\frac{E_J(T_{can}-25)}{298R(T_{can}+273)}} \frac{1 + e^{f(25)}}{1 + e^{f(T_{can})}} \quad (\mu mol(e^-)m^{-2}s^{-1}) \tag{17.24}$$

in which $J^{MAX}_{25,can}$ is the maximum rate of electron transport at 25°C, E_J is the activation energy ($Jmol^{-1}$), T_{can} is the temperature of the canopy (°C), R is the gas constant ($Jmol^{-1}K^{-1}$) and $f(T) = \frac{S(T+273)-H}{R(T+273)}$ where S is the entropy term ($Jmol^{-1}K^{-1}$) and H is the deactivation energy ($Jmol^{-1}$).

Photosynthesis is directly dependent on the PAR intercepted by the vegetative layer, which in turn depends strongly upon the density of vegetation. It is common practice to describe the dependence of the PAR directly absorbed by the canopy on LAI with a negative exponential decay, i.e.:

$$PAR_{can} = PAR(1 - \rho_v)(1 - \exp(-k\, LAI)) \quad (\mu mol\,(photons)\, m^{-2}s^{-1}) \tag{17.25}$$

where *PAR* is the incident photosynthetically active radiation which is dependent on the light source, whether solar or artificial, ρ_v is the reflection coefficient for the vegetation and k is the extinction coefficient.

The net rate at which photosynthesis causes CO_2 to be removed from the air may then be calculated from the following:

$$MC_{i \to v} = m_{CO2}\, h\, (P-R)\, \frac{A_v}{V} \qquad (kgm^{-3}s^{-1}) \qquad (17.26)$$

where m_{CO2} is the molar mass of CO_2 ($kg\, \mu mol^{-1}$) and h is an inhibition factor representing the limit to CH_2O storage in the plant.

Plants also contribute CO_2 to the air as a product of growth and maintenance respiration. Growth respiration is linearly proportional to the flow of carbohydrate to the plant organs and is fairly independent of temperature (Heuvelink 1996). For tomato plant leaves and stem, the growth respiration may be estimated from the following:

$$MC_{v \to i} = \frac{m_{CO2}}{m_{CH2O}} (h_{buf}\, h_{Tcan24}\, g_{Tcan24}\, rg_{org})\, \frac{A_v}{V} \qquad (kgm^{-3}s^{-1}) \qquad (17.27)$$

where m_{CH2O} is the molar mass of carbohydrate and h_{buf} and h_{Tcan24} are inhibition factors ($0 < h < 1$) which represent the impact of insufficient carbohydrates in the buffer and nonoptimal 24-h canopy temperature, respectively. The growth rate coefficients rg_{org} ($kg(CH_2O)m^{-2}s^{-1}$) are independent of temperature, and hence the dimensionless parameter g_{Tcan24} is included in order to represent the effect of mean daily temperatures on growth.

Maintenance respiration, by comparison, is strongly dependent on temperature and varies according to the plant organ under consideration. It may be simulated following the approach proposed for tomato plants by Heuvelink (1996) and implemented by Vanthoor et al. (2011) in which

$$MC_{m \to i} = \frac{m_{CO2}}{m_{CH2O}} (c_m\, Q_{10}^{0.1(T_{can24}-25)})\, C_{org}(1 - e^{-c_{RGR}\, RGR}))\, \frac{A_v}{V} \qquad (kgm^{-3}s^{-1}) \qquad (17.28)$$

where c_m is the maintenance respiration coefficient (s^{-1}), Q_{10} governs the temperature dependence, T_{can24} is the mean 24 h canopy temperature ($°C$), C_{org} is the carbohydrate weight of the plant organ, RGR is the net relative growth rate (s^{-1}) and c_{RGR} is a regression coefficient.

Incorporating supplied CO_2, MC_{supp} and losses due to ventilation, $MC_{i \to e}$, the corresponding CO_2 levels in the internal air, MC, can be calculated from the following:

$$\frac{dMC}{dt} = MC_{supp} - MC_{i \to e} - MC_{i \to v} + \sum_{org} MC_{m \to i} + \sum_{org} MC_{v \to i} \qquad (kgm^{-3}s^{-1}) \qquad (17.29)$$

17.5.2.2 Plant Growth and Crop Development

Once the CO_2 is converted into carbohydrates, these are used by the plant for growth and fruit development. Clearly this is crop-specific, and many different models exist for dry matter production, dry matter partitioning and individual fruit growth where appropriate, for crops such as tomatoes (Stanghellini 1987), cucumber (Marcelis 1994), sweet pepper (Marcelis et al. 2006) and lettuce (van Henten 1994). From a simulation viewpoint, it is useful to consider this as essentially a post-processing exercise as the dry matter partitioning has little impact on the balance of environmental parameters within the greenhouse.

The tomato model implemented by Vanthoor uses a complex "boxcar" method in which the various developmental stages of the fruit are simulated and the number of fruits in each stage at any time is calculated. By comparison, for a non-fruiting plant, the situation may be simpler. For example, in the lettuce model of van Henten (1994), the evolution of dry matter, D, is related to the gross photosynthesis rate, P, and air temperature T_a by the following equation:

$$\frac{dD}{dt} = c_{\alpha\beta}P - c_{D,c_a}D2^{(0.1T_a - 2.5)} \quad (kgm^{-2}s^{-1}) \qquad (17.30)$$

where $c_{\alpha\beta}$ and c_{D,c_a} are empirical constants, representing a yield factor and a respiration rate factor (s^{-1}), respectively.

17.5.3 Growing Underground: An Example Application of the Simulation Model to an Urban Farm

As an example of an urban-integrated farm, consider the case of Growing Underground, a hydroponics farm developed by Zero Carbon Food within a derelict space in London, UK. This innovative farm has made use of an abandoned World War II air raid shelter 33 m below ground level to grow microgreens closer to their location of use. The farm caters to hotels, restaurants and supermarkets and produces a range of crops including peashoots, a variety of basil, coriander, parsley, wasabi and red mustard, rocket salad and celery.

The farm has been set up in two parallel tunnels which have been divided into upper and lower sections and which span approximately 400 m in total length. The tunnels were excavated in London Clay and are lined with concrete and steel liners. The growing area of the farm occupies around 46 m of one tunnel and consists of two aisles of vertically stacked hydroponic trays with an ebb and flow watering system and LED lights to each tray. There are four levels of trays, and each aisle consists of 23 trays of 2 m length, corresponding to a total of 368 m^2 of growing space.

Environmental management of the tunnels depends on adequate ventilation via the installed extraction fans and supplementary air movement fans, together with heating supplied by the LED lights. The extraction fans aim to provide complete air exchange four times per hour, while the purpose of the recirculation fans is to increase air speeds local to the trays and to improve air mixing in the farm, thereby minimising the potential for crop damage due to local air stagnation. The LED lights provide heat while supplying PAR for plant growth. No additional heat is supplied as the depth of the tunnel ensures that relatively stable temperatures are maintained. The lights are switched on for 16 h per day, typically from midnight to 4pm to ensure that heat is supplied at the time when the incoming air is at its coldest. This also helps to maintain stable temperatures.

17.5.3.1 Model Description

A simple model has been developed which represents a 1D slice through the farm and incorporates the heat and mass exchange mechanisms following the methodology described in the previous section. Assuming that the aisles are symmetrical, the model represents a single aisle of trays in half the volume of the space. The layers in the model consist of the floor; the internal air, one layer each of tray, mat and vegetation representing the lumped performance of the four trays; the plastic lining within the tunnel; the concrete tunnel; and three layers representing the ground conditions surrounding the tunnel. The boundary conditions of the model are the temperature and moisture content of the external air, the deep soil temperature and the temperature at the underside of the floor which has been initially assumed to be a constant 17°C based on monitored temperature data. It has been assumed that the temperature and moisture content of the incoming air are the same as those of the nearest weather station in St James' Park, London, approximately 4 miles to the North of the farm location. The deep soil temperature has been assumed to be 14°C. The model requires a number of parameters characterising material properties and operational conditions. The values used for this example simulation are detailed in Table 17.2. For the plants themselves, indicative values have been used for the wide variety of crops grown.

17.5.3.2 Verification and Outcomes

In any simulation it is important to establish that the model can adequately simulate observed conditions in order to give legitimacy to model predictions of unobserved conditions. Monitored data are essential for this. A programme of monitoring temperature, moisture content and CO_2 levels in the tunnel has been carried out in order to both provide additional useful information to the farm managers and also to facilitate validation of the model. The monitors were located as shown in Fig. 17.12. For comparison against the model predictions, it is particularly useful to explore the values recorded at the three in-farm monitors located at column 18 (Fig. 17.13).

The model simulates mean temperature and moisture content in the tunnel; Fig. 17.14 shows the model predictions in light blue for an 18 -month period,

Table 17.2 Parameters for 1D simulation model

Parameter	Value	Parameter	Value
Operational parameters			
Ventilation rate	4 ACH	PAR	77 Wm^{-2}
Internal air velocity	0.5 ms^{-1}	Light operating temperature	27.5°C
Material parameters			
Internal lining thickness	0.005 m	Growing medium thickness	0.01 m
Internal lining density	1500 kgm^{-3}	Growing medium density	511 kgm^{-3}
Internal lining emissivity	0.9	Growing medium emissivity	0.95
Internal lining conductivity	0.03 Wm^{-1}K^{-1}	Growing medium conductivity	0.32 Wm^{-1}K^{-1}
Internal lining-specific heat	1670 Jkg^{-1}K^{-1}	Growing medium-specific heat	1250 Jkg^{-1}K^{-1}
Concrete tunnel lining thickness	0.15 m	Tray thickness	0.005 m
Concrete density	2400 kgm^{-3}	Tray density	1500 kgm^{-3}
Concrete conductivity	0.8 Wm^{-1}K^{-1}	Tray conductivity	0.25 Wm^{-1}K^{-1}
Concrete-specific heat	840 Jkg^{-1}K^{-1}	Tray-specific heat	1670 Jkg^{-1}K^{-1}
Concrete floor thickness	0.15 m		
Plant parameters			
Vegetation-specific heat	3500 Jkg^{-1}K^{-1}	Emissivity	0.9
Extinction coefficient	0.94	Maximum LAI	2.4

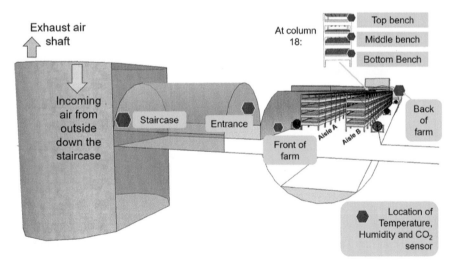

Fig. 17.12 Location of the sensors in the farm measuring temperature, CO$_2$ and RH

Fig. 17.13 Location of three sensors in the "middle" of the farm. The sensors used are Advanticsys IAQM-THCO2

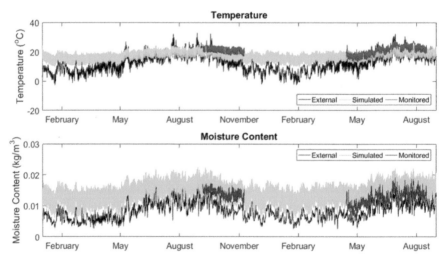

Fig. 17.14 Comparison of simulation against external conditions and monitored data

compared against the external conditions in black and the mean monitored data at column 18 in red. The top plot shows the temperature, and it is clear that the tunnel location has a "buffering" effect on the external temperature; while there is correlation between external and internal temperatures, the simulation predicts lower peak and higher minimum temperatures throughout the year than experienced externally. This result is corroborated by the monitored temperatures for two different periods

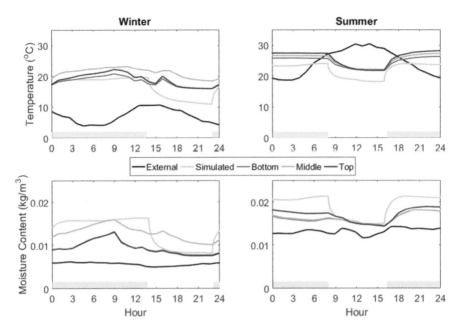

Fig. 17.15 Comparison of simulated and monitored temperatures and moisture content for a typical winter and summer day

shown in red which show that the tunnel is substantially warmer than outside during the winter, and it is less hot during the warmest summer days.

The second graph in the figure shows the simulation results for the moisture content of the air. Here the moisture content is predicted to be higher than that outside throughout the year, to be expected owing to the humid environment of the greenhouse. Again, the simulation results are corroborated by the monitored data (dark blue).

Figure 17.14 illustrates the simulated long-term performance of the tunnel, and it is also interesting to look at the results for a sample day. Sample monitored data are compared for a typical summer and a typical winter day against the predicted mean air temperature and moisture content in Fig. 17.15. The yellow bars at the bottom of each graph indicate the times at which the LED lights are switched on in the farm. The top two graphs illustrate the temperature comparison for a typical winter day on the left and a summer day on the right, with the external temperature variation shown in black. In the simulation model, the different growing trays in the farm are not represented as separate nodes, and the temperature predictions are indicative of average conditions in the tunnel, with the assumption that the air inside the tunnel is fully mixed. The monitored data clearly indicates there is some local temperature stratification with lower temperatures at the bottom tray near the floor. The monitored data also shows local fluctuations not simulated by the model, in all likelihood due to the actions of farm workers in their daily management of the tunnel. Model predictions for the temperature are consistently lower than the monitored data, especially during the periods when the LED lights are switched off.

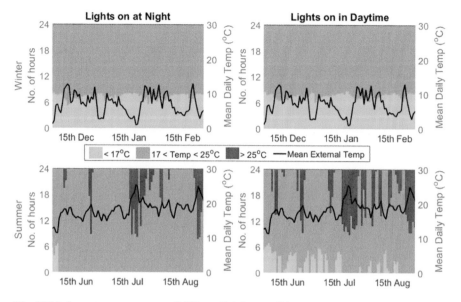

Fig. 17.16 Impact on temperature of different lighting conditions

The two graphs at the bottom of Fig. 17.15 show the corresponding moisture content, with the moisture content of outside air indicated in black. Again there is some local stratification along the height of the trays and an obvious reduction in moisture content of the inside air during the time when the LED lights are off. In summary, the model under-predicts temperature during the time when LED lights are switched off and over-predicts the moisture content of the internal air when the LED lights are switched on. The former is likely due to unaccounted heat gains from farm workers and ambient lighting in the farm during "normal" working hours, and the latter is due to over-prediction of transpiration rates when the LED lights are on. This could be due to an overestimation of the leaf area index (*LAI*), as not all plants are at the same stage of growth at a given time.

17.5.3.3 Simulation to Optimise Environmental Conditions

A validated model may be used to explore the possible variability of indoor environmental conditions that arise as a result of changes to the boundary conditions – extremes of weather, for example – or of changes to operational procedures.

One benefit of operating underground is the ability to control the lighting – and hence the heating – conditions to maintain a stable environment. The LED lights are required for around 16 h per day, dependent on the crop, but the timing may be varied. Results in Figs. 17.16, 17.17, and 17.18 show the impact of changing the timing of the photoperiod with respect to clock time; in Fig. 17.16 the plots show the number of hours in each day that the temperature falls below a specified minimum of

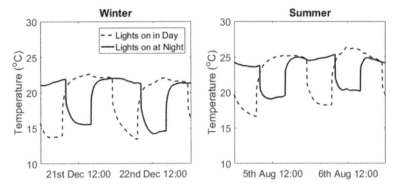

Fig. 17.17 Predicted temperature under different lighting regimes for winter and summer conditions

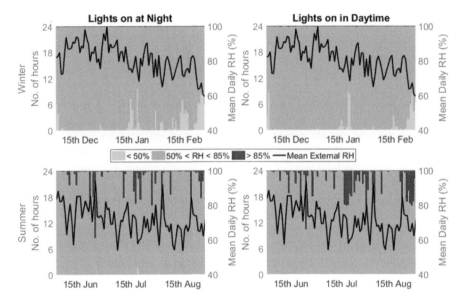

Fig. 17.18 Impact on relative humidity of different lighting conditions

17 °C (in light blue), or above the maximum 25 °C (in red), with the number of hours the environmental conditions lie within these optimal boundaries shown in green. The top two plots in Fig. 17.16 show the difference for the winter period of having the lights on at night, on the left, compared against having the lights on during the daytime, on the right. It is clear that in terms of the total number of hours below optimal, the timing of the lighting is not significant in the winter. The solid black line shows the daily mean external temperature, which during the winter is consistently below the minimum optimal temperature of 17 °C for the farm; the minimum indoor air temperature predicted by the model is lower for the case with LED lighting on

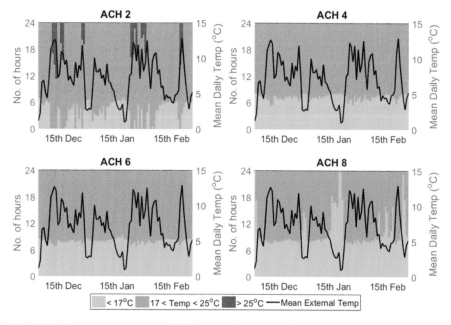

Fig. 17.19 Impact on temperature of different ventilation conditions (winter)

during the daytime and off at night when the external temperatures are coldest, as illustrated in Fig. 17.17. However, provided the lights are switched on for the same number of hours, the total number of suboptimal hours remains the same.

In the summer, however, the situation is different, as illustrated in the bottom two plots of Fig. 17.16. If the lighting is switched on during the daytime as shown in the plot on the right, the indoor air temperatures lie outside the optimal range for a greater proportion of the time, being both hotter during the day and colder at night than desired. This supports the operational procedure of using the LED lighting at night to help maintain a stable environment. The impact on relative humidity is similar (Fig. 17.18); here optimal conditions are assumed to be a relative humidity between 50% and 85%, and varying the timing of the photoperiod is predicted to have a greater impact in the summer months.

In addition to the schedule of LED lights, the farm operators manipulate ventilation rates to control the indoor environment, and here again even a simple simulation can help identify the optimal strategy. The results above illustrated how difficult it is to maintain optimal temperatures in winter, and to counteract this, it is tempting to reduce influx of cold external air by reducing the ventilation rate. Figure 17.19 shows the impact of changing ventilation rates on the temperature, and Fig. 17.20 shows the corresponding impact on the relative humidity. Ventilation rates have been changed from an air exchange rate of 2 air changes per hour (ACH) to 4, 6 and 8 ACH. The top left plot shows that on the coldest days, it is difficult to avoid having a number of hours at suboptimal temperatures even at a very low ventilation rate. However, when the external temperature is above 12 °C, it is

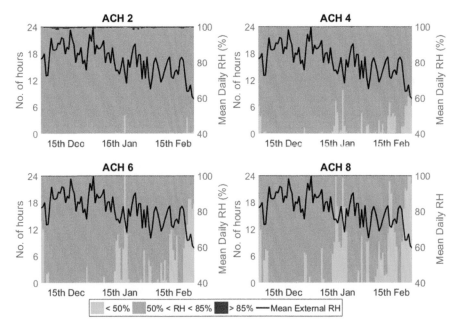

Fig. 17.20 Impact on relative humidity of different ventilation conditions (winter)

possible to maintain adequate temperatures, and the environment may even become too warm. By contrast, at high ventilation rates, on the coldest days, the temperature in the farm may never get to an optimal level; this is clearly a situation to be avoided.

The relative humidity shows a similar story, with reduced ventilation providing better conditions in the winter period. The simulation results suggest that the opposite is the case for the hot days of summer. Figures 17.21 and 17.22 show significant levels of overheating and excessive humidity if the ventilation levels are low, around 2 ACH, with improving environmental conditions as ventilation rates are increased.

Finally, the model may be used to investigate the impact of changes to the local environment on the farm operating conditions. Figure 17.23 shows the impact on the tunnel temperature if the deep soil temperature is different from the assumed value of 14 °C. Such changes could occur as a result of changes to the operating conditions of the nearby London Underground tunnels, or more localised interventions, such as installation of local ground source heat pumps. As before, the plots show the number of hours the temperature falls below the specified minimum of 17 °C (in light blue), or above the maximum 25 °C (in red), for an assumed deep soil temperature of 10 °C (top row), 14 °C (middle row) and 22 °C (bottom row) in the winter (left column) and summer (right column). Substantial reduction of deep soil temperature could lead to more heat being required to maintain optimal conditions in both winter and summer, whereas if the ground warms significantly, overheating may become a problem in summer.

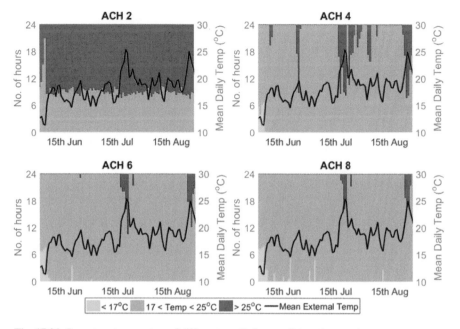

Fig. 17.21 Impact on temperature of different ventilation conditions (summer)

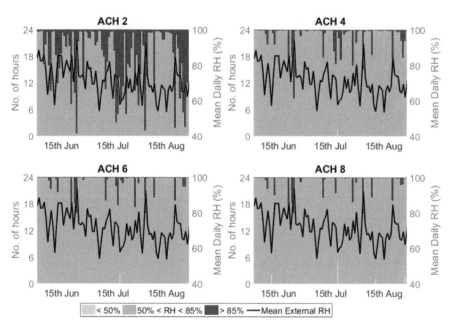

Fig. 17.22 Impact on relative humidity of different ventilation conditions (summer)

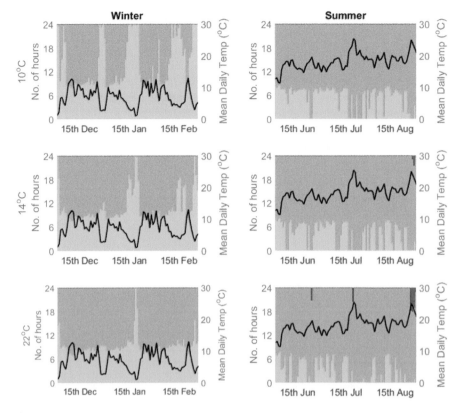

Fig. 17.23 Impact on temperature of changes to deep soil temperature

17.5.4 Implications and Future Work

The simulation model presented in this chapter may be used to investigate various mechanisms of reusing waste resources in cities for urban farming, especially waste heat, light and CO_2 for a wide range of sites. This would allow quantification of the resources required to create suitable and synergistic growing conditions for plants. Moreover, the tool developed could be useful to city planners for understanding the interaction between small-scale urban farms and their immediate environment in a city.

In conclusion, this chapter presented the development and application of a simulation model to investigate the environment of an urban-integrated farm. Applying the model to Growing Underground, an innovative hydroponic farm in previously derelict London tunnels, enabled the testing of different scenarios to optimise farm operations. With appropriate adaptations and co-simulation of plants and building, the same model could be used to investigate time dynamic synergies and trade-offs of building-integrated farms. This would indeed be necessary both from a point of view of human and plant comfort, especially where plants and human occupy the same space or share resources across adjoining spaces.

Finally, simulation science of controlled environment is yet to mature. Testing models against data from more urban farm typologies will continue to improve the response of the model under different scenarios. Ultimately, this will enable an improved understanding of the city-scale environmental impact of integrating urban farming within cities.

Acknowledgements We are very grateful to Growing Underground for their cooperation and for giving us the opportunity to analyse such an interesting case study. In particular, we would like to thank Richard Ballard and Daniel Negoita for their assistance in installing and maintaining the sensor network. Our gratitude is further extended to Paul Fidler, for his invaluable contribution to designing the wireless sensors used for monitoring.

References

ASHRAE (2010) ANSI/ASHRAE Standard 62, Ventilation for Acceptable Air Quality
ASHRAE (2013) Fundamentals handbook. Technical report, American Society for Heating, refrigerating, and air conditioning engineers. ASHRAE, Atlanta
Barthel S, Folke C, Colding J (2010) Social-ecological memory in urban gardens – retaining the capacity for management of ecosystem services. Glob Environ Chang 20:255–265. https://doi.org/10.1016/j.gloenvcha.2010.01.001
Benis K, Ferrao P (2017) Potential mitigation of the environmental impacts of food systems through urban and peri-urban agriculture (UPA) a life cycle assessment approach. J Clean Prod 140:784–795. https://doi.org/10.1016/J.JCLEPRO.2016.05.176, http://www.sciencedirect.com/science/article/pii/S0959652616306552#sec5
Benis K, Reinhart C, Ferrao P (2017) Development of a simulation-based decision support workflow for the implementation of Building-Integrated Agriculture (BIA) in urban contexts. J Clean Prod 147:589–602. https://doi.org/10.1016/j.jclepro.2017.01.130, http://www.sciencedirect.com/science/article/pii/S0959652617301452
Bjorkman O (1968) Carboxydismutase activity in shade-adapted and sun-adapted species of higher plants. Physiol Plant 21(1):1–10. http://doi.wiley.com/10.1111/j.1399-3054.1968.tb07225.x
Bot GPA (1983) Greenhouse climate: from physical processes to a dynamic model. PhD thesis, Wageningen University
Boulard T, Wang S, Haxaire R (2000) Mean and turbulent air flows and microclimatic patterns in an empty greenhouse tunnel. Agric For Meteorol 100(2–3):169–181. https://doi.org/10.1016/S0168-1923(99)00136-7
Boulard T, Roy JC, Pouillard JB, Fatnassi H, Grisey A (2017) Modelling of micrometeorology, canopy transpiration and photosynthesis in a closed greenhouse using computational fluid dynamics. Biosyst Eng 158:110–133. https://doi.org/10.1016/j.biosystemseng.2017.04.001
Castilla N (2013) Greenhouse technology and management, 1st edn. CABI, Oxford
Cerón-Palma I, Sanyé-Mengual E, Oliver-Solà J, Montero JI, Rieradevall J (2012) Barriers and opportunities regarding the implementation of rooftop eco.greenhouses (RTEG) in Mediterranean cities of Europe. J Urban Technol 19(4):1–17. https://doi.org/10.1080/10630732.2012.717685
Chaudhary DD, Nayse SP, Waghmare LM (2011) Application of wireless sensor networks for greenhouse parameter control in precision agriculture. Int J Wirel Mob Netw (IJWMN) 3(1). https://doi.org/10.5121/ijwmn.2011.3113, https://pdfs.semanticscholar.org/5f71/20d409a0eb7228f282c7898b50166762438e.pdf

Crawley DB, Crawley DB, Pedersen CO, Lawrie LK, Winkelmann FC (2000) EnergyPlus: energy simulation program. ASHRAE J 42:49–56. http://citeseerx.ist.psu.edu/viewdoc/summary?doi=10.1.1.122.6852

Custot J, Dubbeling M, Getz-Escudero A, Padgham J, Tuts R, Wabbes S (2012) Resilient food systems for resilient cities. In: Otto-Zimmermman K (ed) Resilient cities 2: cities and adaptation to climate change – proceedings of global forum 2011, 2nd edn. Springer, Bonn, p 436. http://www.springerlink.com/index/10.1007/978-94-007-4223-9_14

Damour G, Simonneau T, Cochard H, Urban L (2010) An overview of models of stomatal conductance at the leaf level. Plant Cell Environ 33(9):1419–1438. https://doi.org/10.1111/j.1365-3040.2010.02181.x

Davis P (2015) Lighting: the principles, Technical guide. Technical report. Stockbridge Technology Centre, Stockbridge. https://horticulture.ahdb.org.uk/sites/default/files/u3089/Lighting_The-principles.pdf

Delor M (2011) Current state of building-integrated agriculture, its energy benefits and comparison with green roofs – Summary. Technical report, University of Sheffield, Sheffield

Department of Agriculture Food and Rural Affairs (2016) Agriculture in the United Kingdom – 2015. Technical report, Department for Environment, Food and Rural Affairs, London

Druma AM (1998) Dynamic climate model of a greenhouse. Technical report, The United Nations University, Reykjavik. http://www.os.is/gogn/unu-gtp-report/UNU-GTP-1998-03.pdf

Earth System Research Laboratory (2017) ESRL global monitoring division – global greenhouse gas reference network. https://www.esrl.noaa.gov/gmd/ccgg/trends/global.html

El-Sharkawy MA (2011) Overview: early history of crop growth and photosynthesis modeling. BioSystems 103(2):205–211. https://doi.org/10.1016/j.biosystems.2010.08.004

Farquhar GD, Von Caemmerer S, Berry JA (1980) A biochemical model of photosynthetic CO_2 assimilation in leaves of C_3 Species. Planta 149:78–90

Ferentinos KP, Albright LD, Ramani DV (2000) Optimal light integral and carbon dioxide concentration combinations for lettuce in ventilated greenhouses. J Agric Engng Res 77(3):309–315. https://doi.org/10.1006, http://www.idealibrary.com

Fernández JE, Bailey BJ (1992) Measurement and prediction of greenhouse ventilation rates. Agric For Meteorol 58(3–4):229–245. https://doi.org/10.1016/0168-1923(92)90063-A, http://linkinghub.elsevier.com/retrieve/pii/016819239290063A

Fitz-Rodríguez E, Kubota C, Giacomelli GA, Tignor ME, Wilson SB, McMahon M (2010) Dynamic modeling and simulation of greenhouse environments under several scenarios: a web-based application. Comput Electron Agri 70(1):105–116. https://doi.org/10.1016/j.compag.2009.09.010

Gijzen H, Vegter GJ, Nederhoff EM (1990) Simulation of greenhouse crop photosynthesis: validation with cucumber, sweet pepper and tomato. Acta Horticulturae 268:71–80

Graamans L (2015) VERTICAL – the re-development of vacant urban structures into viable food production centres utilising agricultural production techniques. PhD thesis, TU Delft. https://repository.tudelft.nl/islandora/object/uuid:f9dd86ce-22a9-4dfe-b66e-ef55230e3856/?collection=research

Graamans L, van den Dobbelsteen A, Meinen E, Stanghellini C (2017) Plant factories; crop transpiration and energy balance. Agri Syst 153:138–147. https://doi.org/10.1016/j.agsy.2017.01.003, http://linkinghub.elsevier.com/retrieve/pii/S0308521X16306515

Grewal SS, Grewal PS (2012) Can cities become self-reliant in food? Cities 29(1):1–11. http://dx.doi.org/10.1016/j.cities.2011.06.003

Grewal HS, Maheshwari B, Parks SE (2011) Water and nutrient use efficiency of a low-cost hydroponic greenhouse for a cucumber crop: an Australian case study. Agri Water Manag 98(5):841–846. https://doi.org/10.1016/j.agwat.2010.12.010

Hartz TK, Baameur A, Holt DB (1991) Carbon dioxide enrichment of high-value crops under tunnel culture 116(6):970–973

Heuvelink E (1996) Dry matter partitioning in tomato: validation of a dynamic simulation model. Ann Bot 77(1):71–80. https://academic.oup.com/aob/article-lookup/doi/10.1006/anbo.1996.0009

House of Commons (1998) The United Kingdom parliament, select committee on environmental, transport, and regional affairs. Technical report, Fifth Report to the House of Commons, London

Jarvis PG (1976) The interpretation of the variations in leaf water potential and stomatal conductance found in canopies in the field. Philos Trans R Soc B Biol Sci 273(927):593–610. http://rstb.royalsocietypublishing.org/cgi/doi/10.1098/rstb.1976.0035

Jolliet O, Bailey BJ (1992) The effect of climate on tomato transpiration in greenhouses: measurements and models comparison. Agri Meteor 58(1–2):43–62. https://doi.org/10.1016/0168-1923(92)90110-P

Kacira M (2016) Controlled environment agriculture research group at the University of Arizona. http://cals.arizona.edu/abe/content/cea

Katsoulas N, Kittas C (2011) Greenhouse crop transpiration modelling,chap 16. In: Evapotranspiration – from measurements to agricultural and environmental applications, INTECH Open, pp 312–328. https://doi.org/10.5772/991

Katsoulas N, Baille A, Kittas C (2001) Effect of misting on transpiration and conductances of a greenhouse rose canopy. Agri Meteor 106(3):233–247. https://doi.org/10.1016/S0168-1923(00)00211-2

Kikegawa Y, Genchi Y, Kondo H, Hanaki K (2006) Impacts of city-block-scale countermeasures against urban heat-island phenomena upon a building's energy-consumption for air-conditioning. Appl Energy 83:649–668. https://doi.org/10.1016/j.apenergy.2005.06.001, www.elsevier.com/locate/apenergy

Kindelan M (1980) Dynamic modelling of greenhouse environment. Trans ASAE 23(5):1232–1239. https://doi.org/10.13031/2013.34752

Kitaya Y, Tsuruyama J, Kawai M, Shibuya T, Kiyota M (2000) Effects of air current on transpiration and net photosynthetic rates of plants in a closed plant production system. In: Kubota C, Chun C (eds) Transplant production in the 21st century. Springer Netherlands, Dordrecht, pp 83–90. http://link.springer.com/10.1007/978-94-015-9371-7_13

Kittas C, Katsoulas K, Baille A (1999) Transpiration and canopy resistance of greenhouse soilless roses: measurements and modelling. ASHS Acta Horticulturae 507:61–68. https://doi.org/10.17660/ActaHortic.1999.507.6

Klein S (2017) TRNSYS 18: a transient system simulation program. http://sel.me.wisc.edu/trnsys

Kolokotroni M, Ren X, Davies M, Mavrogianni A (2012) London's urban heat island: impact on current and future energy consumption in office buildings. Energy Build 47:302–311

Körner O, Hansen JB (2012) An on-line tool for optimising greenhouse crop production. Acta Horticulturae. https://doi.org/10.17660/ActaHortic.2012.957.16

Kozai T, Niu G, Takagaki M (eds) (2015) Plant factory – an indoor vertical farming system for efficient quality food production. Elsevier, Oxford

Law R, Harvey A, Reay D (2013) Opportunities for low-grade heat recovery in the UK food processing industry. Appl Therm Eng 53(2):188–196. https://doi.org/10.1016/J.APPLTHERMALENG.2012.03.024, http://www.sciencedirect.com/science/article/pii/S1359431112002086

Marcelis L (1994) A simulation model for dry matter partitioning in cucumber. Ann Bot 74(1):43–52

Marcelis L, Elings A, Bakker M, Brajeul E, Dieleman J, de Visser P, Heuvelink E (2006) Modelling dry matter production and partitioning in sweet pepper. In: ISHS Acta Horticulturae 718: III international symposium on models for plant growth, environmental control and farm management in protected cultivation (HortiModel 2006), pp 121–128. https://doi.org/10.17660/ActaHortic.2006.718.13

Martineau V, Lefsrud M, Naznin MT, Kopsell DA (2012) Comparison of light-emitting diode and high-pressure sodium light treatments for hydroponics growth of Boston lettuce. HortScience 47(4):477–482. https://doi.org/10.1017/CBO9781107415324.004

McLean D (2017) Integrated environmental solutions Ltd. www.iesve.com

Monteith JL (1966) Photosynthesis and transpiration of crops. Exp Agric 2:1–14

Morris W (1964) The heating and ventilation of greenhouses. Technical report, National institute of agricultural engineering. Silsoe, Bedfordshire

Moss KJ (2007) Heat and mass transfer in buildings, 2nd edn. Taylor & Francis

Nadal A, Llorach-Massana P, Cuerva E, López-Capel E, Montero JI, Josa A, Rieradevall J, Royapoor M (2017) Building-integrated rooftop greenhouses: an energy and environmental assessment in the Mediterranean context. Appl Energy 187:338–351. https://doi.org/10.1016/j.apenergy.2016.11.051, http://www.sciencedirect.com/science/article/pii/S0306261916316361

Nelkin J, Caplow T (2008) Sustainable controlled environment agriculture for urban areas. Acta Horticulturae 801 Part 1:449–455. https://doi.org/10.17660/ActaHortic.2008.801.48

Ogren E (1993) Convexity of the photosynthetic light-response curve in relation to intensity and direction of light during growth. Plant Phys 101:1013–1019. https://www.ncbi.nlm.nih.gov/pmc/articles/PMC158720/pdf/1011013.pdf

Opitz I, Berges R, Piorr A, Krikser T (2016) Contributing to food security in urban areas: differences between urban agriculture and peri-urban agriculture in the Global North. Agri Human Values 33(2):341–358. http://link.springer.com/10.1007/s10460-015-9610-2

Orsini F, Kahane R, Nono-Womdim R, Gianquinto G (2013) Urban agriculture in the developing world: a review. Agron Sustain Dev 33(4):695–720. http://link.springer.com/10.1007/s13593-013-0143-z

Parker T, Kiessling A (2016) Low-grade heat recycling for system synergies between waste heat and food production, a case study at the European Spallation Source. Energy Sci Eng 4(2):153–165. http://doi.wiley.com/10.1002/ese3.113

Penman HL (1948) Natural evaporation from open water, bare soil and grass. In: Proceedings of the royal society

Pieters J, Deltour J (1999) Modelling solar energy input in greenhouses. Solar Energy 67(1–3):119–130. https://doi.org/10.1016/S0038-092X(00)00054-2

Pollet S, Bleyaert P, Lemeur R (2000) Application of the Penman-Monteith model to calculate the evapotranspiration of head lettuce in glasshouse conditions. Acta Horticulturae 519:151–162

Pons O, Nadal A, Sanyé-Mengual E, Llorach-Massana P, Cuerva E, Sanjuan-Delmàs D, Muñoz P, Oliver-Solà J, Planas C, Rovira MR (2015) Roofs of the future: rooftop greenhouses to improve buildings metabolism. Procedia Eng 123:441–448. https://doi.org/10.1016/j.proeng.2015.10.084, http://linkinghub.elsevier.com/retrieve/pii/S1877705815031859

Portis AR (1982) Photosynthesis volume II: development, carbon metabolism, and plant productivity. Academic Press, London

Secretariat of the Convention on Biological Diversity (2012) Cities and biodiversity outlook. Action and policy, a global assessment of the links between urbanization, biodiversity, and ecosystem services. Technical report, Montreal. http://www.cbd.int/en/subnational/partners-and-initiatives/cbo

Seginer I (1984) On the night transpiration of greenhouse roses under glass or plastic cover. Agric Meteorol 30:257–268. Elsevier Science Publishers BV

Skelhorn CP, Levermore G, Lindley SJ (2016) Impacts on cooling energy consumption due to the UHI and vegetation changes in Manchester. Energy Build 122:150–159. https://doi.org/10.1016/j.enbuild.2016.01.035, http://www.sciencedirect.com/science/article/pii/S0378778816300354

Specht K, Siebert R, Hartmann I, Freisinger UB, Sawicka M, Werner A, Thomaier S, Henckel D, Walk H, Dierich A (2013) Urban agriculture of the future: an overview of sustainability aspects of food production in and on buildings. Agri Human Values 1–19. https://doi.org/10.1007/s10460-013-9448-4

Stanghellini C (1987) Transpiration of greenhouse crops – an aid to climate management. PhD thesis, University of Agriculture, Wageningen. http://edepot.wur.nl/202121

Stine SW, Song I, Choi CY, Gerba CP (2005) Effect of relative humidity on preharvest survival of bacterial and viral pathogens on the surface of cantaloupe, lettuce, and bell peppers. J Food Prot 68(7):1352–1358

Studio Glowacka (2013) RIBA Forgotten Spaces 2013 — Studio Glowacka. http://www.studio-glowacka.com/riba-forgotten-spaces-2013/

Takakura T, Jordan T, Boyd L (1971) Dynamic simulation of plant growth and environment in the greenhouse. Trans ASAE 14(5):964–971. https://doi.org/10.13031/2013.38432

Tei F, Scaife A, Aikman DP (1996) Growth of lettuce, onion, and red beet. 1. Growth analysis, light interception, and radiation use efficiency. Ann Bot 78(5):633–643. https://academic.oup.com/aob/article-lookup/doi/10.1006/anbo.1996.0171

Teitel M, Atias M, Barak M (2010) Gradients of temperature, humidity and CO2 along a fan-ventilated greenhouse. Biosyst Eng 106(2):166–174. https://doi.org/10.1016/j.biosystemseng.2010.03.007, http://linkinghub.elsevier.com/retrieve/pii/S1537511010000590

Thompson HC, Langhans RW (1998) Shoot and root temperature effects on lettuce growth in a floating hydroponic system. J Amer Soc Hortic Sci 123(3):361–364. http://journal.ashspublications.org/content/123/3/361.full.pdf

Touliatos D, Dodd IC, McAinsh M (2016) Vertical farming increases lettuce yield per unit area compared to conventional horizontal hydroponics. Food and Energy Secur. http://doi.wiley.com/10.1002/fes3.83

UrbanFarmers AG (2016) UF de Schilde – The "Times Square of Urban Farming". https://urbanfarmers.com/projects/the-hague/

van Henten E (1994) Greenhouse climate management: an optimal control approach. PhD thesis, Wageningen University

Van Iersel MW (2003) Carbon use efficiency depends on growth respiration, maintenance respiration, and relative growth rate. A case study with lettuce. Plant Cell Environ 26(9):1441–1449. https://doi.org/10.1046/j.0016-8025.2003.01067.x

Vanthoor BHE (2011) A model based greenhouse design method. PhD thesis, Wageningen University

Vanthoor BHE, Stanghellini C, van Henten E, De Visser P (2011) A methodology for model-based greenhouse design: part 1, a greenhouse climate model for a broad range of designs and climates. Biosyst Eng 110(4):363–377. http://dx.doi.org/10.1016/j.biosystemseng.2011.06.001

Viljoen A, Bohn K, Tomkins M, Denny G (2010) Places for people, places for plants: evolving thoughts on continuous productive urban landscape. Acta Horticulturae 881:57–65

Von Caemmerer S (2013) Steady-state models of photosynthesis. Plant Cell Environ. https://doi.org/10.1111/pce.12098

Wang J, Lu W, Tong Y, Yang Q (2016) Leaf morphology, photosynthetic performance, chlorophyll fluorescence, stomatal development of lettuce (Lactuca sativa L.) Exposed to different ratios of red light to blue light. Front Plant Sci 7:250. https://doi.org/10.3389/fpls.2016.00250, http://www.ncbi.nlm.nih.gov/pubmed/27014285, http://www.pubmedcentral.nih.gov/articlerender.fcgi?artid=PMC4785143

Wareing PF, Khalifa MM, Treharne KJ (1968) Rate-limiting processes in photosynthesis at saturating light intensities. Nature 220(5166):453–457

Weber CL, Matthews HS (2008) Food-miles and the relative climate impacts of food choices in the United States. Environ Sci Technol 42(10):3508–3513. http://www.ncbi.nlm.nih.gov/pubmed/18546681. http://pubs.acs.org/doi/abs/10.1021/es702969f

Wheeler RM, Mackowiak CL, Sager JC, Yorio NC, Knott WM, Berry WL (1994) Growth and gas exchange by lettuce stands in a closed, controlled environment. J Amer Soc Hortic Sci 119(3):610–615

World Bank (2016) Data. http://data.worldbank.org/

Xu F, Bao HX, Li H, Kwan MP, Huang X (2016) Land use policy and spatiotemporal changes in the water area of an arid region. Land Use Policy 54:366–377. https://doi.org/10.1016/j.landusepol.2016.02.027

Yang X, Short TH, Fox RD, Bauerle WL (1990) Transpiration, leaf temperature and stomatal resistance of a greenhouse cucumber crop. Agri For Meteor. https://doi.org/10.1016/0168-1923(90)90108-I

Zahmatkesh Z, Burian SJ, Karamouz M, Tavakol-Davani H, Goharian E (2015) Low-impact development practices to mitigate climate change effects on urban stormwater runoff: case study of New York City. J Irrig Drain Eng 141(1):04014043. https://doi.org/10.1061/(ASCE)IR.1943-4774.0000770

Chapter 18
Applications for Breeding and High-Wire Tomato Production in Plant Factory

Marc Kreuger, Lianne Meeuws, and Gertjan Meeuws

Abstract For indoor farming to become the next step in agriculture, it is essential that most nutritious and healthy crops can be grown indoors. Some fruits and vegetables currently grown in (high-tech) greenhouses should be suitable from an economic point of view; cost price is relatively high as are the produced volumes. One example for such a crop is the high-wire cultivated varieties of tomatoes. Breeding companies can take the advantage of this technology. Breeding for taste and nutrition only and not having to focus on (a)biotic stress is now possible. This chapter describes the growth and production of a tomato crop in an indoor farm. Growth, development, and fruit yield at least equal greenhouse performance showing the opportunities ahead for these types of crops.

Keywords Tomato · Vine crops · Indoor farming · Fruits · Yield

18.1 Introduction

Indoor, urban, and vertical farming has got a lot of attention the last years. Many local initiatives, start-up companies, and even multinationals move into this area for food production. The concept of local, safe, and fresh food appeals to a lot of producers and consumers. Initially the focus crops for many if not most farms were lettuce, microgreens, and herbs. This makes sense as they are small, fast rotating crops and ideal for an indoor farming setup. However, for a balanced diet produced in an indoor farm, a larger variety of crops needs to be produced. Though field crops like wheat, barley, rice, and corn are not suitable yet due to their low-cost prices, there are enough alternatives available. Some fruits and vegetables currently grown in (high-tech) greenhouses should be suitable from an economic point of view; cost price is relatively high as are the produced volumes. One example for such a crop is the high-wire cultivated varieties of tomatoes. In Europe, medium-sized varieties are sold for 1–3 € per Kg and are reported to yield 60–80 Kg per m^2 per

M. Kreuger (✉) · L. Meeuws · G. Meeuws
Seven Steps To Heaven B.V, Zwaanstraat 31U, 5651 CA, Eindhoven, The Netherlands

year, while cherry tomatoes can be sold for 10 € per Kg while yielding 30–40 kg per m² per year, when cultivated in greenhouses. However, the quality and yields vary throughout the year as environmental conditions change with seasons. Indoor farms can ensure the same environmental condition through the whole year and as result will allow a constant and predicted yield. Moreover, quality will be improved as optimal condition for growth and development can be obtained in such environments.

As the suitable crop list expands, the application for breeding companies should become more attractive. Being able to grow the same variety all over the world together with the benefits of the closed controlled systems enables breeders to refocus their breeding efforts. No longer pests and diseases, drought, and other abiotic stresses are the main targets in breeding. From now on taste, nutrition, and health should receive more attention.

Moreover, indoor farming technology can be a great help in breeding acceleration. Many steps in breeding processes are already performed in closed environments. But growing indoors from seed to seed is now possible offering great opportunities.

This chapter describes how a tomato crop was grown in the indoor environment described in Chap. 8 and what breeding application is possible in such environment.

18.2 Applications in Breeding

For breeding companies speed to market of new traits is key. Breeding cycles in field and greenhouses are bound to the seasons, thereby limiting speed of inbreeding. Breeding for hybrids requires inbreeding of parent lines for a number of cycles, typically 5–10. Creating uniform and homozygous parents is essential to result in heterozygous and uniform hybrids in which heterosis is fully expressed. Selection of the right parent lines may easily take 10 years, and in the case of biannuals like cabbage, it may take even 20 years. Breeders therefore want to speed up this process. There are several options to achieve this.

The first would be to take the inbreeding process to an indoor farm. Working more independent from the climate and seasons enables off-season pollination and seed set. Specific stages in plant development can be mimicked in growth chambers, for instance, vernalization (the cold treatment plants get in winter to induce flowering (Sung and Amasino 2005)). To have full plant cycles in closed environments is more complicated as it requires optimal flowering and seed set. The indoor farming technology as described in Chap. 8 can accelerate the plant life cycle. A cycle time reduction of at least 30% is likely possible, assuming plants follow their normal developmental cycle as they would outdoors.

A second alternative would be to find shortcuts in those cycles. Plant follows programs that lead them through the juvenile phase and maturity to reach flowering and seed set. Those phases can be shortened, or completely different routes can be taken. A good example is described by Evans (1987). This paper describes that winter wheat can flower without vernalization which it normally needs. The way to achieve this is to apply short days during the juvenile phase. By doing so the plant chooses a different route and induces flowers. Some would call this "stress-induced flowering," but the result for the breeder is the same, seeds. It shows the plasticity of plants in their developmental routes, even though normally they would use mostly the same route.

Acceleration of breeding by shortening cycles or applying shortcuts will decrease time to market for new traits and hybrids. Being first on the market is very important for many traits like disease resistance. Especially with those types of traits, a lot of energy and investment need to be done to identify the traits. For these special trials are organized in which the phenotype is assessed in controlled essays. For instance, a batch of young plants is treated with a suspension of spores of a certain disease. Often this trial is done in closed environments, but the plants are grown in a greenhouse. Obviously doing everything indoors will enhance the quality and consistency of the phenotyping essay and therefore speed up the identification of the right trait.

To make indoor farming technology an integrated tool for breeding companies, basically all crops need to be able to grow from seed to seed in such an environment. For some crops, this may be easier than for others. Tomato is known to be difficult to grow indoors. The indoor farming solution as described in Chap. 8 was used to grow tomato varieties. The purpose was to show this crop can be grown from seed to seed in an indoor environment while reaching at least the same quality and yield as they would in a high-tech greenhouse.

18.3 High-Wire Tomato Production

18.3.1 Introduction

To make indoor farming commercially attractive, cost of goods, yields, and sales prices need to be in balance. Since indoor farming is more expensive than conventional greenhouses due to the high starter investments, profitability is important. With the Plant Balance Model (see Chap. 8), yields can be calculated and predicted which should feed into the business case of the farm. The choice of crops and varieties will ultimately drive the profitability of the indoor farm. One of the more profitable crops for indoor growing is tomato. Especially cherry tomato combines relatively high sales price and high yields. For comparison, herbs have a high sales price and yield, but a lower market volume. Lettuce on the other hand has a high yield and a low price but a high market volume.

The capability to grow tomato indoors was recently developed (see Chap. 8). In the first trials, plant quality, yield, and fruit quality were assessed. Overall the difficulty of growing tomato indoors lies in the early stages of development. Before flowering the young tomato plants may have an imbalance in source and sinks (Li et al. 2015), meaning the plant produces too much energy (sugars) for the small number of sinks. It is our hypothesis that if the sugar levels get too high in the leaves, edema will occur. Cells on the leaf surface swell and even burst due to osmotic pressure, thereby damaging the leaf. Restoration and repair is difficult if not impossible and will impact yields. Some varieties are more susceptible than others, and clearly there is a genetic component.

Some reports describe that edema can be prevented by far-red light, white light, or other treatments (Williams et al. 2016). Reduction in source-sink ratio will prevent edema. There are multiple ways to achieve this. Changing light spectrum will reduce sugar production, in the case of white light compared to only blue and red light. As white light is less effective in photosynthesis, with equal intensities in micromoles. $m^{-2}.s^{-1}$, it will result in less sugar production. In the case of far-red light, auxin production and action is induced. As a result, sugar is stored in the vacuole in stem cells during growth and development, resulting in cell elongation. This extra storage of sugar and accumulation of biomass of stems reduce the source-sink ratio, thereby preventing edema. An alternative method is the control of sugar production by adapting light intensities and/or control of the transport of sugar through the plant by controlling evaporation. Though the xylem is the main channel for evaporation, the movement of sugar through the phloem depends on water movement as well. An increase of evaporation while keeping temperature and light intensities the same will in general prevent further development of edema. The ability to control evaporation, light, and temperature independently is key and can be achieved by using a laminar airflow, blue and red light-emitting diodes (LEDs), and humidity control based on vapor pressure deficit (see Chap. 8). Whether or not the hypotheses stated above are true can now be proven.

In a trial conducted in such an environment, growth was measured to validate the Plant Balance Model. Since the model was developed in greenhouses, an assessment in an indoor farm environment was needed. For this plants of two varieties were cultured. Every week complete plants were sampled, and fresh and dry matters were determined of leaves, stems, and fruits. Additionally, leaf area index (LAI) was determined (more details in Schramm 2017).

18.3.2 Results

Lights were mounted on the 4-m high ceiling. Only red and blue LEDs were used in a ratio of 7:3. Consequently the light levels on gutter level were lower than just below the light. Intensities were measured on three levels and showed an increase with increased height (Fig. 18.1). In the figure colors indicate the intensities on the three levels but also indicate the uniformity in the horizontal plane. Due to the limited size of the module (4 by 8 m), a large edge effect was present resulting in limited uniformity.

At the gutter level, the intensity can be as low as 116 $\mu mol.m^{-2}.s^{-1}$; at 2 m above gutter level, it can reach to 250 $\mu mol.m^{-2}.s^{-1}$. The average light intensity 2 m above gutter level is 35% higher than at the gutter level and 19% higher 1 m above gutter

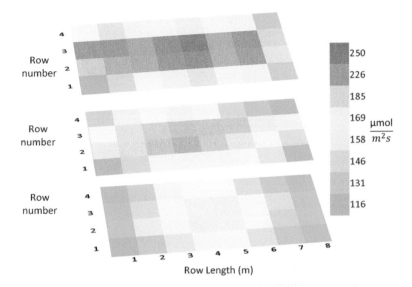

Fig. 18.1 Light intensities on gutter level (bottom) and 1 m (middle) and 2 m above gutter (top). The Y-axis represents the row number and the X-axis represents the length of the gutter. Each box represents an average of three measurements of light intensity measured in $\mu mol.m^{-2}.s^{-1}$

level. Moreover, in every level the light is not equally distributed as the middle rows (2 and 3) experience on average higher light intensities than the two side rows (1 and 4). At the gutter level, this difference is on average 10%, at 1 m above it is 25%, and at 2 m above it is 45%. The first and last 1.5 m (1–2 and 7–8) at every level also experience lower light intensities than the center parts (3–6). At gutter level and 1 m above it is lower in an average of 25%, while 2 m above gutter is only 10% lower.

There is no significant difference in temperature between the inner part, rows 2 and 3, and the outer part, rows 1 and 4 (data not shown). However, 2 m above the gutter level, the temperature is significantly higher than at the gutter level and 1 m above the gutter level (Fig. 18.2). At longer distance from the air inlet, the temperature also increases on this level. This is due to the heat generated by the LEDs mounted on the ceiling. It also shows that at plant level, at least up to 2-m tall, temperature over a single gutter is constant. Uniformity of temperature within the module therefore guarantees uniform growth from that aspect.

Tomato (*Solanum lycopersicum* L.) plants from two different cultivars, a cherry tomato "Axiany" (Cultivar A) and a cluster medium-sized tomato "Axiradius" (Cultivar B), were supplied by Axia Vegetable Seeds (Naaldwijk, the Netherlands). The vine crops have an indeterminate growth habit, meaning on average every week, a new section is being produced consisting of three leaves, three internodes, and a truss at the applied temperature. Total fresh weight (TFW) and total dry weight (TDW) per plant showed linear growth during the trial (Fig. 18.3). Linear regression resulted in a R^2 larger than 0.95 for both cultivars in both TFW and TDW. Stem length also showed a linear growth (Fig. 18.4) for both cultivars. The total leaf area

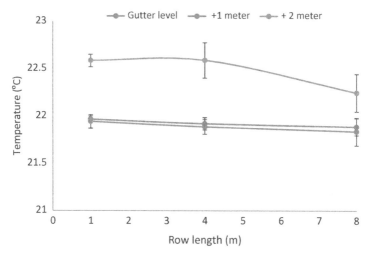

Fig. 18.2 Temperature distribution within the module while lights were on. Dots are the averages of three measurements conducted at three different rows. Bars represent the standard error ($\alpha < 5\%$). Airflow was from right to left

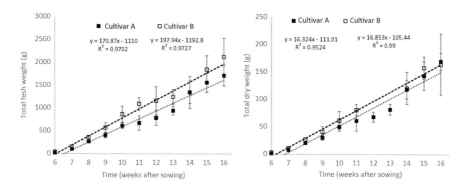

Fig. 18.3 Average of total fresh weight (left) and total dry weight (right) over time and per plant. Average was calculated from a sample of four plants taken once a week from each cultivar. Bars represent the standard deviation ($\alpha < 5\%$) of each sample

showed a somewhat fluctuating growth since some leaves were removed during the trial around 11 weeks after sowing. Overall this shows the crop has a consistent growth rate in biomass, length, and leaves.

Leaf area index (LAI, m^2 of leaves per m^2 of floor area) is known to impact the absorption of light. In general, an LAI of 3 captures most of the sunlight. There is a good correlation between leaf number and LAI in both cultivars (Fig. 18.5), which can allow to determine the number of leaves needed to obtain a certain LAI. To obtain an LAI of 3, cultivar A needs around 20 leaves, while cultivar B needs around 15. To reach an LAI of 3, both cultivars needed a period of 9–10 weeks from sowing

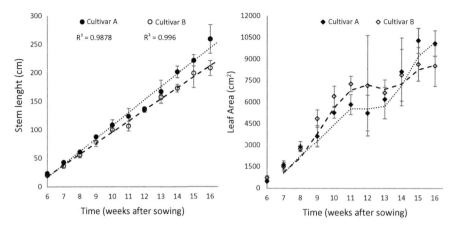

Fig. 18.4 Plant vegetative growth over time of the two tomato cultivars. Left graph shows stem length, right graph leaf area per plant. Bars represent the standard deviation ($\alpha < 5\%$) that was calculated for each sample

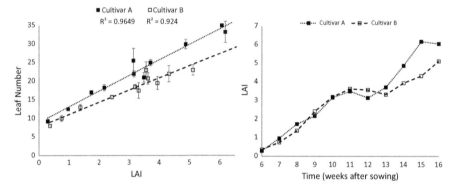

Fig. 18.5 In the left graph the correlation between leaf number and leaf area index (LAI) is shown. Bars represent the standard deviation of the measurements ($\alpha < 5\%$). On the right LAI development over time is shown

(Fig. 18.5). This holds only for this particular trial where each plant had only one stem and was grown at a density of six plants per m^2.

The dry matter partitioning over time is shown in Fig. 18.6. While the stem portion remains rather stable, leaf partitioning seems to decrease. As older leaves die, while stems do not, the leaf partitioning will decrease. As fruits develop their relative portion of the dry matter increases. The realized percentage eventually is the harvest index which can be used in the Plant Balance Model of Chap. 8. By using the linear extrapolation from the fresh weight measurements (Fig. 18.3) and truss weight measurements (not shown), it is possible to estimate the total fresh weight of a single plant in 1 year of growth. The total fruit yield per year can be calculated taking

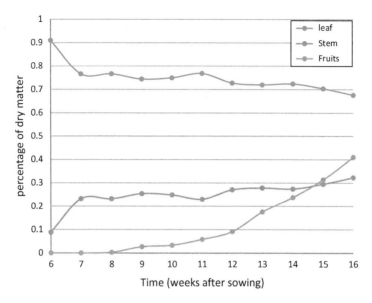

Fig. 18.6 Dry matter partitioning of cultivar A over time

Table 18.1 Calculation of total fresh weight (TFW), total fresh fruit weight (FFW), and harvest index (HI) for a single plant of both varieties

	Cultivar A weight (g)	Cultivar B weight (g)
TFW	7775	9100
FFW	3542	7553
HI	45%	82%

into consideration 16 weeks have passed from sowing till the first truss was harvested, leaving 36 weeks of harvesting. Finally harvest index can be calculated by dividing the yearly fruit yield over total fresh weight (Table 18.1). So far, the data are consistent with greenhouse measurements (Li et al. 2015) indicating the growth and development of tomato plants can be similar in greenhouses and indoor farms. Figure 18.7 shows a tomato crop (three different varieties) grown in the module in the same setup as the described trial.

18.3.3 Conclusions

The data presented demonstrate the ability to grow tomato plants in an indoor environment as it was created. The laminar airflow, red and blue light spectrum, and a constant temperature enable fruit production at commercial level and quality. Though final yields could be calculated only from extrapolated data, further improvements in crop management and climate should result in increased yields. Since this is still in development, this report should be regarded as a first step. It is

Fig. 18.7 Tomato crop grown in the module. Three different varieties were used but in the same setup as the trial described

expected that within the coming years, indoor farms producing tomato fruits of commercial value will show the validity of this technology.

18.3.4 Materials and Methods

The experiment was conducted in the module as described in Chap. 8, located at the Priva company site in De Lier, the Netherlands. The module is essentially a climate room of 4 m × 10 m × 4 m.

18.3.4.1 Environmental Conditions and Irrigation Control

The climate conditions were set to a constant day and night 22 °C with a photoperiod of 20/4 h (day/night). CO_2 was set at 1200 ppm constantly, and relative humidity (RH) was set to 70%–80%. Lights were supplied by red and blue LEDs (red/blue LED ratio of 7:3) in an average light intensity of 132 $\mu mol.m^{-2}.s^{-1}$ on the gutter level and 180 $\mu mol.m^{-2}.s^{-1}$ at 2 m above gutter. The irrigation system consisted of a closed loop hydroponic nutrient film technique (NFT) system supplying a nutrient solution with a recipe specially made for tomatoes. The solution was maintained with an automated irrigation system measuring EC (electrical conductivity) and pH (indicating acidity). EC was maintained on 2.6–2.8 $mS.cm^{-1}$ by dosing a concentrated nutrient solution from A and B tanks. pH was maintained on 5.6–5.8 by dosing a nitric acid from a concentrated acid tank. Once a week a nutrient analysis was conducted by Groen Agro Control (Delft, the Netherlands) on a water sample taken from the mixing tank. The analysis was used to correct the A and B tank solution compositions to optimize the nutrient supply for the plants.

18.3.4.2 Experimental Layout

The experiment included four gutters with a distance of 120 cm between them. Each gutter was 8-m long, and plants were planted in a density of 6 plants per m^2 resulting in 192 plants in total. The first and the last 1.5 m of each gutter were considered border plants. This division resulted in a sample size of 30 plants per gutter. Two different cultivars were used and were planted two gutters per cultivar resulting in a 60 plant samples per cultivar.

Tomato *(Solanum lycopersicum L.)* plants from two different cultivars, a cherry tomato "Axiany" (Cultivar A) and a cluster medium-sized tomato "Axiradius" (Cultivar B), were supplied by Axia Vegetable Seeds (Naaldwijk, the Netherlands). The seeds were germinated and grown till 5 weeks old at Axia greenhouse compartment. Afterwards they were transplanted into 10 cm^2 rock wool blocks and transferred into the module. Pollination was done manually twice a week using an electric pollinator to vibrate the flowering trusses separately. From transplanting once a week, a sample of four plants per cultivar was harvested. Measurements of leaf area and fresh and dry weight were conducted. Stem fresh and dry weight were measured. Moreover, fruit numbers and fresh and dry weight per truss were recorded. All samples were oven dried for 72 h at 80 °C for dry weight measurements.

From the collected data, the model parameters (LAI, HI, and DM (dry matter)) were calculated. LAI was calculated by dividing the average leaf area measured per plant in a specific week multiplied by 6 (six plants per m^2) and divided by 1 m. Dry matter content was obtained by dividing the fruit's dry weight over the fruit's fresh weight.

Alongside the plant measurements, environmental factors such as light intensities and air temperature were also conducted. Light measurements within the photosynthetic active radiation (PAR) were conducted with a quantum sensor (LI-190R Quantum Sensor, LI-COR Inc.), at three different heights above gutter: 0–10 cm above gutter, 1–110 cm above gutter, and 2–210 cm above gutter. The total gutter length is 8 m, and in every meter three measurements were conducted. Finally, an average per meter gutter was calculated. Temperature measurements (P-300 Thermometer, Dostmann Electronic) were taken in the center of each path at three points, meter 1, meter 4, and meter 8, at three different heights.

For light and temperature measurements, statistical tests of variance (ANOVA) and significance of the differences were evaluated using the F-test with $\alpha = 0.05$ on the means. The analysis was conducted using RStudio. Plant measurements standard deviation and standard errors were calculated using Excel.

References

Evans L (1987) Short day induction of inflorescence initiation in some winter wheat varieties. Aust J Pl Phys 14:277–286

Li T, Heuvelink E, Marcelis L (2015) Quantifying the source–sink balance and carbohydrate content in three tomato cultivars. Fr Pl Sc 6:416

Schramm G (2017) MSc internship thesis. Wageningen University, The Netherlands

Sung S, Amasino R (2005) Remembering winter: toward a molecular understanding of vernalization. Ann Rev Plant Biol 56:491–508

Williams K, Miller C, Craver J (2016) Light quality effects on intumescence (Oedema) on plant leaves. In: Kozai T, Fujiwara K, Runkle E (eds) LED lighting for urban agriculture. Springer, Singapore, pp 275–286

Chapter 19
Molecular Breeding for Plant Factory: Strategies and Technology

Richalynn Leong and Daisuke Urano

Abstract Breeding of crops has aimed to improve grain yields, biomass, stress resistance as well as nutritional compositions and other commercial values. Plant factories with artificial lighting (PFALs) provide artificial environmental conditions customized for individual crop species. Light flux, light spectrum, photoperiod, moisture, carbon dioxide concentration and nutrient composition can be optimized and monitored overtime without being influenced by weather. Existing crop cultivars were however selected therefore more suited to the outdoor cultivation. The objective of this chapter is to review the developmental and physiological traits suited to needs of PFALs along with the conventional and emerging biotechnologies in crop breeding. Recent advances in molecular genetics and genomics bridged the knowledge gaps between genetic variations and phenotypic consequences in plants. The genetic information, integrated with a CRISPR (clustered regularly interspaced short palindromic repeats)/Cas9 (CRISPR-associated protein 9) genome-editing technique, enables crop breeding without labour intensive conventional screenings.

Keywords Transgene · CRISPR/Cas9 · Phylogenetic tree · Sequence similarity network · High-value crops · Tipburn · QTL

19.1 Introduction

Plant factories with artificial lighting (PFALs) provide a unique and controlled environment for the optimal growing of plants that allows consistent production all year around. The technology is multidisciplinary, encompassing engineering, biology, crop science and agriculture, and it is constantly being developed and improved to provide to the needs of consumers – high-quality, high-value, affordable and accessible crops. However, there are still indefinitely many ways to

---

R. Leong · D. Urano (✉)
Temasek Life Sciences Laboratory, 1 Research Link, National University of Singapore, Singapore, Singapore
e-mail: richalynn@tll.org.sg; daisuke@tll.org.sg

improve plant growth and make PFALs more efficient, i.e. maximizing yields while minimizing inputs and waste products. This chapter describes current approaches used to improve plant yield and quality, reduce cost of production and produce high-value crops, followed by the new approaches to take. Genetic and metabolic engineering in plants are achieved through the modern technique CRISPR/Cas9. Genes are chosen and targeted through data mining and computational biology and the use of evolutionary analysis of protein sequences and model plant systems to obtain genome and proteome information. The leafy vegetable, lettuce, will be used as an example and model for current and future techniques to improve growth and value. The ideas described below aim to encourage and guide the next generation of scientists to take on new approaches to crop improvement and complement them with PFALs technology. We will also discuss the future trend that is increasingly prominent in all areas of farming and agriculture.

19.2 Desired Plant Traits for an Artificially Controlled Environment

PFALs are suitable to increase vegetable yields efficiently, decrease usage of chemicals and pesticides in farming and reduce transportation of produce from farmlands to urban distributors. PFALs provide an ideal platform for urban farming, producing specialised crops, leafy vegetables and medicinal plants, and greatly reduce the carbon footprint that conventional open farm systems would produce. With technological innovation, use of information technology and computerized systems, the workflow process from environmental control to cultivation to harvesting and transport are optimally controlled. These continuous efforts have increased plant growth in PFALs more than twice faster than grown in outfields, but the growth rate can be improved even more by designing new crops with physiological traits optimal to the multilayered vertical cultivation systems.

The selection criteria under PFAL environments include (1) short stature growing to about 30 cm in height, (2) preference for low light flux (100–150 μmol m^{-2} s^{-1} on leaf surface), (3) tolerance to high-density planting, (4) short cultivation time from seedling stage to harvest, (5) efficient nutrient uptake from the soil-free hydroponics, (5) preference for CO_2 supplementation and (6) a high value-per-weight ratio (Fig. 19.1). The cultivation period from germination to seedling stage is not essential for reducing the overall cost of production, because of limited requirements for cultivation space at the early developmental stage. Similarly, preference for higher light flux and continuous lighting would not be essential, because the lighting practices increase initial and running costs. Elite crop varieties for outdoor cultivation need to possess a resilience to environmental stresses and resistance to harmful organisms according to individual climate and soil conditions. The requirements contrast to the breeding principle for PFALs that provide perfectly controlled environments with the minimum risks of pathogen and insect infestations. Released

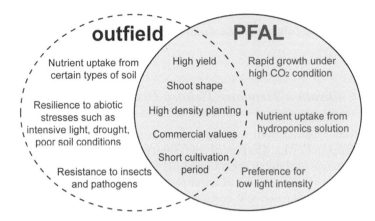

Fig. 19.1 Crop-breeding strategies for outfield and PFAL conditions

from an unavoidable trade-off between plant growth and stress resistance, PFAL-specific varieties that attain a maximal growth rate can be selected efficiently.

19.2.1 Plant Height and Leaf Area

As the distance between light source and planting bed increases, the amount of light reaching the leaf surface drops. Thus a typical vertical farming system places light-emitting diode (LED) light 40–50 cm from cultivation bed. Similarly, high-density planting at the early developmental stage improves light-use efficiency by increasing leaf area index (LAI = leaf area per unit ground surface area $[m^2 \cdot m^{-2}]$). Designing plant cultivars with short stature and larger leaf area with appropriate phyllotactic patterns is the primary breeding strategy to maximize the light-use efficiency in PFALs.

In outside environment, the shape and complexity of vegetative tissues are influenced by growing conditions, allowing plants to adapt to the surrounding environments. These developmental strategies are controlled by combinational actions of five hormones: auxin, gibberellic acid (GA), brassinosteroid (BR), cytokinin and strigolactone. Genetic mutations that decrease the amounts of those hormones or disrupt hormone perceptions resulted in dwarfisms in *Arabidopsis*, rice and other plants (Clouse and Sasse 1998). GA-insensitive mutant alleles in wheat (reduced height-1) exhibit the short stature phenotype (Peng et al. 1999). Another semidwarfing gene (*SD-1*) in rice encodes an enzyme GA 20-oxidase 2 that is essential for GA biosynthesis (Sasaki et al. 2002). The *sd-1* mutant rice has been selected by independent breeding practices in various areas, demonstrating the importance of the *SD-1* gene. *Arabidopsis rotundifolia 3* (*rot3*) mutant decreases leaf length while increasing leaf width, due to the loss of P450 enzyme (*CYP90C1*) that catalyses brassinosteroid production (Ohnishi et al. 2006). Genetic or non-genetic manipulations of those hormonal actions would enable to shape plants to fit to multilayered cultivation systems in PFALs.

19.2.2 Cultivation Time and Tipburn Occurrence

Short cultivation period from seedling to harvest increases annual production of leafy vegetables in PFALs. The typical cultivation time for different lettuce varieties has greatly decreased through optimizing LED light intensities, photoperiods and carbon dioxide concentrations for efficient photosynthesis and plant growth (Kozai 2013). However, production still needs to be slowed down to a certain extent to minimize the occurrence of tipburn. Tipburn is a physiological disorder commonly found in leafy greens such as lettuce and kale. This common necrosis occurrence at the leaf edges is caused by a lack of calcium ions in the plant which results in inefficient cell wall assembly as the plant grows. Causes of calcium deficiency in leaves are inhibition of calcium ion absorption at the root zone, inhibition of the transfer of calcium ions from root to shoot due to inefficient transpiration and immobility of calcium ions once it is fixed to the cell walls (Barta and Tibbitts 2000). Tipburn occurrences also increase as light intensity and duration increase (Tibbitts TW and Rao Rao Rama 1968). In Sect. 19.6, tipburn occurrence in leafy vegetables will be further discussed. Various molecular players in plants will be introduced and ideas proposed to reduce cultivation periods and improve plant quality through genome editing.

19.2.3 Cell Death

Reducing produce loss contributes to higher yields and overall minimizing costs. Produce loss is contributed by unsellable and visually displeasing vegetable crops due to browning and rotting of leaves, shoots and fruits. PFAL is known to markedly reduce unsellable parts by providing optimal growth conditions and minimal invasion by pathogens. To further reduce agricultural waste, produce loss could be reduced by targeting genes that control cell death (Sakuraba et al. 2012). A well-known gene is stay-green (*SGR*), which drives senescence in plants by destabilizing chlorophyll-protein complexes, allowing enzymes to break down chlorophyll. Knockout mutants of *SGR* delayed senescence (Hörtensteiner 2009). Functional *sgr* mutant vegetables which retain photosynthetic activity are valuable targets to be used in PFALs to increase yields while reducing input of environmental factors like light, water and carbon dioxide. Examples of transgenic crops that have decreased SGR activity are alfalfa (Zhou et al. 2011) and rice (Jiang et al. 2007).

19.2.4 Preference for Low Light Intensity

Different types of plants require different light intensities for maximum photosynthesis and growth rates. The light intensity of direct sunlight reaches over 1000 μmol m^{-2} s^{-1}, while a modern PFAL system employs a lower light intensity (100–150 μmol m^{-2} s^{-1}) to balance the energy input and the growth rate of plants. While sun plants possess multiple protective mechanisms against intense light stress, the light-preferring plants demonstrate etiolation and yellowish coloration under low light circumstances. Oppositely, some shade-loving plants naturally have an effective light-harvesting system to capture enough numbers of photons from dim light for photosynthesis, although are damaged under direct sunlight. Chloroplasts serve as a site for light harvesting, electron transport and carbon fixation. The light-use efficiency by plants arises from various morphological, subcellular and molecular regulations. For example, the amount of flavonoids and other polyphenols, which absorb photosynthetically active radiation, negatively influence the light-use efficiency at low light intensity. Successful genetic manipulation of these photosynthetic systems would raise the light-use efficiency of plants grown in PFALs.

19.3 Genetic Engineering and Crop Breeding

The physiological and morphological traits mentioned in 19.2 can be controlled through molecular breeding methods such as genome editing to introduce certain mutations or transgenes into the genome (Fig. 19.2). Naturally occurring mutations provide the genetic variations of crops that are heritable to next generations. Over many centuries since the dawn of agriculture, farmers, breeders and biologists have been selecting and producing crops with favourable traits. This was done by phenotypically selecting, combining and propagating through repeated crossings over many plant generations. It was a long and arduous process that took up to 15 years to isolate new crop varieties with desired traits. The identification of genetic material in the 1940s has allowed breeders to induce genetic mutations to crop genomes commonly by radiation to deoxyribonucleic acid (DNA). This technique allowed a quicker selection of plants with advantageous traits that increased yield or made them tolerant to environmental stresses. This became the green revolution through to the 1970s.

Despite the success of classical genetics and crossing of varieties to get high yielding and tolerant crops, it still took a long time, and success rates were dependent on chance. In the 1980s and 1990s, emerging genetic engineering and agricultural biotechnology encompassed the identification of specific genes linking to phenotypes, the precise introduction of specific genes into crops from the same or different species and the efficient and precise plant breeding by the use of genetic markers or quantitative trait loci (QTL) (Collard and Mackill 2008). With increasing genetic knowledge, scientists and agricultural biologists are able to breed crops with

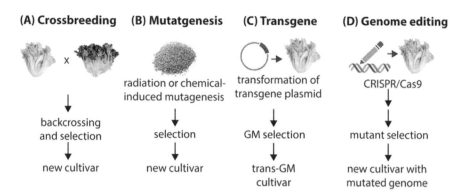

Fig. 19.2 Diagram of conventional and modern plant-breeding strategies. (**a**) Crossbreeding of two existing varieties. A new variety can be isolated after several generations of selective breeding. (**b**) Random mutagenesis by irradiation or chemicals. Mutations associated with a phenotype of interest can be isolated by backcrossing and subsequent genome sequencing. (**c**) Genetically modified transgenic crops. A transgene from an unrelated organism can be cloned and transferred to the genome of crops. (**d**) Genome manipulation by CRISPR/Cas9. A crop genome can be manipulated at the single nucleotide resolution with CRISPR/Cas9 technology. This technique enables breeders to introduce multiplex mutations into the genome of any plant species

advantageous traits at an accelerated rate. In 1997, Calgene introduced the genetically modified tomatoes Flavr Savr which have a longer shelf life than conventional tomatoes. They did this by introducing an antisense gene that blocks the production of the enzyme polygalacturonase which breaks down pectin in cell walls, resulting in decreased softening of fruit (Krieger et al. 2008). Ribonucleic acid interference (RNAi) is another up and rising biotechnology that is applied in crop science to improve nutritional contents and tolerance in stress conditions (Saurabh and Vidyarthi 2014). In genetic engineering, the barrier of sexual incompatibility between plant species is also overcome. Additionally, it is shown that genetically modified organisms (GMOs) are more environmentally friendly as more water, soil and energy are conserved following decreased wastage of destructed crops due to abiotic stress such as drought and biotic stress such as pests.

19.3.1 Current Breeding Practices of Lettuce

Breeding in lettuce for field or greenhouse farming aims to develop lettuce cultivars that produce economically sustainable yields and quality with less input of water and nutrients. Conventional breeding approaches in lettuce include pedigree and backcrossing methods. Lettuce gene pools are often generated from cultivated lettuce *Lactuca sativa* and the wild species *L. serriola* and *L. saligna*. The application of marker-assisted selection, QTLs and computational research in molecular genetics and genomics has greatly advanced lettuce crop breeding. QTLs, which are chromosomal regions directly related to certain traits of a crop, are extensively elucidated and used in crops for precise selection of favourable traits. Many genetic studies done on lettuce were to understand resistances in various lettuce diseases such as *Verticillium* wilt and to cultivate disease resistance crops (Lebeda et al. 2014). In pedigree methods for self-pollinating crops such as lettuce, the ideal plants are selected through phenotypic scans early in the F2 (second filial) generation which are then reselected in subsequent generations until homozygosity is achieved. The whole process takes about 5–10 years. Pedigree breeding has also led to various varieties of crisphead, butterhead and loose leaf. Varieties were selected based on their resistances to early bolting and tipburn.

In backcrossing breeding methods, several generations are backcrossed with one of the parent line until homozygosity is reached. Backcrossing is usually applied to incorporate new genes to an elite variety. In lettuce, backcrossing method was used between wild lettuce *L. saligna*, which was the only species having resistance to lettuce downy mildew, and cultivated species *L. sativa*. Subsequent generations were backcrossed to *L. sativa* parent line. The set of backcrossed inbred lines (BILs) which had chromosomal segments of *L. saligna* in *L. sativa* genetic background covered almost 100% of wild lettuce *L. saligna* genome. Developing backcrossed lines improved QTL studies; in this case having BILs was helpful to study the genetic function of downy mildew resistance in wild lettuce (Jeuken and Lindhout 2004).

Lettuce varieties like crisphead display many physiological disorders such as tipburn, premature bolting and ribbiness. Breeding crisphead lettuce for resistance to these physiological disorders is a long-term strategy to achieve economically sustainable yields and quality. To identify QTL resulting in these physiological disorders, recombinant inbred lines (RILs) are derived by crossing resistant parent line and susceptible parent line. Subsequent generations are intercrossed to form RILs, which are then phenotypically and genetically analysed using single nucleotide polymorphisms (SNPs) as genome markers. Through computational analysis and comparisons between phenotype and genotype, a genetic linkage map is formed, and QTL related to a certain physiological disorder is identified (Jenni et al. 2013). Further gene identification from the identified QTL can be tested for resistance in multiple genetic backgrounds of lettuce and across various environmental types.

Consumption of fruits and vegetables rich in phytonutrients protects against a wide range of human diseases. These types of high-value crops are especially important and suitable for making a profitable indoor farming system that produces crops with higher value than field grown crops. QTLs can also be identified for beneficial phytonutrients such as phenolic compounds including phenolic acids and flavonoids. Studies have identified QTLs and genes that function in biosynthetic pathways of phytonutrients production. Similarly, lettuce RILs are derived through crossing between parent wild lettuce *L. serriola* and cultivated lettuce *L. sativa* and were analysed for their phytonutrients and antioxidant status. Downstream analysis of identified gene targets has shown differential expression between wild lettuce and cultivated lettuce, suggesting a strong correlation between phytonutrients synthesis and availability in lettuce (Damerum et al. 2015).

With the recent advances in QTL analysis with breeding techniques, it will be exciting to identify the genetic basis for favourable traits in crops. Further tests and genetic engineering techniques will produce commercial varieties suitable for PFALs.

19.4 Target Genes for the Genetic Modification of Plants

19.4.1 The Model Plant, Arabidopsis

Arabidopsis, a weed in the mustard family, is the first plant for which genome was sequenced in year 2000. It is the model plant for many crops. Its genome is relatively small consisting of a set of genes that controls plant processes from growth and development to flowering and reproduction to defence mechanisms against biotic and abiotic stresses. *Arabidopsis* is commonly used as a reference for comparative genomics and gene functions in other plants. Researchers for various crops and vegetables tap on public databases of *Arabidopsis* and increasingly other species such as rice to develop more focused molecular tools for gene discovery and engineering. *Arabidopsis* shows close genetic relationship to leafy vegetables *Brassica*, thus making it a very suitable model plant to study gene functions in leafy vegetables which are the common crops cultivated in PFALs. As various vegetable crops are getting sequenced, genetic and functional information are unravelled by comparison to the model plant. Using the gene ontology database and genome and proteome databanks of *Arabidopsis*, identification of functional homologues and comparative analysis of the target protein can be carried out.

19.4.2 Evolutionary Relationships of Leafy Vegetables

Brassica are commonly used for diet, as catch and cover crops, and for biofuel production. Important vegetables include the *B. rapa* (Chinese cabbage, pak choi and turnip), *B. oleracea* (cabbage, broccoli, kale, brussels sprout) and *B. napus* (rutabaga and Hanover kale). Some *Brassica* species are also harvested for their oilseeds (commonly known as canola), making them the third leading source of vegetable oil in the world, after soybean and palm oil. It would be of great commercial value to expand the *Brassica* species in high-density PFALs. Lettuce is another leafy vegetable commonly grown in indoor farming systems. Many of PFALs have established optimal environments for lettuce varieties. Thus, lettuce would be a "model" vegetable to investigate the effects and enhancements of genetic engineering in leafy vegetables in addition to the already optimal conditions for growth.

Leafy vegetable species do not form a monophyletic clade. While majority of leafy vegetable species lies in the *Brassica* lineage, other vegetables are distributed

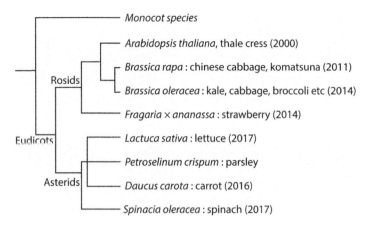

Fig. 19.3 Evolutionary relationships of crop species grown in PFALs. The phylogenetic tree shows relative relationships of vegetables, herbs and medicinal plants with the model plant *Arabidopsis*. Two major Eudicot clades Rosids and Asterids were separated from each other approximately 120 million years ago. The binomial names and representative crops are indicated at each edge. Numbers in parentheses indicate the year the genome paper was released

across subclades within eudicots (Fig. 19.3). The *Arabidopsis* and *Brassica* lineages split only about 20 million years ago (MYA), while other vegetables such as lettuce and spinach are separated earlier from *Arabidopsis* about 120 MYA (Murat et al. 2015). It has been empirically known that species with longer evolutionary distance less conserves mutant phenotypes of homologous genes. Nevertheless it would still be helpful for many features of genetics and functional studies as their genomes are sequenced fully.

19.4.3 Informatics Pipeline to Identify Target Genes for Genetic Engineering

Technological advances in generating and analysing large-scale biological data have opened a new era for crop breeding practices in the last decade. Bioinformatics method integrates multidisciplinary data with statistical methods and helps researchers decide on taking the best breeding strategy. *Arabidopsis* geneticists have identified various genetic mutations that result in morphological changes and environmental responses as described earlier. The accumulated knowledge can be integrated by bioinformatics pipeline to aid designing new vegetable varieties. Basic Local Alignment Search Tool (BLAST) search followed by phylogenetic analysis can be used to identify gene targets in the sequenced crops using the *Arabidopsis* homologue as a query sequence (Fig. 19.4a).

It has been empirically known that proteins with similar sequences possess similar functions across species. This evolutionary theory is commonly used to

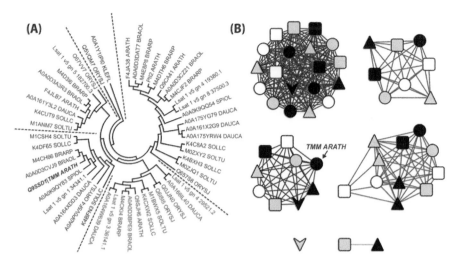

Fig. 19.4 Protein sequence analyses for the identification of target genes for molecular breeding. (a) A phylogenetic tree of *Arabidopsis* TMM (too many mouths) and its homologues in leafy vegetables, carrot, potato, tomato and rice. Protein sequences homologous to *Arabidopsis* TMM were searched by protein BLAST programme from their proteome databases. Multiple sequence alignment for all retrieved proteomic sequences was built using the MAFFT algorithm for the construction of a maximum likelihood tree showing the phylogeny. A green alga, *Klebsormidium*, was used as outgroup. (b) Sequence similarity network of TMM homologues was generated using the same sequence data as in Fig.3A. Each node represents protein sequence. Nodes are connected with edges if two sequences exhibited pairwise sequence similarity score greater than a predefined threshold value. Nodes in the same shape and in the same colour represent same plant species. These analyses of evolutionary relationships of proteins facilitate the intuitive identification of target genes for gene editing. True homologues and protein redundancy can be identified from the tree and network

predict protein functions in non-model species (Lee et al. 2007). Phylogenetic tree summarizes an evolutionary trajectory of a certain gene family using tree topology and branch lengths – the point when genes split from others and how rapidly they evolved. The algorithms for tree construction normally cluster orthologous sequences in the same monophyletic clade in the tree. Orthologues represent homologous sequences in different species that evolved from a common ancestral sequence by a speciation event; hence they are frequently involved in the same or relevant biological processes in different organisms. On the other hand, paralogues – two similar sequences located on different positons on genome often derived by gene duplication – evolve under less selection pressure hence have a higher chance to acquire additional functions apart from the original. The tree topology allows us to determine orthologous sequences from multiple homologous sequences in species. If single orthologous gene was retained in genome, the gene would be the target gene to modify the gene function. There is nevertheless a substantial limitation of the in silico approach. Paralogous genes in the same gene family may exhibit functional overlaps in crops, possibly resulting in functional redundancy in their mutant phenotypes.

While phylogeny is an effective way to cluster orthologous sequences in the same monophyletic branch, it is a computationally expensive approach hence not suitable to process a large number of sequence data. Sequence similarity network (SSN) is an alternative approach to identify functional orthologues in different species (Gerlt 2017). This method calculates pairwise similarity scores between all pairs of sequences in the multiple sequence alignment and then visualizes them with vertices and edges in a graph (Fig. 19.4b). Each vertex represents an individual protein sequence, and edges are drawn between vertices of which similarity score is greater than a user-defined threshold value. SSN requires less computational power compared with the construction of phylogenetic trees. The less computational time is adequate to recapitulate sequence homologies in a large set therefore to cope with a prospective increase in the number of vegetable genomes. As identification of target genes is the first step towards genetic engineering of crops, the in silico process should be thoroughly carried out and assessed with multiple theoretical and experimental approaches.

19.5 Development of an Integrated Pipeline for Genetic Engineering

19.5.1 Transgene and Its Disadvantage

Transgenics is one of the most commonly used approaches for genetic engineering of crop species. This method enables the transfer of a gene or other DNA fragments to plant genomes usually from unrelated organisms. The successful introduction of the transgene confers new abilities to the host plants – production of oral vaccines, antibodies, vitamins and pharmaceuticals or resistance to herbicidal chemicals. One type of transgene crop is the Roundup Ready crops (RR crops), which are engineered to be tolerant to the herbicide glyphosate. RR crops have improved agricultural efficiency worldwide. RR crops carry the gene coding for a glyphosate-insensitive enzyme, rendering them tolerant to the herbicide application, while weeds are killed off efficiently. The gene was obtained from *Agrobacterium* sp. strain CP4 (Funke et al. 2006). However, these types of transgenic crops received much controversial feedback from critics and consumers as while consuming the GM crop itself is safe, the massive amounts of herbicide glyphosate sprayed onto the crops may be harmful to humans if not regulated. Super weeds may also arise as weeds obtain resistance to the overspraying of the herbicide, which makes the need to produce newly engineered crops. Although it is shown that consuming GMOs is no riskier than consuming the non-GM equivalents (Zeljenková et al. 2016), the trend of genetic engineering is moving towards gene editing by CRISPR/Cas9 which is deemed safer and more effective.

19.5.2 Genome Editing by CRISPR/Cas9 Technology

The CRISPR/Cas9 technology is the latest gene-editing method that precisely and efficiently targets and edits a gene or genes of interest. Clustered regularly interspaced short palindromic repeats (CRISPR) was identified in a wide range of bacteria as an adaptive immunity defence mechanism. Small fragments of DNA that resemble invading DNA of viruses are incorporated into a CRISPR region amidst a series of short repeats of 20 base pairs. Near the CRISPR region is the DNA fragment that codes for an enzyme called CRISPR-associated protein, Cas9. Upon translation, Cas9 forms a complex with CRISPR RNA to recognise and cleave viral DNA, thus providing a defence against invading viruses.

Modern biotechnology has enabled customizable CRISPR fragments, called guide RNA (gRNA), to direct Cas9 to induce site-specific genome modifications located next to protospacer adjacent motif (PAM) sequence by complementary base pairing. Mutations are then introduced through DNA cleavage and error-prone non-homologous end joining (NHEJ) or homology-directed repair (HDR) of the target site (Hsu et al. 2014). The CRISPR/Cas9 technology is being widely tested in common crops such as rice and wheat, to create new varieties with resistance to biotic and abiotic environmental stresses in this changing landscape and climate in the world.

CRISPR/Cas9 technology has proved successful in various plants including *Arabidopsis*, tobacco, sorghum and rice (Jiang et al. 2013; Xu et al. 2016; Chilcoat et al. 2017). It has thus far produced few commercialized crop for field and greenhouse farming but none for indoor plant factory cultivation. Examples of CRISPR/Cas9-edited crops include the mushroom and potato which resists browning (Waltz 2016a). The effect was achieved by knocking out the gene coding for the enzyme polyphenol oxidase which causes browning. Being able to induce numerous genomic modifications at one go, the technology has great potential in producing crop varieties that are resistant to biotic and abiotic stresses and also able to generate higher yields. The agronomy company DuPont Pioneer aims to showcase the strength of this technology in its upcoming CRISPR/Cas9 corn that is genetically modified to be waxy, a trait that has been highly valued by the food industries and materials industry since its discovery in the 1900s. This CRISPR crop will be launched around 2020 (Waltz 2016b). The lengthy breeding process by backcrossing is bypassed with this technology. Along with the genetic and functional knowledge of what makes a plant with favourable traits, CRISPR/Cas9 is able to produce ideal crops at a highly efficient rate.

19.5.3 Future Use of CRISPR/Cas9 Technology for Indoor Farming

CRISPR/Cas9 technology can be used as a breeding approach in future PFALs. It would be much simpler to breed a new cultivar with high yields and high quality by removing genes related to stress tolerance in a crop. Without tolerance, the crop would still be able to thrive in indoor conditions optimized for the plant. CRISPR can enable such "impractical" breeding approach to achieve successful and efficient PFALs. In production of high-value crops by targeting metabolic pathways, CRISPR/Cas9 system can be modified or customized to target sites specifically. Several genes may need to be targeted and gene expression levels at different plant tissues or different developmental stages. Customizations include making specific Cas9 or gRNA expression promoters, modifying Cas9 for different nuclease activity, or making combinations of different Cas9 variants. There are numerous ways to alter and develop the CRISPR/Cas9 system to meet our needs in genetic and metabolic engineering to create new crops (Murovec et al. 2017; Yin et al. 2017)

The first step in carrying out CRISPR/Cas9 editing inside plants is to engineer the gRNA, usually 20 base pairs long, and the Cas9 gene into transformation vectors (Fig. 19.5). The gRNA specifically binds to a DNA sequence in a target gene and guides Cas9 to the target site to make a cleavage. The vector containing the gRNA and Cas9 gene is then transformed into *Agrobacterium*. *Agrobacterium tumefaciens* is a plant-infecting bacterium that causes tumours in plants when its transfer DNA (T-DNA) plasmid is inserted into the plant genome. In biotechnology, scientists have modified the plasmid into transformation vectors by removing the tumour-inducing genes and inserting genes of interest in their place. A selectable marker such as antibiotic resistance is also inserted into the vector for selection of plants that have successfully incorporated the vector into their genome.

In the case of CRISPR/Cas9, *Agrobacterium* containing the gRNA and Cas9 are used to infect the target plant crop. Protoplasts or leaf discs of the plants are used to incubate with the *Agrobacterium*, and whole plants are regenerated using plant tissue culture upon successful transformation of vector into the plant genome. Another common method of infection used in *Arabidopsis* is the floral dip method, which does not require plant tissue culture or regeneration (Clough and Bent 1998). Since vegetable species such as *Brassica* are phylogenetically close to *Arabidopsis*, the floral dip method is likely to succeed, further reducing time by bypassing the time-consuming preparation of tissue cuttings or regeneration of whole plants.

Successfully regenerated plants can then be screened through polymerase chain reaction (PCR) and sequenced to check for gene editing. Plants with positive gene editing at target sites are then crossed with wild-type plants to produce a T2 generation that has Cas9-free gene-edited plants. These plants are then crossed to form T3 generation which can be used for experiment to test for their effectiveness in better growth and yields in PFALs.

A DNA-free CRISPR/Cas9 method is now gaining popularity in crop science as there is no foreign DNA introduced into the plants which may alleviate regulatory

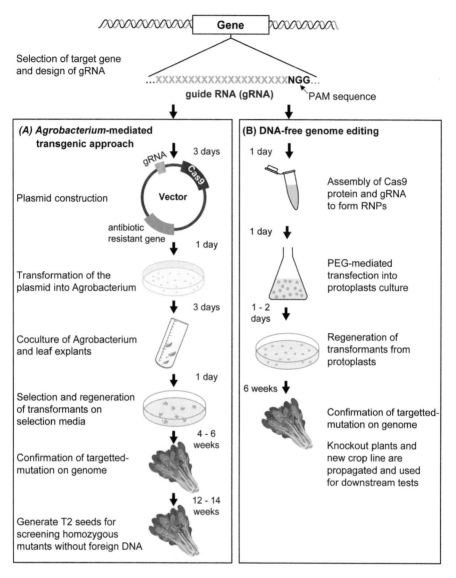

Fig. 19.5 CRISPR/Cas9-mediated genome-editing strategies for generating heritable lettuce mutants. Both approaches begin with the selection of the genomic target based on phylogenetic and bioinformatics analysis, followed by the design of guide RNA (gRNA). gRNA complements the genomic target sequence following a 3-nucleotide sequence NGG protospacer adjacent motif (PAM). (**a**) The agrobacterium-mediated transgenic approach relies on a vector construct containing Cas9 and gRNA sequence which will be transformed into *Agrobacterium*, which in turn infects cotyledons, enabling DNA integration and expression of Cas9 and gRNA. In the plant cell, gRNA guides Cas9 to genomic target sequence and induces double-stranded breaks of DNA. Successful mutants grow on antibiotic selection media and are induced by hormones to regenerate into whole plants. T1 plants are then backcrossed to wild-type parent line to eliminate CRISPR/Cas9 vector construct in T2. (**b**) The DNA-free CRISPR/Cas9 introduces mutations in targeted regions of

concerns and society may be more accepting of these non-GMO crops. Preassembled complexes of purified Cas9 and gRNA are transfected into plant protoplasts. Successful target mutagenesis by the DNA-free method has been seen in *Arabidopsis* and crop plants like lettuce, tobacco and rice (Woo et al. 2015).

The whole process from gene targeting to successful CRISPR/Cas9 editing and testing of generated "mutant" plants in PFALs would take about 1 year, much shorter than conventional breeding methods such as mutagenesis or hybridization. As compared to other genome-editing techniques such as transcription activator-like effector nuclease (TALENs) and zinc finger nuclease (ZFNs), CRISPR/Cas9 is more cost-effective and has less technical problems (Xu et al. 2016; Malzahn et al. 2017).

19.6 Cultivating High-Quality Crops at High Density

One of the main goals of PFALs is to cultivate high-density and high-quality crops at the least cost; thus it is beneficial to produce crops that have decreased cultivation periods and enhanced quality in terms of being fresh and green. A search for candidate genes and proteins to improve vegetable yield in PFALs begins by identifying the causes of physiological failures in cultivation. Tipburn, the browning and necrosis of leaf edges, is a common physiological disorder that occurs in leafy vegetables and fruits such as lettuce, Chinese cabbage and strawberries. This disorder has resulted in massive economic loss and agricultural wastage due to lengthened cultivation periods and reduced sellable fresh and green crops. Tipburn in lettuce is known to be caused by a lack of calcium in leaves, commonly the younger leaves enclosed in a developing head (Barta and Tibbitts 2000).

Regulating calcium uptake and distribution in leaves to decrease tipburn is beneficial for plant health. Calcium transporter genes in the leaf and root systems include the calcium/proton exchanger (CAX) and autoinhibited calcium ATPase (ACA). These calcium transporters are found in tonoplasts, plasma membrane and endoplasmic reticulum of plant cells. The functions of some of these calcium transporters have been elucidated and proven to regulate calcium concentration in cytosolic and vacuole compartments in the plant cells (Fig. 19.6a). Expression levels of CAX1, ACA4 and ACA11 in cabbage are found to vary in tipburn-sensitive and tipburn-resistant lines and also in various abiotic stress conditions (Lee et al. 2013). CAX1 downregulation in *Arabidopsis* led to decreased accumulation of calcium in

Fig. 19.5 (continued) genome without the use of an external DNA vector construct. Cas9 and protein-gRNA ribonucleoproteins (RNPs) are preassembled and transfected into plant protoplasts via the PEG-mediated technique. Successfully transfected mutants are sequenced and regenerated into whole plants

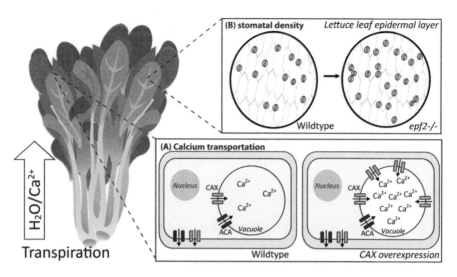

Fig. 19.6 Areas to be genetically targeted to decrease tipburn occurrence. (**a**) Calcium transporter CAX and ACA found in vacuolar and cellular membranes of plant cells facilitates calcium transport from cytosol to vacuoles and apoplastic regions. In tipburn areas of the leaves, vacuolar and apoplastic concentration of calcium is significantly lower. CAX and ACA genes can be engineered to increase calcium concentration, reducing tipburn occurrence. (**b**) Transpiration efficiency affects calcium transport to leaves. Increasing transpiration increases calcium availability in leaves. Modifying genes that control stomatal numbers to increase transpiration rates would aid in increasing calcium concentration in leaves, thus decreasing tipburn

vacuoles of mesophyll cells. The plant grew slowly and had decreased transpiration and carbon dioxide assimilation (Gilliham et al. 2011). Within plant leaves, calcium concentration in vacuoles was present in a decreasing fashion from the leaf base to the leaf apex (Lee et al. 2013). Increasing vacuole concentration of calcium may aid increasing availability of calcium in the leaves. This may be achieved by overexpressing CAX1 gene in the plant. Overexpressing CAX1 in rice increased its nutritional value (Kim et al. 2005). In the case of tipburn, tissue death and slow growth of vegetables may be avoided by overexpression of CAX1.

Transpiration is the driving force for transport of water and nutrients in plants (Nilson and Assmann 2007). Decreased transport of calcium ions to young growing leaves via transpiration is one of the indirect causes of tipburn. Transpiration efficiency is controlled by the opening and closing and distribution of stomata. Targeting genes that regulate stomatal opening or density to increase transpiration rates would be advantageous in decreasing tipburn (Fig. 19.6b). Information of these genes with known functions can be searched through published literature on the model plant *Arabidopsis*. Stomata-specific genes include *open stomata1* (OST1), *too many mouths* (TMM) and *epidermal patterning factor* (EPF) (Dow et al. 2017). Overexpression of EPF2 in *Arabidopsis* decreased stomatal density by inhibiting the progression of protodermal cell to meristemoid mother cell in stomatal development (Hara et al. 2009). With the knowledge of gene sequence, target gene knockout or site-specific expression through targeting promoter can increase expression and increase number of stomata on the leaves. Gene expression level editing can be achieved through a variant of Cas9, one that does not possess the nuclease activity and acts as an RNA-guided DNA-binding protein (La Russa and Qi 2015). Ultimately, with increased transpiration, calcium ions are enabled to be transported efficiently to leaves, especially the younger leaves which have a lower probability of efficient transpiration due to its smaller surface area. Through stomatal gene analysis and editing to increase transpiration rates and reduce tipburn, maximal growth conditions in PFAL could be applied, thus increasing harvest rates and quality.

19.7 Cultivating High-Value Crops

Production cost is the primary issue of plant factory; hence building a profitable business with PFALs requires products with higher value than the outfield-grown crops. High-value crops include leafy vegetables and fruits that have functional components and medicinal components and plants that can produce additional pharmaceuticals that include functional proteins and secondary metabolites. A lot of research has been done to manipulate the nutrient composition, light intensity and light radiation to attain certain characteristics in plants to suit consumer needs. For example, low-potassium vegetables were cultivated for patients suffering from kidney diseases by removing potassium and sodium ions from the nutrient solution while maintaining magnesium at 5%. Low-nitrate vegetables are also being produced by replacing nitrate ions with ammonia in the nutrient solutions. Low-nitrate vegetables are an important source of food as nitrate is accumulated in the plant, and overconsumption of nitrates are toxic to the human body when nitrate is converted to nitrite in the gut. Nitrate accumulation in plants is also decreased by increasing light intensity, as nitrate reductase activity increases with increasing light intensity in certain plants.

19.7.1 Genetically Modified Plants for Pharmaceuticals

Genetic engineering has enabled plants to produce pharmaceuticals such as oral vaccines, antimicrobial agents and secondary metabolites. Researchers have introduced functional protein-encoding genes in rice, potato, soybean, lettuce, tomato and strawberry plants. Conventionally, biological systems such as *E. coli*, yeast, mammalian cells and insect cells are used to produce vaccines. Plants have been proposed as a cheap, efficient and clean system to produce these vaccines and functional proteins (Thomas et al. 2011). By using plants, there is decreased risk of contamination with animal and human pathogens. Moreover, growing these GM plants in PFALs not only enables optimal growing conditions for the stable cultivation of the plants, it also ensures low risk of gene diffusion which brings about contamination of the food chain.

One major drawback of vaccine-producing plants is the difficulty in standardizing the vaccine dose. Different plant parts such as the fruit, leaves and roots express genes at different levels and at different rates, making it hard to standardize. Such problems are faced by GM plants modified by transgenic engineering techniques, in another words by overexpression of certain genes. Overexpression of genes in plants has the following two disadvantages: (1) the possibility of having secondary effects due to gene expression or disruption of the plants' natural genes and metabolic pathways and (2) misconception by society that transgenics are unnatural and have negative side effects on human health and the environment (Key et al. 2008).

19.7.2 Genetic Engineering for Vitamin Biofortification

Consumers tend to choose foods based on their nutritional and other benefits on human health as well as their price. Providing vegetables with increased nutritional content is therefore an effective way to bring additional value to plants grown in PFALs. Lack of important vitamins including vitamin A, B, C (L-ascorbic acid) and E causes a range of mild to severe diseases (Blancquaert et al. 2017). PFALs can play an important role in making high nutrient crops accessible to the margins of developed countries and places where there is no access to fresh foods or supermarkets.

Transgene approach successfully generated new varieties of crops, especially in rice, maize and wheat with enhanced nutritional contents such as vitamins A, B2, B9 and C (Farre et al., 2014). These studies targeted metabolic pathways for the synthesis and degradation of vitamin compounds and genetically modified the expression levels of key enzymes in the pathways. A common strategy is to augment the metabolic flux by overexpressing the rate-limiting enzymes in the pathway (Fig. 19.7a). While the chemical synthesis involves sequential actions of multiple enzymes, there are a small number of rate-limiting enzymes of which activity affects the amount of desired product drastically. Another effective way is to prevent further conversion of the desired product to another inactive or unavailable product (Fig. 19.7b). Blocking a by-product pathway or a negative feedback regulation also increases the metabolic flux towards the desired product (Fig. 19.7c, d). We introduce a successful example below that generated a transgenic rice line conferred with an enhanced vitamin B9 (folate) content.

The biosynthesis of folate involves 11 steps of enzymatic reactions at three different subcellular locations. Dihydroneopterin triphosphate (HMDHP) and p-aminobenzoate (pABA) are the two major precursors of folate that are synthesized in cytosol and chloroplasts. Mitochondria serve as the location for the latter steps of folate biosynthesis (Fig. 19.8). Guanosine triphosphate cyclohydrolase I (GTPCHI) or aminodeoxychorismate synthase (ADCS) is a key enzyme that catalyses the rate-limiting steps of HMDHP or pABA synthesis, respectively. In rice and tomato, a

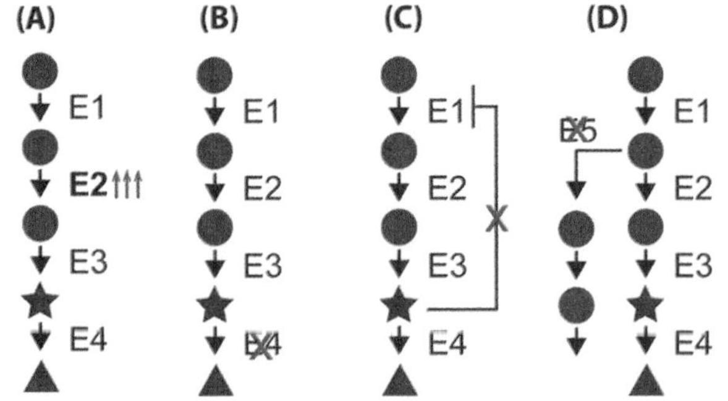

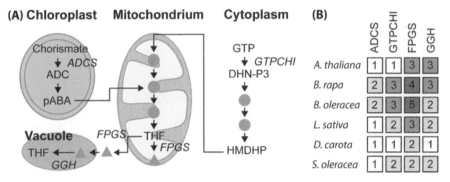

Fig. 19.7 Strategies for the metabolic engineering of secondary metabolite pathways. (**a**) Enhancement of the metabolic flux by overexpressing rate-limiting enzymes. (**b**) Stabilization of the desired product. (**c**) Attenuation of inhibitory feedback regulation. (**d**) Disruption of a by-product pathway. Circles, stars or triangles represent intermediates, desired products or less active compound converted from the desired product

Fig. 19.8 Enzymes for the biosynthesis of vitamin B9 in leafy vegetables. (**a**) Folate biosynthesis pathways. Italicized names represent key enzymes of which expression levels would affect the folate content in the crops of interest. Abbreviations of compound and enzyme names are GTP, guanosine triphosphate; DHN-P3, dihydroneopterin triphosphate; HMDHP, 6-hydroxymethyl-dihydropterin; ADC, aminodeoxychorismate; pABA, para-aminobenzoic acid; THF, tetrahydrofolate; GTPCHI, GTP cyclohydrolase; ADCS, aminodeoxychorismate synthase; FPGS, folylpolyglutamyl synthase; GGH, gamma-glutamyl hydrolase. (**b**) Folate biosynthesis enzymes encoded in *Arabidopsis* and vegetable genomes

simultaneous overexpression of GTPCHI and ADCS succeeded to increase folate contents by up to 25- and 100-fold due to the increased supply of the two folate precursors (Díaz de la Garza et al. 2007; Storozhenko et al. 2007). A lettuce with GTPCHI overexpression resulted in significant but modest enhancement on folate content (Nunes et al. 2009), likely because the supply of pABA was not enough in the transgenic lettuce. Species-specific bottlenecks may exist and limit an effective biofortification. As an example, the overexpression of the two enzymes enhanced the folate production in potato tuber only up to threefold (Blancquaert et al. 2013). This may be due to the presence of other rate-limiting factors or feedback regulations in the folate synthesis pathway in potato. Although there are still certain problems to be solved to accurately control the metabolic flux, engineering of biosynthetic pathways of multiple vitamins successfully provided a new corn cultivar containing 169-, 6- and 2-fold higher contents of vitamin A, B9 and C compared with its parental cultivar (Naqvi et al. 2009).

Changing environmental factors would influence the composition of vitamins, pigments and other secondary metabolites. Each type of plant has its own set of "optimal" light and nutrient composition to maximize the production of compounds of interest. This type of environmental regulations should be carried out along with the genetic engineering method to increase the nutritional and other benefits.

19.7.3 Plants with High Antioxidant Contents

Phenolic flavonoids scavenge reactive oxygen species generated by photosynthesis activity and protect cellular components from oxidative damage in plants. Because of their strong anti-oxidative activity, the flavonoids represent important phytonutrients for anti-oxidation and anti-inflammation in the human body. Production of the botanical antioxidants per unit of weight is highly influenced by environmental and nutritional factors. For instance, subjecting plants to different ratios of LED red and blue light affects their production of phenolic flavonoids. In the form of stress, subjecting plants like red leaf lettuce and herbs to harmful ultraviolet (UV) rays (UV-B and UV-A) increases anthocyanin content, a type of flavonoid. The anthocyanin content in plants can be increased by metabolic engineering method. While there are scarce knowledge on key genes and proteins of the pathways and how their expression profile is affected by UV radiation, a technique called RNA sequencing can be used to produce large numbers of sequences that can be mined to identify candidate genes (Farré et al. 2014). By comparing the whole genome gene expression of crops with and without subjecting them to supplemental UV radiation, we can identify the gene network that stimulates anthocyanin production in plants. With increasing knowledge of metabolic networks in anthocyanin and other antioxidants, scientists can enhance availability and accumulation of the products in common plants that grow in PFALs such as lettuce and kale.

19.8 Future Perspective

PFALs provide optimal conditions for cultivating any plants as there is full control over environmental factors such as radiation, temperature, carbon dioxide concentration and composition of the nutrient solutions. Moreover, there is no need for spraying herbicide or pesticide in the enclosed areas because of near zero contamination with weeds or pests. The use of molecular genetics to breed new crop varieties has remained unexplored and thus would further deliver a substantial improvement in urban agriculture business with PFALs.

Transgenic crop breeding method has achieved a considerable yield increase in the past decade. However, the term "genetically modified organisms" brought with it some controversy when it came to food, because of the unnatural genetic transfer from unrelated organisms. CRISPR/Cas9 technology offers an easier and precise way to carry out crop engineering without using foreign DNA. With this modern biotechnology, we can design custom-made crops that overcome physiological limitations such as tipburn and cultivation time, as well as producing favourable proteins as required. The CRISPR/Cas9 crops would not be needed to pass through regulatory processes as no foreign DNA is retained in the genome of the plant (Waltz 2016a), hence more welcomed by the public. The genome-editing technology should be unified with accumulated knowledge on plant genes and molecular systems. It would be very promising to produce to high yielding, healthy and fresh vegetables and high-value crops and to expand the types of crops that can be grown in PFALs in the near future.

References

Barta DJ, Tibbitts TW (2000) Calcium localization and tipburn development in lettuce leaves during early enlargement. J Am Soc Hortic Sci 125:294–298

Blancquaert D, Storozhenko S, Van Daele J et al (2013) Enhancing pterin and Para-aminobenzoate content is not sufficient to successfully biofortify potato tubers and Arabidopsis thaliana plants with folate. J Exp Bot 64:3899–3909. https://doi.org/10.1093/jxb/ert224

Blancquaert D, De Steur H, Gellynck X, Van Der Straeten D (2017) Metabolic engineering of micronutrients in crop plants. Ann N Y Acad Sci 1390:59–73. https://doi.org/10.1111/nyas.13274

Chilcoat D, Bin LZ, Sander J (2017) Use of CRISPR/Cas9 for crop improvement in maize and soybean. Prog Mol Biol Transl Sci 149:27–46. https://doi.org/10.1016/bs.pmbts.2017.04.005

Clough SJ, Bent AF (1998) Floral dip: a simplified method for agrobacterium-mediated transformation of Arabidopsis thaliana. Plant J 16:735–743. https://doi.org/10.1046/j.1365-313X.1998.00343.x

Clouse SD, Sasse JM (1998) BRASSINOSTEROIDS: essential regulators of plant growth and development. Annu Rev Plant Physiol Plant Mol Biol 49:427–451. https://doi.org/10.1146/annurev.arplant.49.1.427

Collard BCY, Mackill DJ (2008) Marker-assisted selection: an approach for precision plant breeding in the twenty-first century. Philos Trans R Soc Lond Biol Sci 363:557–572. https://doi.org/10.1098/rstb.2007.2170

Damerum A, Selmes SL, Biggi GF et al (2015) Elucidating the genetic basis of antioxidant status in lettuce (Lactuca sativa). Hortic Res 2:15055. https://doi.org/10.1038/hortres.2015.55

Díaz de la Garza RI, Gregory JF, Hanson AD (2007) Folate biofortification of tomato fruit. Proc Natl Acad Sci U S A 104:4218–4222. https://doi.org/10.1073/pnas.0700409104

Dow GJ, Berry JA, Bergmann DC (2017) Disruption of stomatal lineage signaling or transcriptional regulators has differential effects on mesophyll development, but maintains coordination of gas exchange. New Phytol 216:69–75. https://doi.org/10.1111/nph.14746

Farré G, Blancquaert D, Capell T et al (2014) Engineering complex metabolic pathways in plants. Annu Rev Plant Biol 65:187–223. https://doi.org/10.1146/annurev-arplant-050213-035825

Funke T, Han H, Healy-Fried ML et al (2006) Molecular basis for the herbicide resistance of roundup ready crops. Proc Natl Acad Sci 103:13010–13015. https://doi.org/10.1073/pnas.0603638103

Gerlt JA (2017) Genomic enzymology: web tools for leveraging protein family sequence-function space and genome context to discover novel functions. Biochemistry 56:4293–4308. https://doi.org/10.1021/acs.biochem.7b00614

Gilliham M, Dayod M, Hocking BJ et al (2011) Calcium delivery and storage in plant leaves: exploring the link with water flow. J Exp Bot 62:2233–2250. https://doi.org/10.1093/jxb/err111

Hara K, Yokoo T, Kajita R et al (2009) Epidermal cell density is autoregulated via a secretory peptide, EPIDERMAL PATTERNING FACTOR 2 in arabidopsis leaves. Plant Cell Physiol 50:1019–1031. https://doi.org/10.1093/pcp/pcp068

Hörtensteiner S (2009) Stay-green regulates chlorophyll and chlorophyll-binding protein degradation during senescence. Trends Plant Sci 14:155–162. https://doi.org/10.1016/j.tplants.2009.01.002

Hsu PD, Lander ES, Zhang F et al (2014) Development and applications of CRISPR-Cas9 for genome engineering. Cell 159:313–319. https://doi.org/10.1186/s40779-015-0038-1

Jenni S, Truco MJ, Michelmore RW (2013) Quantitative trait loci associated with tipburn, heat stress-induced physiological disorders, and maturity traits in crisphead lettuce. Theor Appl Genet 126:3065–3079. https://doi.org/10.1007/s00122-013-2193-7

Jeuken MJW, Lindhout P (2004) The development of lettuce backcross inbred lines (BILs) for exploitation of the Lactuca saligna (wild lettuce) germplasm. Theor Appl Genet 109:394. https://doi.org/10.1007/s00122-004-1643-7

Jiang H, Li M, Liang N et al (2007) Molecular cloning and function analysis of the stay green gene in rice. Plant J 52:197–209. https://doi.org/10.1111/j.1365-313X.2007.03221.x

Jiang W, Zhou H, Bi H et al (2013) Demonstration of CRISPR/Cas9/sgRNA-mediated targeted gene modification in Arabidopsis, tobacco, sorghum and rice. Nucleic Acids Res 41:e188. https://doi.org/10.1093/nar/gkt780

Key S, Ma JKC, Drake PMW (2008) Genetically modified plants and human health. J R Soc Med 101:290–298. https://doi.org/10.1258/jrsm.2008.070372

Kim KM, Park YH, Kim CK et al (2005) Development of transgenic rice plants overexpressing the Arabidopsis H +/Ca2+ antiporter CAX1 gene. Plant Cell Rep 23:678–682. https://doi.org/10.1007/s00299-004-0861-4

Kozai T (2013) Resource use efficiency of closed plant production system with artificial light: concept, estimation and application to plant factory. Proc Jpn Acad Ser B Phys Biol Sci 89:447–461. https://doi.org/10.2183/pjab.89.447

Krieger EK, Allen E, Gilbertson LA et al (2008) The Flavr Savr tomato, an early example of RNAi technology. Hortscience 43:962–964

La Russa MF, Qi S (2015) The new state of the art : Cas9 for gene activation and repression. Mol Cellular Biol 35:3800–3809. https://doi.org/10.1128/MCB.00512-15.Address

Lebeda A, Křístková E, Kitner M et al (2014) Wild Lactuca species, their genetic diversity, resistance to diseases and pests, and exploitation in lettuce breeding. Eur J Plant Pathol 138:597–640. https://doi.org/10.1007/s10658-013-0254-z

Lee D, Redfern O, Orengo C et al (2007) Predicting protein function from sequence and structure. Nat Rev Mol Cell Biol 8:995–1005. https://doi.org/10.1038/nrm2281

Lee J, Park I, Lee ZW et al (2013) Regulation of the major vacuolar Ca2+ transporter genes, by intercellular Ca2+ concentration and abiotic stresses, in tip-burn resistant Brassica oleracea. Mol Biol Rep 40:177–188. https://doi.org/10.1007/s11033-012-2047-4

Malzahn A, Lowder L, Qi Y (2017) Plant genome editing with TALEN and CRISPR. Cell Biosci 7:21. https://doi.org/10.1186/s13578-017-0148-4

Murat F, Louis A, Maumus F et al (2015) Understanding Brassicaceae evolution through ancestral genome reconstruction. Genome Biol 16:262. https://doi.org/10.1186/s13059-015-0814-y

Murovec J, Pirc Ž, Yang B (2017) New variants of CRISPR RNA-guided genome editing enzymes. Plant Biotechnol J 15:917–926. https://doi.org/10.1111/pbi.12736

Naqvi S, Zhu C, Farre G et al (2009) Transgenic multivitamin corn through biofortification of endosperm with three vitamins representing three distinct metabolic pathways. Proc Natl Acad Sci 106:7762–7767. https://doi.org/10.1073/pnas.0901412106

Nilson SE, Assmann SM (2007) The control of transpiration. Insights from Arabidopsis. Plant Physiol 143:19–27. https://doi.org/10.1104/pp.106.093161

Nunes ACS, Kalkmann DC, Aragão FJL (2009) Folate biofortification of lettuce by expression of a codon optimized chicken GTP cyclohydrolase I gene. Transgenic Res 18:661–667. https://doi.org/10.1007/s11248-009-9256-1

Ohnishi T, Szatmari A-M, Watanabe B et al (2006) C-23 hydroxylation by Arabidopsis CYP90C1 and CYP90D1 reveals a novel shortcut in Brassinosteroid biosynthesis. Plant Cell Online 18.3275–3288. https://doi.org/10.1105/tpc.106.045443

Peng JR, Richards DE, Hartley NM et al (1999) "Green revolution" genes encode mutant gibberellin response modulators. Nature 400:256–261. https://doi.org/10.1038/22307

Sakuraba Y, Schelbert S, Park S-Y et al (2012) STAY-GREEN and chlorophyll catabolic enzymes interact at light-harvesting complex II for chlorophyll detoxification during leaf senescence in Arabidopsis. Plant Cell Online 24:507–518. https://doi.org/10.1105/tpc.111.089474

Sasaki A, Ashikari M, Ueguchi-Tanaka M et al (2002) Green revolution: a mutant gibberellin-synthesis gene in rice. Nature 416:701–702. https://doi.org/10.1038/416701a

Saurabh S, Vidyarthi AS (2014) RNA interference : concept to reality in crop improvement, pp 543–564. https://doi.org/10.1007/s00425-013-2019-5

Storozhenko S, De Brouwer V, Volckaert M et al (2007) Folate fortification of rice by metabolic engineering. Nat Biotechnol 25:1277–1279. https://doi.org/10.1038/nbt1351

Thomas DR, Penney CA, Majumder A, Walmsley AM (2011) Evolution of plant-made pharmaceuticals. Int J Mol Sci 12:3220–3236. https://doi.org/10.3390/ijms12053220

Tibbitts TW, Rama RR (1968) Light intensity and duration in the development of lettuce tipburn. Am Soc Hortic Sci 93:454–461

Waltz E (2016a) Gene-edited CRISPR mushroom escapes US regulation. Nature 532:293. https://doi.org/10.1038/nature.2016.19754

Waltz E (2016b) CRISPR-edited crops free to enter market, skip regulation. Nat Biotechnol 34:582–582. https://doi.org/10.1038/nbt0616-582

Woo JW, Kim J, Il KS et al (2015) DNA-free genome editing in plants with preassembled CRISPR-Cas9 ribonucleoproteins. Nat Biotechnol 33:1162–1164. https://doi.org/10.1038/nbt.3389

Xu R-F, Li HHH, Qin R-Y et al (2016) CRISPR/Cas9: a powerful tool for crop genome editing. Nat Biotechnol 7:1–12. https://doi.org/10.1093/mp/sst121

Yin K, Gao C, Qiu J-L (2017) Progress and prospects in plant genome editing. Nat Plants 3:17107. https://doi.org/10.1038/nplants.2017.107

Zeljenková D, Aláčová R, Ondrejková J et al (2016) One-year oral toxicity study on a genetically modified maize MON810 variety in Wistar Han RCC rats (EU 7th framework Programme project GRACE). Arch Toxicol 90:2531–2562. https://doi.org/10.1007/s00204-016-1798-4

Zhou C, Han L, Pislariu C et al (2011) From model to crop: functional analysis of a STAY-GREEN gene in the model legume Medicago truncatula and effective use of the gene for alfalfa improvement. Plant Physiol 157:1483–1496. https://doi.org/10.1104/pp.111.185140

Chapter 20
Production of Value-Added Plants

Shoko Hikosaka

Abstract For decades, humans have cultivated plants in open fields and greenhouses, with the aim of utilizing their primary and secondary metabolites. However, conditions in open fields and greenhouse environments are affected by weather conditions. Recently, due to research and development of plant factory with artificial lighting (PFAL), it has become possible to produce a target substance efficiently by artificially controlling environmental conditions. In this section, we introduce examples of research conducted using PFALs by changing various environmental conditions, such as light and temperature, to increase production efficiency of useful secondary metabolites in vegetables, medicinal plants, and genetically modified plants. Various functional foods, such as vegetables, fruits, and crops, have been developed in recent years. Additionally, in PFALs, the application of various environmental conditions, such as light intensity, blue/red ratio, ultraviolet (UV) radiation, and low root-zone temperature, can increase the concentrations and content of functional and medicinal compounds. Therefore, PFALs are effective for stable cultivation with accumulation of functional compounds at high concentrations, and they can help in producing plants according to consumer needs (e.g., in terms of shape, color, and characteristics of flavors). Completely closed plant production systems (CCPPS) are considered key technologies and advanced facilities for the efficient and stable production of plant-made pharmaceuticals (PMPs) by transgenic plants. To achieve stable and high-quality production of PMPs, optimal procedures of environmental control in CCPPS should be established for each transgenic plant. Nowadays, plants are considered smart cells, which can produce several secondary metabolites, including useful materials for humans. Owing to the recent development of high-throughput gene and metabolite analysis, and bioinformatics technology, it is expected that these developments would facilitate rapid selection of environmental conditions suitable for controlling the concentration of target secondary metabolites.

S. Hikosaka (✉)
Laboratory of Environmental Control Engineering, Graduate School of Horticulture, Chiba University, Chiba, Japan
e-mail: s-hikosaka@faculty.chiba-u.jp

Keywords Bioactive compounds · Medicinal plants · Plant-made pharmaceuticals · Secondary metabolites · Transgenic plants

20.1 Introduction

In a plant factory with artificial lighting (PFAL), it is possible to subject plants to environmental conditions that are different from the natural conditions, and the plants may efficiently produce substances that are useful to humans. Additionally, we can apply the optimal environment conditions, which were found in growth chamber experiments at small scales. This suggests that control of environmental conditions can enable the cultivation of plants that could not be cultivated easily under natural or greenhouse conditions until date. Moreover, it is possible to expand the range of cultivated plants, their target organs, and compounds to a greater extent than before. In this chapter, I will provide the outline of environmental control adopted in the PFAL for value-added plant production, including (1) functional compounds in horticultural crops, (2) bioactive compounds in medicinal plants, and (3) pharmaceutical and functional protein production by transgenic plants. Finally, I will introduce the "Smart Cell Project," which is currently being promoted by the Japanese government and aims to produce active substances in plants with high efficiency in an artificial environment.

20.2 Classification of Value-Added Plants

We classified the target plants as shown in Fig. 20.1. The daily demand of plants varies from relatively high levels in horticultural crops to low levels in medicinal plants. In contrast, bioactive compounds production in medicinal plants is relatively high. Furthermore, the cultivation methods are known, genetic variation is small, and knowledge and information are plentiful in vegetables, whereas those in medicinal plants are unknown, wide, and limited, respectively. The same plant (e.g., red perilla) is occasionally classified into different categories, e.g., herbal and medicinal, depending on the variety. Moreover, the genetic variation in horticultural crops is smaller than that in herbal or medicinal plants.

However, the highly value-added transgenic plants, which mainly produce proteins, such as vaccines and enzymes, usually do not exhibit variation of target protein concentration. The cultivation methods for high yield of target protein in transgenic plants are unknown, genetic variation is small, and knowledge and information are limited. Therefore, the category of transgenic plants was separated from that of non-transgenic plants in Fig. 20.1.

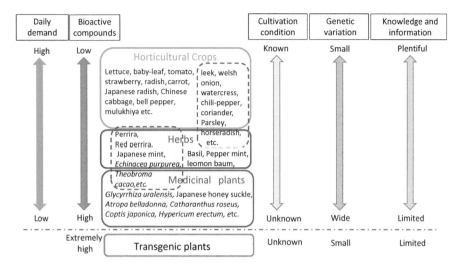

Fig. 20.1 Classification of value-added plants

20.3 Functional Compounds (Phytochemicals) in Horticultural Crops

20.3.1 Background

Although rapid urbanization has improved the quality of human life globally, the number of patients with lifestyle diseases is increasing in many developed countries owing to malnutrition or unbalanced diet consumption. Many epidemiological studies report that fruits and vegetables are closely related to the reduced risk of cardiovascular disease, cancer, diabetes, Alzheimer's disease, cataracts, and age-related functional decline (Liu 2003). To reduce medical costs, governments are promoting many campaigns encouraging people to improve their diets. Therefore, people tends to select high-quality foods and functional foods. However, to maintain a diet suitable for the maintenance of good health, it is necessary to consume adequate quantities of health-promoting functional compounds while properly controlling the total food intake. Therefore, the production of cereals, vegetables, and fruits containing large quantities of functional compounds is required.

20.3.2 Functional Compounds in Plants

To prevent the accumulation of reactive oxygen species (ROS) generated within cells, plants have developed a complex antioxidant defense system. Numerous molecules involved in this system belong to the category of compounds termed as

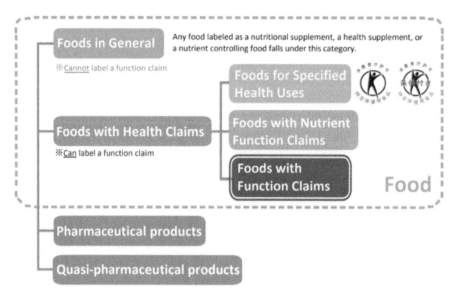

Fig. 20.2 Introduction of the new labeling system of "Foods with Function Claims" by the Consumer Affairs Agency of Japan. (http://www.caa.go.jp/foods/pdf/151224_1.pdf)

secondary metabolites. Secondary metabolites generally include pigments and phenolics (Bartwal et al. 2013). The Ministry of Education, Culture, Sports, Science and Technology (MEXT) of the Japanese government aims to systematize non-nutritive functional substances produced by plants, listed in the table of food compounds and contents and functions of flavonoids/polyphenols, terpenoids/carotenoids, sulfur-containing compounds, volatile ingredients, and spices (Hibino 2004). These functional compounds are not present in high concentrations in vegetables and fruits as compared with medicated plants. Therefore, daily or frequent consumption of these compounds is recommended, and the consumed amount and market size of functional vegetables and fruits are overwhelmingly larger than those of medicinal plants.

20.3.3 Target Concentrations of Functional Compounds

To cultivate value-added functional plants in a PFAL, the concentration of functional compounds should be significantly higher and more effective than that of other cultivars, and these value additions should appeal directly to consumers. In Japan, a new system of "Foods with Health Claims" called "Foods with Function Claims" was introduced in April 2015 (Fig. 20.2, Consumer Affairs Agency, Government of Japan) to make a large number of products clearly labeled with certain nutritional or health functions available and to enable consumers make more informed choices. To support this system, Agriculture, Forestry and Fisheries Research Council (AFFRC) introduced technical information necessary for appropriate display to farmers (http://www.affrc.maff.go.jp/kinousei/gijyututekitaio.htm).

Although labeled vegetables and fruits reflect high value, farmers have to maintain stable concentrations of functional compounds in plants throughout the year. Therefore, technological developments in PFAL are required for the stable accumulation of functional compounds in plants at high concentrations. Because the content of plant secondary metabolites vary greatly with environmental conditions during cultivation, stable production in the PFAL is desirable. Furthermore, because a PFAL creates environmental conditions that cannot be created in open fields or greenhouses, the concentrations of functional compounds are expected to be higher than those in plants cultivated in fields or greenhouses.

20.3.4 Environmental Conditions

For the stable accumulation of functional compounds in plants at high concentrations, it is first necessary to identify suitable varieties with high concentrations of the relevant functional compounds. Li et al. (2011) reported that the concentrations of β-carotene and lutein in spinach depended strongly on the cultivars. Moreover, it is necessary to develop environmental control technology to achieve the desired form, texture, and concentration of functional compounds for the selected variety. Because functional compounds serve as defense agents against various environmental stress conditions, abiotic and biotic stresses induce the production of functional compounds in plants (Bartwal et al. 2013; Dixon and Paiva 1995).

Light manipulation is a key environmental variable that can be controlled for increasing the concentrations of functional compounds (phytochemicals) in plants under controlled environment conditions, such as greenhouses and plant factories (Goto et al. 2016). There are many reports that functional compound concentrations and antioxidant capacity were enhanced by light of high intensity (Li et al. 2011), blue light (Johkan et al. 2010; Son and Oh 2013), UV light (Lee et al. 2014; Goto 2012, Goto et al. 2016), or blue and UV light in combination (Ebisawa et al. 2008a, b). According to many reports, light of high intensity and short wavelength (blue and UV light) tends to increase the concentrations of functional compounds (phytochemicals) to protect the tissue from excessive light energy (Fig. 20.3). In contrast, light of long wavelength—red and far-red light—tends to increase biomass and leaf area, respectively (Holopainen, et al. 2017). Therefore, the combination of light quality, strength, and irradiation period can help in promoting growth and increasing the concentration of functional compounds in the PFAL.

However, it is well known that air or root-zone temperatures, particularly low temperatures, increase the concentrations of ascorbic acid (Ito et al. 2013), flavonoids, including anthocyanin (Solecka et al. 1999), and phenylpropanoids (Janasa et al. 2002). Generally, environmental stress, such as low temperature, water deficit, light of high intensity, and UV light, increases in the generation of ROS in tissues. Therefore, ROS induces biosynthesis and accumulation of bioactive compounds

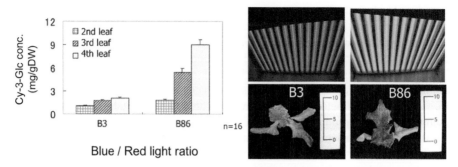

Fig. 20.3 Effects of blue/red light ratio on anthocyanin concentration (Cy-3-glu: cyanidin-3-glucoside) in lettuce leaves (*Lactuca sativa* L. "Red Fire")

with antioxidative effects. Although the long-term treatment of plants at low air or root-zone temperatures inhibits growth, short-term treatment often increases the concentrations of functional compounds in plants without inhibiting growth. Additionally, the responses of plants and harvested fruits to controlled air or gas exposure were reported in many studies (Sharmaa and Davisa 1997). They are absorbed by the entire plant, and it is expected to induce responses in the entire plant leaves. These environmental stimuli and combinations are expected to improve the quality of plants in the PFAL.

20.3.5 Conclusion

Various functional foods, such as cereals, vegetables, fruit, and tea, have been developed in recent years. Additionally, in a PFAL, the application of various environmental factors, such as light intensity, blue/red ratio, UV, low root-zone temperature, and controlled air, can increase the concentrations and content of functional compounds. Therefore, a PFAL is effective for stable cultivation with accumulation of functional compounds at high concentrations, and it can help in producing plants according to consumer needs (e.g., shape, color, and flavor).

20.4 Medicinal Plant Production

20.4.1 World Market and the Current Condition of the Medicinal Plant Industry

Medicinal and herbal plants are raw biological materials for cosmetics, aroma, medicinal products, and functional foods. Their use by traditional healthcare practitioners and traditional healers, as well as at the household level, has contributed to their increased demand. According to BCC Research, the global market for botanical and plant-derived drugs was valued at $24.4 billion in 2014. This total market is expected to reach nearly $35.4 billion in 2020, with a compound annual growth rate of 6.6% from 2015 to 2020 (The International Trade Centre 2017). This market is also expected to show significant growth in the Asia-Pacific region during the forecast period, i.e., between 2017 and 2027. For the further development of this market, manufacturers need to invest more in research and development of the products for improving their product portfolio (Future Market Insights Global & Consulting Pvt. Ltd. 2017).

However, because the global demand for medicinal plants has increased, the price of crude drugs imported from China has also increased recently due to farmed and natural products increasing in price (Kang 2008, 2011). In addition, labor costs have increased because aging and urbanization have caused a shortage of laborers (Kang 2008, 2011).

Additionally, the excessive harvesting of wild medicinal plants has increased desertification in a number of major medicinal plant-producing countries, and the increased demand for crude drugs in Asian and European countries has caused a more severe shortage of crude drug resources (Kang 2008; Koike et al. 2012). Moreover, China, which is the main country producing medicinal plants, has restricted the export of medicinal plants. Therefore, the supply of medicinal plants to importing countries, including Japan, is insufficient. In addition, the residual agricultural chemicals contained in medicinal plants have caused problems. In Japan, 90% of the domestic consumption of medicinal plants are imported from overseas (Fig. 20.4, Japan Kampo Medicines Manufacturers Association), and the

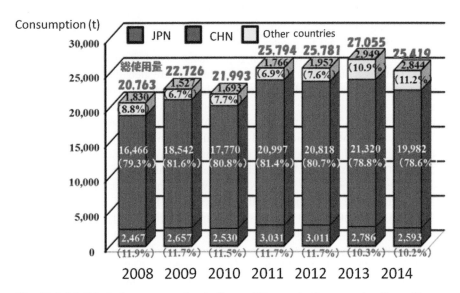

Fig. 20.4 Medicinal plant consumption in Japan, China, and other countries (Japan Kampo Medicines Manufacturers Association)

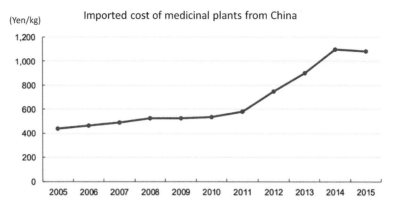

Fig. 20.5 Imported cost of medicinal plants from China (Cabinet Secretariat, Office of Healthcare Policy 2016)

import cost has increased (Fig. 20.5). Therefore, technical developments toward safe and high-quality production of medicinal plants in a greenhouse or plant factory with artificial lighting (PFAL) are required in the near future.

20.4.2 Merits and Demerits of Medicinal Plant Production in PFALs

In a PFAL, it is possible to produce high-quality medicinal plants (e.g., in terms of ideal shape and concentration of medicinal compound) annually with no agrochemicals (e.g., pesticides and fungicides). Thus, PFALs seem to be suitable cultivation systems for solving the abovementioned problems related to medicinal plant production. However, outfield or greenhouse productions are globally popular methods for growing medicinal plants, because most of them are harvested once annually, and dried plants traditionally have been consumed (changes of composition and concentration of medicinal compound have not been considered so far). There are few existing facilities for the commercial production of medicinal plants and a few studies on medicinal plant production under artificial environment conditions.

Many of the existing PFALs (excluding research and exhibition facilities, of which there are reported to be approximately 191 units in Japan as of March 2016) produce leafy vegetables, mainly lettuce (Japan Facilities and Horticultural Association). Presently, plant factories, including artificial light types and sunlight types (greenhouses), are used to supply horticultural crops annually that contain unstable nutrients required constantly by humans, such as vitamins. However, because medicinal plants are ingested only when needed, many kinds of medicinal plants contain bioactive compounds in concentrations that are high enough to be sufficient in small portions only. Therefore, it is often sufficient to collect a wild plant or harvest a cultivated variety once a year and store the collection for a long time. Thus, medicinal plant production in a PFAL is not realistic when producing traditional Kampo medicine, which has a limited price value (prescribed), or assuming a market size that accepts long-term preserved products.

20.4.3 Superiority of Medicinal Plant Production in PFALs

Presently, many commercial PFALs with profitable management are producing leafy vegetables and seedlings. The production of other fruits vegetables, root vegetables, cereals, and medicinal plants by PFALs is lesser than that of leafy vegetables and seedlings, owing to problems in terms of cultivation technology and costs. In order to establish the commercial production method and cultivation technology for medicinal plants, the following problems need to be addressed:

1. There are various types of medicinal herb, and the type suitable for cultivation in PFALs needs to be selected.
2. The target organs and growth stages with high accumulations of medicinal compounds need to be selected.
3. Some medicinal plants are large in size, which makes cultivation in a PFAL difficult.
4. The breeding of medicinal plants is not progressing rapidly; genetic variation is large, and plant shapes and growth rates can differ.
5. Knowledge of fundamental ecological characteristics is insufficient; bloom and dormancy of some medicinal plants are unknown.
6. The environmental conditions and cultivation methods most favorable for photosynthesis are unknown.
7. Depending on the target organ (e.g., root), cultivation can last for several years.
8. The conditions, which would enlarge the target organ and increase the concentration of the target compound, are unknown.
9. The market size is small, and the economic efficiencies (profitability) of medicinal plants are unknown as compared with those of horticultural crops.

Although it is not easy to solve these problems, suitable cultivar-specialized research and development for medicinal plant production in PFALs is expected in the near future. For instance, small-sized medicinal plants with high concentrations of bioactive compounds, having an expected high price or annual demand, should be nominated for commercial production in PFALs. In particular, it appears that there is an advantage in utilizing PFALs for the annual production of medicinal plants for use in manufacturing over-the-counter (OTC) drugs, supplements, cosmetic raw materials, or high-demand drugs (e.g., anticancer, anti-Alzheimer's disease) because their prices are not affected by governmental regulation of drug prices.

Because PFALs can control environmental factors, we can apply optimal environmental conditions for plant and target organ growth (Malayeri et al. 2010; Mosaleeyanon et al. 2006; Zobayed et al. 2005a) and apply the environmental stress locally, such as the amount of ultraviolet (UV) radiation and temperature.

Moreover, PFALs are suitable for the large-scale annual seedling production of medicinal plants. Even for medicinal plants with a long cultivation period that are not profitable to grow in PFALs, there is a good possibility of obtaining an advantage if seedlings are produced and sold.

20.4.4 Variety

Medicinal plants contain phenylpropanoids, flavonoids, and terpenoids, which are bioactive compounds often found in functional vegetables. In addition, they often contain alkaloids at concentrations higher than those in functional vegetables. Because the concentration of these bioactive compounds varies greatly depending on the variety, it is first necessary to find suitable varieties of medicinal plants in order to produce them efficiently. Thereafter, it is necessary to develop the environment control technology to increase target organ yield in the medicinal plants and to ensure that the target bioactive compound concentration is met. The figures below represent the difference in perillaldehyde concentration between two varieties of red perilla, with one used as medicine and the other as food (Fig. 20.6).

20.4.5 Environmental Conditions

The target organs of medicinal plants include leaves, stems, roots, flowers, buds, and fruits. In addition, the target bioactive compounds include phenylpropanoids, flavonoids, carotenoids, and alkaloids. Many previous studies have focused on the effects

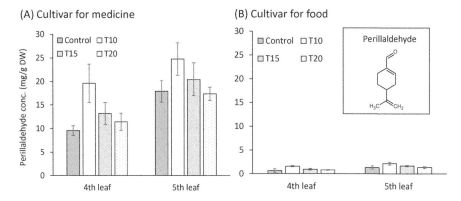

Fig. 20.6 Perillaldehyde concentration in two varieties of the red perilla, *Perilla frutescens* (L.) Britt. var. *acuta* (Thunb.) Kudo, grown for (**a**) medicine and (**b**) food. "T10," "T15," and "T20" indicate the temperature of the nutrient solutions, 10, 15, and 20 °C, respectively, applied to the plants for 6 days (Ogawa unpublished data)

of environmental factors and stress on the growth and bioactive compound concentrations of medicinal plants.

Generally, the yield (i.e., total content) of bioactive compounds increases with the biomass of the target organ. Therefore, the first step toward efficient medicinal plant production includes establishment of optimal conditions for photosynthesis and growth. However, the conditions necessary for increasing plant growth and increasing medicinal compound concentration may be different.

Second, appropriate environmental stimuli should be applied to the plant to increase the target bioactive compound concentration per unit dry weight of the plant. It is effective to provide a treatment that raises the bioactive compound concentration after maximizing the growth of the harvesting organ. Previous studies, including greenhouse experiments, have reported that the concentrations and/or content of bioactive compounds in many kinds of medicinal plants and herbs are increased by UV irradiation (Ebisawa et al. 2008a, b; Hikosaka et al. 2010), low air temperature (Christie et al. 1994; Miura and Iwata 1983), low root-zone temperature (Sakamoto and Suzuki 2015; Voipio and Autio, 1994), salinity (Hichem and Mounir. 2009; Sreenivasulu et al. 2000), drought (Bettaieb et al. 2011; Cheruiyot et al. 2007), and ozone (Booker and Miller 1998; Sudheer et al. 2016). In this section, we introduce the research results of an experiment to increase the concentration of bioactive compounds in medicinal plants by using environmental controls in a PFAL.

20.4.5.1 Light (Photosynthetically Active Radiation)

Light conditions, including light period, light quality, and light intensity, affect the growth rate and morphogenesis of plants, and indirectly affects secondary metabolite production. Once the optimum range of light conditions for the target medicinal plant is determined, it becomes less difficult to promote growth and achieve a high yield of the target organ in a PFAL.

There are many reports stating that a sufficient amount of light increases the total yield of the target organ with or without an increase in the target bioactive compound concentration via plant growth, because all secondary metabolites are derived from photoassimilates. For instance, the number of flower buds (target organ) on Japanese honeysuckle (*Lonicera japonica* Thunb.) (Fig. 20.7) is increased without a decrease in the main bioactive compound concentrations (chlorogenic acid and luteolin), when grown in a winter greenhouse with supplemental lighting (Hikosaka et al. 2017). Higashiguchi et al. (2016) reported that high light intensity treatment with a photosynthetic photon flux density (PPFD) of 142 $\mu mol\ m^{-2}\ s^{-1}$ and a 24/0 h light/dark photoperiod successfully maximized the content of asperuloside iridoids in the plant *Hedyotis diffusa* because these light conditions increased both the plant growth and asperuloside concentration.

As for the concentration of alkaloids, the optimal red light intensity for vindoline (VDL) and catharanthine (CAT) production in leaves of *Catharanthus roseus* (L.) G. Don. was suggested to be a PPFD of 150–300 $\mu mol\ m^{-2}\ s^{-1}$ (Fukuyama et al.

Fig. 20.7 Flower buds of Japanese honeysuckle (*Lonicera japonica*)

2015). These two compounds are used as important and expensive anticancer drugs. Their previous study indicated that the production of VDL and CAT under red light was greater than under blue light, a mixture of red and blue light, and fluorescent lamp-based white light.

20.4.5.2 UV Light

Many studies have concentrated on the effects of environmental factors and stress on the growth and glycyrrhizin (GL) concentration of Chinese licorice (*Glycyrrhiza uralensis*) (Afreen et al. 2005; Hou et al. 2010; Sun et al. 2012 (Fig. 20.8)). Japanese mint (*Mentha arvensis* L. var. *piperascens*) is a plant of the Lamiaceae family, and it is used in herbal medicines. It exhibits increased l-menthol and antioxidant capacity when subjected to UV irradiation (Hikosaka et al. 2010). In PFALs, the UV light is mainly absorbed by the upper leaves of plants, so the influence of UV is more noticeable in the upper leaves than the lower leaves (Fig. 20.9).

20.4.5.3 Temperature

Air and root-zone temperatures are known to cause many physiological and biochemical changes and often alter the secondary metabolite production in plants. In general, the suitable temperature range for increasing plant growth increases the biomass of the target organ and often does not change or decrease the target compound concentration per unit dry weight of the plant.

To increase the bioactive compound concentration, excessive high or low (air or root-zone) temperatures, which reduce the photosynthesis of the plant, are effective. For example, in one study, high-temperature (35 °C) treatment increased the hypericin, pseudohypericin, and hyperforin concentrations in the shoot tissues of St. John's wort (*Hypericum perforatum* L.) (Zobayed et al. 2005b). In perilla [*Perilla frutescens* (L.) Britt. var. *acuta* Kudo], short-term (6 days) exposure to root-zone

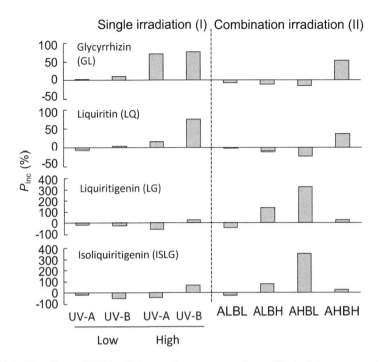

Fig. 20.8 The effects of UV irradiation on the percentage change (P_{Inc}) of the concentrations of four medicinal ingredient as compared with that in the control in the dried main roots of Chinese licorice plants (*Glycyrrhiza uralensis* Fisch). Glycyrrhizin (GL), liquiritin (LQ), liquiritigenin (LG), and isoliquiritigenin (ISLG) are the major flavonoids in licorice roots. (Sun et al. 2012)

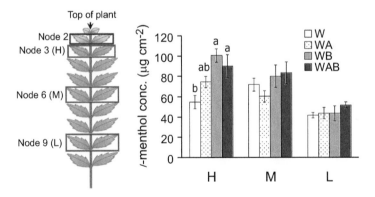

Fig. 20.9 Effects of UV light on the l-menthol concentration of upper (H), middle (M), and lower (L) leaves of Japanese mint (*Mentha arvensis* L. var. *piperascens*) (Hikosaka et al. 2010). W, WA, WB, and WAB indicate the light source as fluorescent lamps (FL) alone, FL + UV-A, FL + UV-B, and FL + UV-A + UV-B, respectively

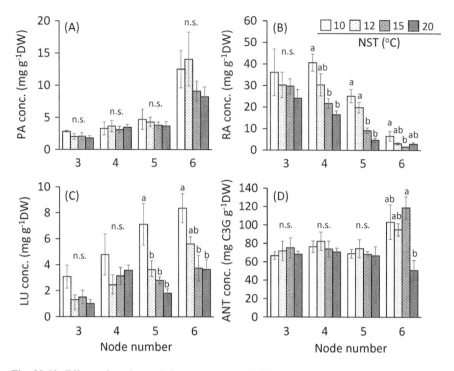

Fig. 20.10 Effects of nutrient solution temperature (NST) on the concentrations on dry weight basis of bioactive compounds (**a**, perillaldehyde (PA); **b**, rosmarinic acid (RA); **c**, luteolin (LU); **d**, anthocyanin (ANT)). Vertical bars indicate SE (n = 6). Different letters indicate significant differences among treatments at P < 0.05, and n.s. indicates no significant difference by Tukey-Kramer's test (Ogawa et al. 2018)

temperatures of 10 °C appeared to be effective for increasing rosmarinic acid and luteolin concentrations (and content), without a decrease in plant growth (Fig. 20.10) (Ogawa et al. 2018).

20.4.6 Conclusion

Medicinal plant production in PFALs is expected to reduce the use of agrochemicals and improve the quality and yield of bioactive compounds, although there are few facilities for the commercial production of medicinal plants at present, globally. PFALs are suitable for the large-scale and annual seedling production of medicinal plants. However, further research and development is required to determine the most effective environmental controls for maximizing the growth efficiency of different varieties and growth stages of medicinal plants and their respective bioactive compounds.

20.5 Pharmaceutical and Functional Protein Production by Transgenic Plants

20.5.1 Methods of Plant-Made Pharmaceutical Production

A few decades passed since the concept of pharmaceutical and functional protein production by transgenic plants was proposed. Plant-made pharmaceuticals (PMPs) obtained from transgenic plants enhance immune functions in mammalian (e.g., human beings and livestock), and they provide many advantages in comparison with the conventional production methods using expression in other bio-based methods (e.g., animals, cells, or microorganisms) (Daniell et al. 2001, 2009; Ma et al. 2003; Sack et al. 2015; Sainsbury and Lomonossoff 2014; Yao et al. 2015). For instance, the risk of contamination with animal pathogens (e.g., prions, viruses, and mycoplasmas) is lower in plant expression systems than in *Escherichia coli*, yeast, or mammalian cell expression methods, thus enhancing safety. Moreover, in general, the cost of plant cultivation and PMPs-derived products is only 0.1% of the cost of mammalian cell culture systems and 2–10% of that of microbial systems (Yao et al. 2015). Presently, commercial cultivation and production of pharmaceutical and functional proteins by transgenic plants are underway in some countries, including Japan.

Of the two major methods of PMPs production (Fig. 20.11), one is the method using stable transgenic plants, which involves traditional deoxyribonucleic acid (DNA) modification for recombinant protein expression and relies on the generation of stable seed propagation. Although regeneration is sometimes time-consuming, the greatest advantage of this method is the stable expression and edibility without intermediate extraction and filtration process when host plants are horticultural crops (e.g., lettuce, strawberry, and rice).

Another method of PMPs production is transient gene expression, involves their production in plant tissues, which generates recombinant proteins rapidly within days without DNA modification of the host plant. In general, this method requires extraction and filtration process for obtaining the final products of the PMPs. Presently, the production can be scaled up to commercially relevant levels. However, the problems in both methods include unstable yields caused by environmental conditions for plant cultivation (Sack et al. 2015), gene silencing, glycosylation (Leuzinger et al. 2013; Sainsbury and Lomonossoff, 2014; Yao et al., 2015), and low yields of the expressed proteins (Ahmad 2014) owing to degradation and effects of phenol in tissues.

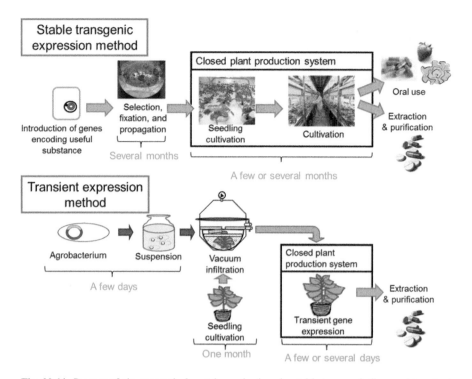

Fig. 20.11 Process of pharmaceutical protein productions by stable transgenic lines and transient gene expression methods (Drawn by Saito and Shiraishi)

20.5.2 Completely Closed Plant Production Systems for PMPs Production

For both PMPs production methods mentioned in Section 20.5.-1, plants should be cultivated in closed plant production systems with some equipment, such as filters and autoclave facilities, to prevent gene diffusion. Therefore, establishment of completely closed plant production systems (CCPPS) (i.e., a specialized PFAL) is a key technology for ensuring success in achieving stable PMPs production.

Particularly, environmental control in CCPPS for stable transgenic plant production is more important because the stable expression in the transgenic plants for edible use is essential for clearing the pharmaceutical standards. Additionally, the environmental control for that is more complicated, because in general, the cultivation period throughout the growth stage of stable transgenic plants is longer than that of plants with transient gene expression. Goto (2011) mentioned that a closed plant production system for stable transgenic plants has been considered an important technology for supporting the stable production of pharmaceutical and functional proteins. As a first practical approach, National Institute of Advanced Industrial

Fig. 20.12 Completely closed plant production systems (CCPPS) in AIST

Fig. 20.13 Interberry alpha developed for use as canine medicine (http://www.hokusan-kk.jp/product/interberry/index.html; https://unit.aist.go.jp/bpri/bpri-pmt/result_e.html) (18 Dec. 2017)

Science and Technology (AIST) in Japan established a CCPPS for PMPs production in 2007 (Fig. 20.12).

Several research groups have succeeded in introducing PMPs-related genes and confirmed the protein accumulation in plants such as lettuce (Matsui et al. 2011), tomato (Sun et al. 2007), rice (Sugita et al. 2005), strawberry (Hikosaka et al. 2013), and soybean (Maruyama et al. 2014). A successful use of the CCPPS was in veterinary drugs, e.g., Interberry alpha produced by stable transgenic everbearing strawberry. To develop this stable transgenic strawberry, which produces canine interferon α, the research group at AIST collaborated with Hokusan Co., Ltd. and Kitasato Institute (Fig. 20.13). It was approved for use in 2013, as the first example, worldwide, of a drug produced using stable transgenic plants.

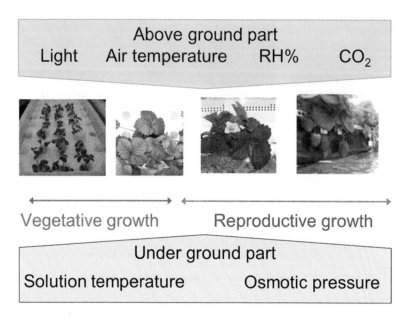

Fig. 20.14 Growth stages of everbearing strawberry and environmental factors for aboveground and underground parts of the plant

20.5.3 Environmental Control for Stable Transgenic Plants

CCPPS is considered an advanced and necessary facility for the efficient and stable commercial production of functional proteins using stable transgenic plants, as mentioned above. To achieve efficient production, environmental conditions should be optimized for maximizing the yield of target organs (e.g., leaf, fruit, and/or root) and accumulation of the target functional protein, while minimizing the cultivation period, and energy and resource consumption, thereby reducing the total cost. For example, in transgenic strawberry and rice, it is essential to optimize each environmental condition for vegetative and reproductive growth stages (Fig. 20.14).

For more than 10 years, our research group has investigated the optimal environmental conditions required for efficient production of PMPs, using several kinds of stable transgenic rice (Kashima et al. 2015) and everbearing strawberry (Hikosaka et al. 2009, 2013). We found that the environmental conditions for the vegetative stage (Miyazawa et al. 2009) and flower initiation in everbearing strawberry were the same between non-transgenic and stable transgenic strawberry. For instance, anthesis of both non-transgenic and transgenic everbearing strawberries was enhanced by continuous lighting (i.e., 24 h photoperiod) and daily light integral. Furthermore, under a 16 h photoperiod, blue light enhanced flowering better in these transgenic strawberries than red light did (Yoshida et al. 2012).

In contrast, there is no information about the suitable growth conditions for target gene expression and protein accumulation in transgenic plants; therefore, it is

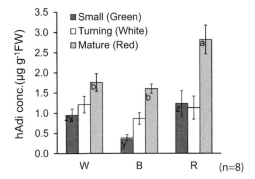

Fig. 20.15 Effects of light quality [white (W), blue (B), and red (R)] on human adiponectin (hAdi) concentration in *hAdi*-expressing everbearing strawberry fruit at different fruit maturity stages (n = 8). Vertical bars indicate standard errors. Different letters indicate significant differences. Red light promoted hAdi accumulation (Hikosaka et al. 2013). FW, fresh weight

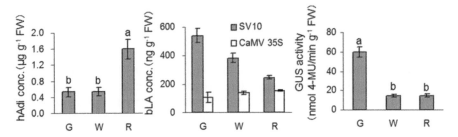

Fig. 20.16 Three types of recombinant proteins (hAdi, human adiponectin; bαLA, bovine α lactarbmine; GUS, marker protein of GUS) and two types of promoters (SV10, strawberry vein banding virus; CaMV 35S, cauliflower mosaic virus 35S) were used, and they were present in green (G), white (W), and red (R) mature fruits of everbearing strawberry (Hikosaka et al. 2011)

necessary to study the conditions for each stable transgenic plant. A study by our research group reported that light quality affected the concentration of human adiponectin protein, which increases human immune activity, in the stable transgenic strawberry fruits, particularly at the mature stage (Hikosaka et al. 2013, Fig. 20.15).

Moreover, according to other results of our group, the stable transgenic everbearing strawberries, which contained different combinations of two promoters (SV10 and 35S promoters) and three target genes, showed different accumulation responses of target proteins [i.e., human adiponectin, bovine alpha-lactalbumin, and beta-glucuronidase (GUS)] in each growth stage (Fig. 20.16). As shown in Fig. 20.16, the accumulation of target proteins in transgenic plants should be checked independently for each transgenic line. Therefore, it remains unclear whether high growth rate and short harvest period of target organs are essential for achieving high expression and high yield of target proteins.

The stable and efficient PMPs production by stable transgenic plants has not been established until date. However, it is expected that progressive technology, such as elucidation of gene constructs and promoters with high expression levels, or perfect environmental control for target transgenic plants in CCPPS will be developed in the near future.

20.5.4 Environmental Control for Transient Gene Expression Method

The most important advantages of the transient gene expression method are rapid PMPs (vaccine and antibody) production against pandemics. Recently, certain companies in the United States, Canada, the EU, and other countries established the PMPs factories (i.e., PFALs for PMPs production) by using transient gene expression method at commercial levels. D'Aoust et al. (2010) in Medicago Inc. group reported the recent developments toward the successful large-scale production of influenza virus-like particles (VLPs).

The most popular host plant for this method is *Nicotiana benthamiana*, which is often used as a model plant in molecular research. A few commercial PMPs factories using the transient gene expression method cultivated *N. benthamiana* at the pre-infiltration stage (before inoculation stage) in a greenhouse for mass production and cost saving at the cultivation stage. However, the environmental conditions in a greenhouse are not as stable as those in CCPPS; therefore, plant and leaf size often vary depending on the greenhouse environment. Eventually, it is possible that the automatic infiltration treatment would involve certain amounts of cost and labor.

Stable expression for PMPs production using transient gene expression methods is not as severe as that achieved using stable transgenic plants for edible use, because PMPs production by transient methods requires the processes of extraction, purification, quality check, and regulation of concentrations. However, as mentioned above, a closed PMPs factory is necessary for all transgenic plants. Therefore, many researchers have studied the optimal environment conditions and harvest timings for ensuring high expression and yield of target proteins. Fujiuchi et al. (2016) reported that the hemagglutinin (HA, antigen protein of influenza virus) accumulation in intact leaves of *N. benthamiana* was affected by environmental conditions. Matsuda et al. (2012) reported that light intensity affects HA accumulation. Patil and Fauquet (2015) showed that both light intensity and temperature have a significant effect on the systemic movement of the silencing signal in studies using agroinfiltration of *N. benthamiana*.

According to previous reports, the optimal environment for gene expression and accumulation of the target proteins in *N. benthamiana* differs, depending on the expressed genes. These include not only the genes of target proteins but also other related genes (e.g., promoters and constructs). It is suggested that certain environmental condition for the post-infiltration stage may not be suitable for the production of other proteins. Although the time for environmental optimization will be shorter in the transient gene expression method than in the stable transgenic plant cultivation method, further studies are required to achieve high expression even for the transient gene expression method.

20.5.5 Conclusions

CCPPS is considered a key technology and an advanced facility for efficient and stable production of PMPs. To achieve high expression levels and stable PMPs production, optimal procedures of environmental control in CCPPS should be established for each transgenic plant. In addition, it is expected that combining molecular biotechnological methods will boost the efficient production of PMPs in CCPPS.

20.6 Additional Information: New Project Involves PFALs

20.6.1 Bioeconomy and Smart Cell Industry Society

In recent years, biotechnology has been expected as the solution to food problems, energy problems, and environmental problems (The Ministry of Economy, Trade and Industry in Japan (METI), 2016). According to the Organisation for Economic Co-operation and Development (OECD), the global bio-market of 2030 has been forecast to constitute 2.7% of the overall GDP (approximately US$ 1.6 trillion or ¥ 200 trillion). OECD (2009) advocates the concept of "bioeconomy" as a market (industry group) where biotechnology can greatly contribute to economic production. In Europe and the United States, the governments have set out policies on bioeconomy since 2011. Moreover, China has established numerous research bases of bioeconomy and started strategic actions in cooperation with the United Kingdom and the United States. In Japan, the domestic bioindustry market was reported to be ¥ 3.1 trillion in 2015, and the market size is expanding every year.

METI (2016) reported the basic research policy toward the realization of a future society produced by biotechnology, termed as the "smart cell industry society." A smart cell is defined as, "a biological cell with highly functional design and controlled expression of function." Plants are considered smart cells, which can produce numerous secondary metabolites, including useful materials for human beings. Many of these secondary metabolites are difficult to produce by chemical synthesis or exceeding productivity by conventional methods. In this section, I will introduce some aspects of the Smart Cell Project related to PFAL.

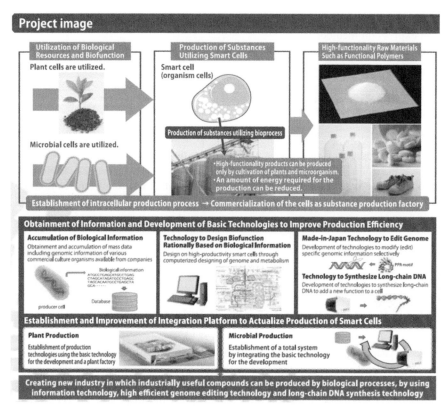

Fig. 20.17 Three key technologies for promotion of efficient production of useful materials for humans using plants as smart cell (NEDO)

20.6.2 Smart Cell Project (Original Work Written by New Energy and Industrial Technology Development Organization (NEDO))

Aiming to clarify Japan's direction in a new trend in biotechnology, NEDO in Japan started the project titled, "Development of Production Techniques for Highly Functional Biomaterials using Smart Cells of Plants and other Organisms" (Smart Cell Project; 2016–2020). This project aims to support the development of innovative biotechnologies for producing industrial materials, such reagents, perfumes, cosmetics, and plastics from plants and microorganisms by leveraging the Japanese original genome editing technology independent of foreign technologies, as well as the latest information technology (IT) and artificial intelligence (AI) technologies (Fig. 20.17) in new trends in biotechnology, with experts from industry, academia, and government (http://www.nedo.go.jp/english/news/AA5en_100149.html).

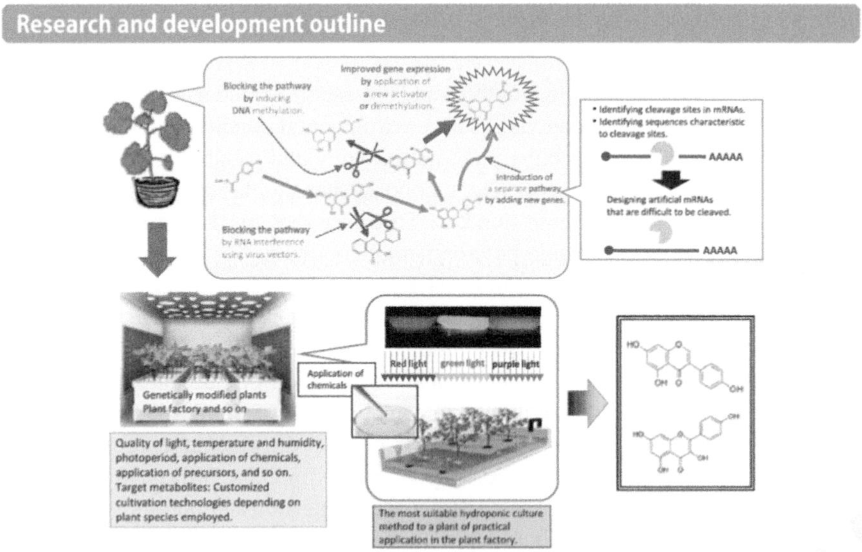

Fig. 20.18 The outline of Smart Cell Project for elucidating genes and pathways of the relevant secondary metabolism by using PFALs (NEDO)

In this project, NEDO will develop technologies for 1) high-speed collection system of precise and large-scale biological information, which is required for gene design, and 2) intracellular process design by genome editing. Finally, NEDO aims to industrialize the production of highly functional substances by controlling/modifying the intracellular process with energy-saving and low-cost techniques. Particularly, this project aims at the establishment and improvement of integration platform to actualize the production of smart cells by plant and microbial production.

In addition to plant genome recombinant technologies for improving production by plants, the objectives of this project are to establish new technologies, such as prevention of DNA methylation, to identify key sequences critical for high gene expression, which are involved in secondary metabolism. For many crop species, few genes involved in secondary metabolite productions are known. Consequently, genetic manipulations for higher production are not possible. However, there is ample evidence that environmental conditions for cultivation could critically affect the yield of secondary metabolites without elucidating genes and pathways of the relevant secondary metabolism. In order to assist in developing strategies for higher production of secondary metabolites by controlling cultivation conditions, changes in gene expression under various cultivating conditions will be examined. By utilizing the combinations of the technologies mentioned above, effective application methods will be evaluated (Fig. 20.18).

20.6.3 Value-Added Plant Production in the Future

As mentioned above, plants contain thousands of useful secondary metabolites, and their metabolite concentrations vary with environmental conditions. However, the environmental conditions affecting plant morphology and metabolite concentration are not completely clear because the secondary metabolites of plants and their metabolic pathways are large in number. Therefore, we must invest time, labor, and money to select the suitable environmental conditions for increasing the production of target compounds even in PFALs.

In recent years, it has become possible to determine the gene expression in plants after environmental stimulation over time, using the next-generation sequencer. PFAL can provide the accurate data of plant growth and production of secondary metabolites under varied environmental conditions. Henceforth, it appears that environmental conditions that will improve production efficiency of secondary metabolites can be estimated in a shorter period by combining informatics technology. PFALs, as smart factories, will facilitate the suggestion and optimization of the environmental condition for efficient production of value-added plants in the near future.

References

Afreen F, Zobayed SMA, Kozai T (2005) Spectral quality and UV-B stress stimulate glycyrrhizin concentration of *Glycyrrhiza uralensis* in hydroponic and pot. Plant Physiol Biochem 43:1074–1081
Ahmad K (2014) Molecular farming: strategies, expression systems and bio-safety considerations. Czech J Genet Plant Breed 50:1–10
Bartwal A, Mall R, Lohani P et al (2013) Role of secondary metabolites and brassinosteroids in plant defense against environmental stresses. J Plant Growth Regul 32:216–232
Bettaieb I, Hamrouni-Sellami I, Bourgou S et al (2011) Drought effects on polyphenol composition and antioxidant activities in aerial parts of *Salvia officinalis* L. Acta Physiol Plant 33:1103–1111
Booker FL, Miller JE (1998) Phenylpropanoid metabolism and phenolic composition of soybean [*Glycine max* (L.) Merr.] leaves following exposure to ozone. J Exp Bot 49:1191–1202
Cabinet Secretariat, Office of Healthcare Policy (2016) http://www.kantei.go.jp/jp/singi/kenkouiryou/
Cheruiyot EK, Mumera LM, Ngetich WK et al (2007) Polyphenols as potential indicators for drought tolerance in tea (*Camellia sinensis* L.). Biosci Biotechnol Biochem 71:2190–2197
Christie PJ, Alfenito MR, Walbot V (1994) Impact of low-temperature stress on general phenylpropanoid and anthocyanin pathways: enhancement of transcript abundance and anthocyanin pigmentation in maize seedlings. Planta 194:541–549
Consumer Affairs Agency of Japan. http://www.caa.go.jp/foods/pdf/151224_1.pdf
Daniell H, Streatfield SJ, Wycoffc K (2001) Medical molecular farming: production of antibodies, biopharmaceuticals and edible vaccines in plants. Trends Plant Sci 6:219–226
Daniell H, Singh ND, Mason H, Streatfield SJ (2009) Plant-made vaccine antigens and biopharmaceuticals. Trends Plant Sci 14:669–679

D'Aoust MA, Couture MM, Charland N, Trépanier S, Landry N, Ors F, Vézina LP (2010) The production of hemagglutinin-based virus-like particles in plants: a rapid, efficient and safe response to pandemic influenza. Plant Biotechnol J Jun 8(5):607–619

Dixon RA, Paiva NL (1995) Stress-induced Phenylpropanoid metabolism. Plant Cell 7:1085–1097

Ebisawa M, Shoji K, Kato M et al (2008a) Supplementary ultraviolet radiation B together with blue light at night increased quercetin content and flavonol synthase gene expression in leaf lettuce (*Lactuca sativa* L). Environ Contr Biol 46:1–11

Ebisawa M, Shoji K, Kato M et al (2008b) Effect of supplementary lighting of UV-B, UV-A, and blue light during the night on growth and coloring in red-leaf lettuce. Shokubutu Kankyo Kogaku 20:158–164

Fujiuchi N, Matoba N, Matsuda R (2016) Environment control to improve recombinant protein yields in plants based on agrobacterium-mediated transient gene expression. Front Bioeng Biotechnol 4:1–6

Fukuyama T, Ohashi-Kaneko K, Watanabe H (2015) Estimation of optimal red light intensity for production of the pharmaceutical drug components, vindoline and catharanthine, contained in *Catharanthus roseus* (L.) G. Don. Environ Control Biol 53:217–220

Future Market Insight Global & Consulting Pvt Ltd (2017) Herbal medicinal products market: homeopathic medicines product type segment to register the highest CAGR of 10.4% during the forecast period: global industry analysis (2012–2016) and opportunity assessment (2017–2027): ReportBuyer. https://www.reportbuyer.com/product/5134483/herbal-medicinal-products-market-homeopathic-medicines-product-type-segment-to-register-the-highest-cagr-of-10-4-during-the-forecast-period-global-industry-analysis-2012-2016-and-opportunity-assessment-2017-2027.html

Goto E (2011) Production of pharmaceutical materials using genetically modified plants grown under artificial lighting. Acta Hort (907):45–52

Goto E (2012) Plant production in a closed plant factory with artificial lighting. Acta Hortic (956):37–49

Goto E, Hayashi K, Furuyama S, Hikosaka S, Ishigami Y (2016) Effect of UV light on phytochemical accumulation and expression of anthocyanin biosynthesis genes in red leaf lettuce. Acta Hortic 1134:179–185

Hibino K (2004) Functional food factors and secondary metabolites of plant. Bull Coll Nagoya Bunri Univ 28:1–15 (in Japanese)

Hichem H, Mounir D (2009) Differential responses of two maize *(Zea mays* L.) varieties to salt stress: changes on polyphenols composition of foliage and oxidative damages. Ind Crop Prod 30:144–151

Higashiguchi K, Uno Y, Kuroki S et al (2016) Effect of light intensity and light/dark period on Iridoids in Hedyotis diffusa Kazuki. Environ Control Biol 54:109–116

Hikosaka S, Sasaki K, Goto E, Aoki T (2009) Effects of in vitro culture methods during the rooting stage and light quality during the seedling stage on the growth of hydroponic everbearing strawberries. Acta Hortic 842:1011–1014

Hikosaka S, Ito K, Goto E (2010) Effects of ultraviolet light on growth, essential oil concentration, and total antioxidant capacity of Japanese mint. Environ Control Biol 48:185–190

Hikosaka S, Yoshida H, Goto E, Matsumura T, Tabayashi N (2011) Target protein concentrations in different mature stages of transgenic strawberry fruits. Hortic Res (Japan) 10-1(Suppl):130 (in Japanese)

Hikosaka S, Yoshida H, Goto E, Tabayashi N, Matsumura T (2013) Effects of light quality on the concentration of human adiponectin in transgenic everbearing strawberry. Environ Control Biol 51:31–33

Hikosaka S, Iwamoto N, Goto E (2017) Effects of supplemental lighting on the growth and medicinal ingredient concentrations of Japanese honeysuckle (*Lonicera japonica* Thunb). Environ Control Biol 55:71–76

Holopainen JK, Kivimäenpää M, Julkunen-Tiitto R (2017) New light for phytochemicals. Trends Biotechnol. https://doi.org/10.1016/j.tibtech.2017.08.009

Hou JL, Li WD, Zheng QY et al (2010) Effect of low light intensity on growth and accumulation of secondary metabolites in roots of *Glycyrrhiza uralensis* Fisch. Biochem Syst Ecol 38:160–168

Ito A, Shimizu H, Hiroki R, Nakashima H, Miyasaka J, Ohdoi K (2013) Effect of different durations of root area chilling on the nutritional quality of spinach. Environ Control Biol 51:187–191

Janasa KM, Cvikrováb M, Pałagiewicza A et al (2002) Constitutive elevated accumulation of phenylpropanoids in soybean roots at low temperature. Plant Sci 163:369–373

Japan Kampo Medicines Manufactures Association. http://www.nikkankyo.org/aboutus/investigation/pdf/shiyouryou-chousa04.pdf

Johkan M, Shoji K, Goto F, Hashida S-n, Yoshihara T (2010) Blue light-emitting diode light irradiation of seedlings improves seedling quality and growth after transplanting in red leaf lettuce. Hortscience 45:1809–1814

Kang D (2008) The problems of the herbal medicines (current situation of medicinal plants). Kampo Med 59:397–425 (in Japanese)

Kang D (2011) The actual condition and the problem of Chinese natural medicines. Yakuyo Shokubutsu Kenkyu 33:32–36 (in Japanese)

Kashima K, Mejima M, Kurokawa S, Kuroda M, Kiyono H, Yuki Y (2015) Comparative whole-genome analyses of selection marker–free rice-based cholera toxin B-subunit vaccine lines and wild-type lines. BMC Genomics 16:48

Koike H, Yoshino Y, Matsumoto K et al (2012) Study on conditions to increase the domestic production of herbal materials by changing crops production from tobacco. Kampo Med 63:238–244 (in Japanese with abstract in English)

Lee MJ, Son JE, Oh MM (2014) Growth and phenolic compounds of Lactuca sativa L. grown in a closed-type plant production system with UV-A, -B, or -C lamp. J Sci Food Agric 94:197–204

Leuzinger K, Dent M, Hurtado J, Stahnke J, Lai H, Zhou X, Chen Q (2013) Efficient Agroinfiltration of plants for high-level transient expression of recombinant proteins. J Vis Exp 77:e50521. https://doi.org/10.3791/50521

Li J, Hikosaka S, Goto E (2011) Effects of light quality and photosynthetic photon flux on growth and carotenoid pigments in spinach (*Spinacia oleracea* L.). Acta Hortic (907):105–110

Liu RH (2003) Health benefits of fruit and vegetables are from additive and synergistic combinations of phytochemicals. Am J Clin Nutr 78:517S–520S

Ma JK-C, Drake PMW, Christou P (2003) The production of recombinant pharmaceutical proteins in plants. Nat Rev Genet 4:794–805

Malayeri SH, Hikosama S, Goto E (2010) Effects of light period and light intensity on essential oil composition of Japanese mint grown in a closed production system. Environ Control Biol 48:141–149

Maruyama N, Fujiwara K, Yokoyama K, Cabanos C, Hasegawa H, Takagi K, Nishizawa K, Uki Y, Kawarabayashi T, Shouji M, Ishimoto M, Terakawa T (2014) Stable accumulation of seed storage proteins containing vaccine peptides in transgenic soybean seeds. J Biosci Bioeng 118:441–447

Matsuda R, Tahara A, Matoba N, Fujiwara K (2012) Virus vector-mediated rapid protein production in *Nicotiana benthamiana*: effects of temperature and photosynthetic photon flux density on hemagglutinin accumulation. Environ Control Biol 50:375–381

Matsui T, Takita E, Sato T, Aizawa M, Ki M, Kadoyama Y, Hirano K, Kinjo S, Asao H, Kawamoto K, Kariya H, Makino S, Hamabata T, Sawada K, Kato K (2011) Production of double repeated B subunit of Shiga toxin 2e at high levels in transgenic lettuce plants as vaccine material for porcine edema disease. Transgenic Res 20:735–748

Miura H, Iwata M (1983) Effect of temperature on anthocyanin synthesis in seedlings of Benitade (*Polygonum hydropiper* L.). J Jpn Soc Horticult Sci 51:412–420

Miyazawa Y, Hikosaka S, Goto E, Aoki T (2009) Effects of light conditions and air temperature on the growth of everbearing strawberry during the vegetative stage. Acta Hortic (842):817–820

Mosaleeyanon K., Zobayed SMA, Afreen F et al (2006) Enhancement of biomass and secondary metabolite production of St. John's Wort *ypericum perforatum* L.) under a controlled environment Environ Contr Biol 44:21–30

Ogawa E, Hikosaka S, Goto E (2018) Effects of nutrient solution temperature on the concentration of major bioactive compounds in red perilla. J Agri Meteor 74(2):71–78

Patil BL, Fauquet CM (2015) Light intensity and temperature affect systemic spread of silencing signal in transient agroinfiltration studies. Mol Plant Pathol 16(5):484–494

Sack M, Rademacher T, Spiegel H, Boes A, Hellwig S, Drossard J, Stoger E, Fischer R (2015) From gene to harvest: insights into upstream process development for the GMP production of a monoclonal antibody in transgenic tobacco plants. Plant Biotechnol J 13:1094–1105

Sainsbury F, Lomonossoff GP (2014) Transient expressions of synthetic biology in plants. Curr Opin Plant Biol 19:1–7

Sakamoto M, Suzuki T (2015) Effect of root-zone temperature on growth and quality of hydroponically grown red leaf lettuce (*Lactuca sativa* L. cv. Red wave). Am J Plant Sci 6:2350–2360

Sharmaa YK, Davisa KR (1997) The effects of ozone on antioxidant responses in plants. Free Radic Biol Med 23:480–488

Solecka D, Boudet A-M, Kacperska A (1999) Phenylpropanoid and anthocyanin changes in low-temperature treated winter oilseed rape leaves. Plant Physiol Biochem 37(6):491–496

Son KH, Oh MM (2013) Leaf shape, growth, and antioxidant phenolic compounds of two lettuce cultivars grown under various combinations of blue and red light-emitting diodes. Hortscience 48:988–995

Sreenivasulu N, Grimm B, Wobus U et al (2000) Differential response of antioxidant compounds to salinity stress in salt-tolerant and salt-sensitive seedlings of foxtail millet (Setaria italica). Physiol Plant 109:435–442

Sudheer S, Yeoh WK, Manickam S, Ali A (2016) Effect of ozone gas as an elicitor to enhance the bioactive compounds in Ganoderma lucidum. Postharvest Biol Technol 117:81–88

Sugita K, Endo-Kasahara S, Tada Y, Lijun Y, Yasuda H, Hayashi Y, Jomori T, Ebinuma H, Takaiwa F (2005) Genetically modified rice seeds accumulating GLP-1 analogue stimulate insulin secretion from a mouse pancreatic beta-cell line. FEBS Lett 579:1085–1088

Sun HJ, Kataoka H, Yano M, Ezura H (2007) Genetically stable expression of functional miraculin, a new type of alternative sweetener, in transgenic tomato plants. Plant Biotechnol J 5(6):768–777

Sun R, Hikosaka S, Goto E et al (2012) Effect of UV irradiation on growth and concentration of four medicinal ingredients in Chinese licorice (*Glycyrrhiza uralensis*). Acta Hortic (956):643–648

The International Trade Centre (ITC) (2017) Medicinal and aromatic plants and extracts. http://www.intracen.org/itc/sectors/medicinal-plants/

The Ministry of Economy, Trade and Industry in Japan (2016). http://www.meti.go.jp/press/2016/07/20160714001/20160714001-1.pdf (in Japanese)

Voipio I, Autio J (1994) Responses of red-leaved lettuce to light intensity, UV-A radiation and root zone temperature. Acta Hortic 399:183–190

Yao J, Weng Y, Dickey A, Wang KY (2015) Plants as factories for human pharmaceuticals: applications and challenges. Int J Mol Sci 16:28549–28565

Yoshida H, Hikosaka S, Goto E, Takasuna H, Kudou T (2012) Effect of light quality and light period on flowering of everbearing strawberry in a closed plant production system. Acta Hortic 956:107–112

Zobayed SMA, Afreen F, Kozai T (2005a) Necessity and production of medicinal plants under controlled environments. Environ Control Biol 43:243–252

Zobayed SMA, Afreen F, Kozai T (2005b) Temperature stress can alter the photosynthetic efficiency and secondary metabolite concentrations in St. John's wort. Plant Physiol Biochem 43:977–984

Chapter 21
Chemical Inquiry into Herbal Medicines and Food Additives

Natsuko Kagawa

Abstract Medicinal and aromatic plants are valuable sources of natural products that contribute to human health. Herbal medicines and some food additives are derived from medicinal and aromatic plants, and to prove their safety and efficacy, chemical analyses of the concentrations of active natural compounds in raw plant materials are essential. This chapter discusses the phytochemistry of medicinal and aromatic plants, the chemical requirements for the production of raw plant materials, and the analytical methods used for quality control. Plant factories offer the use of technology that more efficiently enhances the production of targeted compounds by controlling a plant's metabolism and can provide sustainable harvesting of medicinal and aromatic plants, which will conserve valuable source-plant species for future generations.

Keywords Medicinal and aromatic plant · Herbal medicine · Natural product · Bioactive compound · Chemical analysis · Quality concentration · Sustainable harvest · Conservation of source-plant species

21.1 Introduction

Plant factories with artificial lighting (PFALs) have great potential for the cultivation and sustainable harvesting of medicinal and aromatic plants. Phytochemicals produced by these plants are utilized as pharmaceuticals, herbal medicines, food additives, and cosmetics on the world market. To supply these plants as raw materials for industry and to conserve plant species from the threat of overharvesting, ideas for their cultivation are proliferating. When cultivating medicinal and aromatic plants, however, it is difficult to maintain quality concentrations of phytochemicals. PFALs must develop cultivation methods that will promote both plant growth and chemical accumulation.

N. Kagawa (✉)
Center for Environment, Health and Field Sciences, Chiba University, Kashiwa, Chiba, Japan
e-mail: knatsuko@faculty.chiba-u.jp

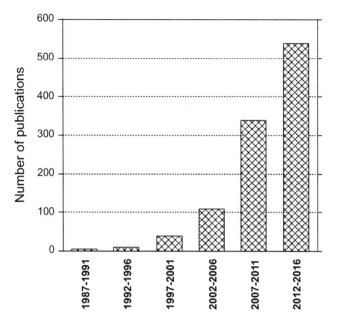

Fig. 21.1 The number of publications in 5-year increments

21.2 Issue of Concern with Medicinal Plants

Global interest in the cultivation of medicinal and aromatic plants is mounting. A survey of the topics of most interest via a search of science websites is revealing. In 5-year increments, Fig. 21.1 tracks the trends in the number of scientific publications that cite both the words "cultivation" and "medicinal plants" as topics that have increased during the past 30 years. In particular, the number of publications listed for the period from 2007 to 2011 reached 339, which is 3 times that of the previous period (2002 to 2006). The number of countries by author affiliation has also increased, and from 2012 to 2016, researchers in more than 100 countries had contributed to studies closely related to the cultivation of medicinal plants.

Figure 21.2 shows another trend whereby the number of research areas in publications related to the cultivation of medicinal plants reached 63 during the latest period (2012 to 2016) and included topics such as plant science, agriculture, pharmacology and pharmacy, biotechnology applied to microbiology, environmental science ecology, chemistry, and food science technology.

Growing pressure to conserve species threatened by over-harvesting is a subplot of these trends. International action is pursuing the achievement of sustainable harvesting since the market for herbal production is worldwide and most countries are involved in the trade of medicinal and aromatic plants as exporting, importing, manufacturing, or consuming countries.

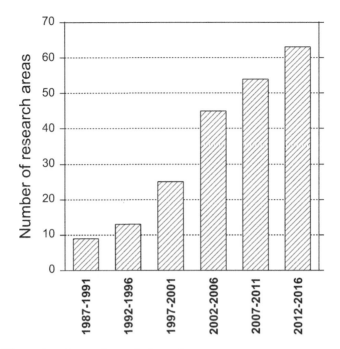

Fig. 21.2 The number of research areas in 5-year increments

21.3 Active Compounds in Herbal Medicines

Medicinal and aromatic plants are used for the production of herbal medicines and food additives. Herbal medicines in Europe are marketed as pharmaceuticals in the form of extracts, teas, tinctures, and capsules obtained from medicinal and aromatic plants (Lubbe and Verpoorte 2011). These are known as botanical drugs in the United States.

In Japan, the raw materials from plants that are used for herbal medicines include traditional medicines such as Kampo, and these are referred to as crude drugs and related drugs in the *Japanese Pharmacopoeia*. Crude drugs and their related products include extracts, powders, tinctures, syrups, spirits, fluidextracts, or suppositories containing crude drugs as the active ingredient and combination preparations containing crude drugs as the principal active ingredient in the *Japanese Pharmacopoeia* (JP17, 2016).

Plants produce compounds that are numerous and diverse and are called natural products, secondary metabolites, phytochemicals, or natural compounds. Plants and their crude extracts are complicated mixtures of compounds and generally contain many hundreds of phytochemicals. Natural compounds in the extracts of medicinal and aromatic plants have been investigated in clinical trials for many herbal medicines for their ability to prevent and treat diseases. In most cases, those efforts have focused on identifying the bioactive compounds that are responsible for the medicinal effects of herbal medicines.

Many people who rely on modern Western scientific medicines tend to prefer a single substance as the most important ingredient in a medicine (Houghton 2001). The active principle concept in medicine is responsible for the progress of modern chemistry, and this has enabled many natural products to be characterized as bioactive molecules when isolated as pure substances from the extracts of medicinal and aromatic plants.

It is important to note that a high percentage of the world's population still uses crude extracts of local plants as traditional medicines, although their bioactive compounds have not been fully elucidated. In contrast to the use of those folk medicines, the use of herbal medicines that are standardized in terms of the active constituents is increasing in industries such as food, cosmetics, and pharmaceuticals.

In the food and cosmetics industries, various constituents extracted from plants are currently being used as flavorings (drinks, foods, confectionaries, perfumes, and cosmetics), sweeteners (drinks, foods, and confectionaries), colorants (drinks, foods, confectionaries, and cosmetics), antioxidants (drinks, foods, cosmetics, and health products), and other functional additives (health products, nutraceuticals, functional foods, beauty care products, and cosmeceuticals). These widespread trends in markets have resulted in a global demand for medicinal and aromatic plants of very high quality.

21.4 Non-sustainable Harvest of Plants

Raw materials from plants for use as herbal products such as herbal medicines, dietary supplements, food additives, and cosmetics are taken from both wild and cultivated plants. Worldwide, approximately two-thirds of all medicinal plants are harvested in the wild rather than from cultivated sources (Vines 2004).

About 15,000 species of medicinal plants are targeted in global concerns about depletion of wild populations, loss of habitat, invasive species, and pollution (Hamilton 2008). Non-sustainable harvesting of plants as sources of raw materials by commercial collectors has become a major cause of local extinctions and degradation of habitat.

There are well-documented examples of wild harvesting that have threatened populations of plant species and their habitats, which include *Arctostaphylos uva-ursi* (bearberry), *Thymus* spp. (wild thyme), *Piper methysticum* (kava kava), *Glycyrrhiza glabra* (liquorice), *Chamaelirium luteum* (false unicorn), *Hydrastis canadensis* (goldenseal), *Panax quinquefolius* (American ginseng), and *Panax ginseng* (Asian ginseng) (Vines 2004). Table 21.1 summarizes the chemical structures of the representative compounds in the popular herbal medicines listed above, along with the bioactivities reported as extracts or single substances, and major uses in industry (Abourashed and Khan 2001; Attele et al. 1999; Briskin 2000; Budzinski et al. 2000; Chan et al. 2000; Chen et al. 2017; Einbond et al. 2017; Gastpar and

21 Chemical Inquiry into Herbal Medicines and Food Additives

Table 21.1 Phytochemicals of popular herbal medicines in the wild

Plant name	Chemical representative		Plant part	Activity	Use
Arctostaphylos uva-ursi (bearberry)	Arbutin CAS: 497-76-7	1,2,3,6 Tetra O galloyl β D glucose CAS: 79886-50-3	Leaf	Antiseptic Astringent	Urogential tract infections
				Tyrosinase inhibitory	In cosmetic: skin-whitening
Thymus spp. (wild thyme)	Thymol CAS: 89-83-8	Carvacrol CAS: 499-75-2	Leaf Stem	Antimicrobial Antifungal Antioxidant	Essential oil Postharvest diseases
Piper methysticum (kava kava)	(+)-Kavain CAS: 500-64-1	(+)-Dihydromethysticin CAS: 19902-91-1	Root Rhizome	Antidepressant	Anxiolytic
Glycyrrhiza glabra (liquorice)	Glycyrrhizin CAS: 1405-86-3	Liquiritin CAS: 551-15-5	Root Stolon	Antiinflammatory Estrogenicity	Expectorant Peptic ulcer
Chamaelirium luteum (false unicorn)	Chamaeliroside A CAS: 1323952-20-0		Root Rhizome	Female reproductive health	
Hydrastis canadensis (goldenseal)	Berberine CAS: 2086-83-1	(−)-Hydrastine CAS: 118-08-1	Root Rhizome	Antiinflammatory Antimicrobial Cytochrome P450 inhibitory	Tonic for appetite and digestion Inflammation Infection
Panax ginseng (Asian ginseng) *Panax quinquefolius* (American ginseng)	Ginsenoside Rg$_1$ CAS: 22427-39-0		Root Rhizome	Antineoplastic Antistress Antioxidant	Cardiovascular CNS Endocrine Immune

Klimm 2003; Hamid et al. 2017; Li et al. 2016; Li et al. 2000; Martin et al. 2014; Van der Voort et al. 2003; Weber et al. 2003).

When medicinal plant species are threatened by wild harvesting, the damage can lead to a loss of genetic diversity and can increase the risk of adulteration from plants of the same genus, which can lead to a reduction, or loss, of the active components in the raw materials (Lubbe and Verpoorte 2011).

Generally, harvesting by local people for their own use is not a conservation problem. Industrial-scale harvesting of wild plants is harmful to sustainability (Hamilton 2008). The demand for medicinal plants as raw materials for use by industries is enormous and growing faster than the supply from the wild harvesting of slow-growing and limited plants in nature, which means over-harvesting by commercial collectors will continue until alternative methods of supply can be proposed.

21.5 Cultivation of Medicinal Plants

Cultivation is one solution to sustainable supply. Species with a large market share in the herbal medicines industry such as *Ginkgo biloba* (ginkgo), *Hypericum perforatum* (St. John's wort), and *Panax ginseng* (Asian ginseng) are now being cultivated by pharmaceutical and herbal companies, and some, such as Asian ginseng, have become rare in the wild (Canter et al. 2005; Vines 2004).

Cultivation has advantages as a method of supplying raw materials for commercial products. First, control of quality and quantity for raw materials is easier. The entirety of the supply chain from germination to harvesting is controlled with cultivated plants, which decreases the risks of adulteration and misidentification of plants. Under uniform conditions, cultivated plants generally have fewer chemical variations in terms of chemical profiles and content when only one variety is dealt with. Secondly, both the price and supply of raw materials tend to be stable. Wild harvests are more vulnerable to environmental problems such as invasive species, pollution, climate change, or trade regulations.

It is expensive to cultivate some plant species since investment and special technologies are needed for large-scale production. With medicinal and aromatic plants, however, the fear of losing a species as a wild resource of raw materials has provided great incentive to develop cultivation processes. For instance, many studies on the cultivation of *Glycyrrhiza* (liquorice) have been performed with Japanese herbal companies and academia to provide a stable supply because 100% of raw materials derived from liquorice in Japanese herbal medicines are imported from producer countries, and almost all of it comes from Chinese wild resources (JKMMA 2016; Akiyama et al. 2017).

One problem is that some herbal medicines are difficult to cultivate using normal processes. Cultivators often encounter phenomena such as low germination rates, slow growth rates, fast flowering, or low content of active compounds. In general, species with relatively small or local habitats must be cultivated with special treatment and/or under special environmental conditions that reproduce an environment similar to their natural habitat.

Environmental conditions for cultivation should be optimized based on the natural habitat because this greatly affects the growth of plants, the biosynthesis of secondary metabolite, and the accumulation of bioactive substances. To overcome low yields and low quality of cultivated plants, the use of some technologies has proven efficient and less labor-intensive.

Hydroponic growth of medicinal plants with underground parts (root, rhizome, stolon) that are used as raw materials has been established and applied to *Glycyrrhiza uralensis* (liquorice), *Atropa belladonna* (belladonna), and *Coptis japonica* (goldthread or canker root) in PFAL for 6 months to 1 year, which has successfully shortened the cultivation times for these slow-growing plants by almost 4 years, as well as promoting sufficient concentrations of active compounds (Yoshimatsu 2012). It generally takes 3 to 5 years of field cultivation or wild growth for the accumulation of targeted active compounds to reach the same levels of concentrations as those reaped by hydroponics cultivation.

21.6 Phytochemistry with Plant Factories

Recently, the high potential that PFALs possess for the cultivation of medicinal and aromatic plants has attracted the herbal product industry. The technology of PFAL can control environmental factors and provide comprehensive conditions in which medicinal plants can be cultivated on a commercial scale, even for plants that have special environmental requirements.

The goals of PFAL are to increase biomass yields along with uniformity in the concentration of phytochemicals based on standards of safety, quality, and efficacy. The chemical components in raw materials depend on the type of herbal products. For herbal medicines, increasing the concentration of bioactive compounds in extracts is important, and reducing the level of toxic ingredients, such as harmful compounds or heavy metals, is also essential. For food additives and cosmetics, beneficial compounds that are used as flavorings, sweeteners, colorants, and other functional additives are selected for increased production.

Again, plants have many hundreds of compounds, and it is very difficult to evaluate the potency of each. Therefore, processes of extraction and separation/purification for target compounds from cultivated plants are almost a necessity in order to standardize the safety, quality, and efficacy of raw materials. Uniformity in the concentrations of compounds in extracts can simplify the processes and make them more predictable while reducing the cost of raw materials. In industry, the desired compounds for specific products are generally the secondary metabolites produced in the plants. Difficulties in the cultivation of medicinal and aromatic plants are not confined to the enhancement of their growth but also include the regulation of secondary metabolites.

Plants produce targeted secondary metabolites through different biosynthetic pathways and accumulate them in different plant parts (roots, stems, leaves). The metabolic pathways by which target compounds are biosynthesized have been well investigated for some plants, and this has shown that secondary metabolites are adaptations to environmental stimuli such as fluctuating temperature and light conditions (antioxidants), stress (proline), infection (flavonoids), or herbivory (alkaloids) (Canter et al. 2005).

Environmental factors (temperature, humidity, light, water supply, minerals, and carbon dioxide) that can be controlled in PFAL influence plant metabolism. In addition, it is important to note that the accumulation of secondary metabolites is independently affected by various environmental conditions such as ultraviolet irradiation, drought, nutrient deficiency, and salinity as a response to abiotic stresses. In other words, secondary metabolites may be controlled intentionally with the addition of abiotic treatments. Therefore, the technology of a plant factory has great potential for easier regulation of secondary metabolites in the concentration of target compounds and for commercial production of raw materials with uniform quality.

An example of engineering changes in the concentrations of medicinal compounds was reported for *Perilla frutescens* grown in PFAL via the application of different environmental conditions (Lu et al. 2017). In that study, the concentration of rosmarinic acid (anti-allergic activity) was affected by light intensity and nutrient concentration in a hydroponic solution, and the highest concentration of rosmarinic acid was achieved with an optimized combination of light intensity and nutrient solution. The elicitor for accumulation of rosmarinic acid was the gap between the highest light intensity and the lowest nutrient solution, which resulted in low nutrient uptake stress to the perilla plant. Interestingly, the concentration of perillaldehyde (antimicrobial) that was produced through the monoterpene biosynthetic pathway was less affected by those stimuli than that of rosmarinic acid synthesized via the phenylpropanoid pathway.

This means that expression of phytochemicals in medicinal plants can be controlled by the fine regulation of every environmental condition and that the efficient production of targeted secondary metabolites can be enhanced in plants cultivated via PFAL technology.

21.7 Chemical Analysis of Medicinal Plants

21.7.1 Introduction

Chemical analysis of phytochemicals produced in medicinal and aromatic plants is conducted to evaluate their potency or function. Each bioactive compound that is responsible for the medicinal effect of an herbal medicine is primarily identified as the active principle compound. Since the potency of herbal medicines relies on the amount of the active compound, its concentration in raw materials must be determined via quantitative analysis. The compounds that add to the desired functions of herbal products are characterized as the main ingredients. Since the value of raw materials as functional additives is measured according to the main ingredients, the quality of raw materials is frequently assessed via their level of contamination.

21.7.2 Organic Molecules

Chemical analysis of secondary metabolites involves processes for extraction, separation, and identification. Every process involving organic molecules has been technologically developed and optimized for the successful analysis of secondary metabolites. With organic molecules, a method for each process is carefully selected by the characteristics and behavior of a target compound for analysis so that a suitable scheme can incorporate both speed and accuracy.

21.7.3 Extraction

Compounds are largely differentiated by their level of solubility in either water or organic solvents. Each secondary metabolite has a unique solubility. For instance, essential oils used as flavoring are mixtures of volatile and hydrophobic compounds produced in plants that are generally not soluble in water.

The aim of extraction is to use a solvent to extract compounds from solid materials and to concentrate the compounds in a solution for a separation process. The choice of a solvent that will dissolve a compound is most important. Nonpolar organic solvents, such as n-hexane, n-pentane, isopentane, and diethyl ether, are the preference for the extraction of nonpolar and volatile compounds, monoterpenoids (C10 unit) and phenylpropanoids, which are the secondary metabolites in essential oils. Hydrophilic compounds of secondary metabolites show low levels of volatility and are extracted using any mixture of polar and water-soluble solvents such as alcohols (e.g., methanol, ethanol), acetonitrile, acetone, and water.

Supercritical fluid extraction (SFE) is a method that is used to extract compounds via a supercritical fluid that is neither a gas nor a liquid. Supercritical fluids have physical properties (density, viscosity, diffusivity) that lie between those of gases and liquids; thus, the solubility of compounds in supercritical fluids can be more enhanced by the properties of supercritical fluids than by solvents. Carbon dioxide is typically used as a substrate of a supercritical fluid. Carbon dioxide is a nonpolar molecule in the chemical structure in a fluid and is suitable for the extraction of relatively nonpolar oils and fats; however, various applications that include nonvolatile triterpenoids (C30 unit) have proven that secondary metabolites can be extracted more efficiently and quicker by SFE either with the use of cosolvents or with the pretreatment of plant materials. SFE is now being utilized in the production of decaffeinated coffee and tea by the selective extraction/removal of caffeine from normal coffee beans and tea leaves.

21.7.4 Separation

Plants contain many hundreds of compounds as secondary metabolites. Extracts from the plants still contain hundreds of compounds even after screening by solubility via extraction. Some methods used to remove contamination from extracts are useful when the contamination may disturb the measurement of target compounds. In addition, those methods enhance the measurement sensitivity of target compounds by promoting detectable concentration levels. These methods include liquid-liquid extraction (also known as solvent extraction), solid phase extraction (SPE), and coagulation with filtration. Such pretreatment of extracts is time-consuming, but it is sometimes required in the preparation of samples intended for mass spectrometry.

21.7.5 Identification

21.7.5.1 Gas Chromatography: GC

Chemical analysis that uses a gas chromatograph is referred to as gas chromatography. The abbreviation "GC" frequently refers to both the process of gas chromatography and a gas chromatograph (JSAC 2011). Gas chromatography is the most dominant method used for the analysis of essential oils extracted from plants because it is specialized for the separation and identification of compounds that can be easily vaporized at temperatures from 250–350 °C. An essential oil extracted from medicinal and aromatic plants contains hundreds of monoterpenes and phenylpropanoids that are semi-volatile with relatively low boiling points.

A gas chromatograph equipped with either a flame ionization detector (FID) or a mass spectrometer is useful for the quantitative analysis of the compounds contained in essential oils. Gas chromatography/mass spectrometry (GC/MS) provides information about the chemical structures of the compounds detected by MS.

Electron impact (EI) is applied to GC/MS as an ionization method since it can effectively ionize gas-phase molecules via bombardment with a high-energy electron beam (Silverstein et al. 2005). The ions generated are recorded as the mass spectrum of ions that separate on the basis of mass/charge (m/z). The gas-phase method is applicable to compounds that vaporize under the operating temperature of a GC and are stable at that temperature. Those compounds typically have a molecular weight < 300. Monoterpenes (C10 unit) frequently show peaks of m/z 79, 93, 107, and 121 as fragment ions on the mass spectra (JSAC 2011).

Each of the compounds isolated by GC/MS shows a mass spectrum that is unique sufficiently for identification of the compounds by computer search that compares the similarities between the mass spectrum of the detected compound and that of known compounds in libraries and databases that currently register the EI mass spectra of more than 200,000 organic compounds from the results of GC/MS. This utility together with the great sensitivity of the EI method makes GC/MS a powerful and popular tool for chemical analysis.

21.7.5.2 High-Performance Liquid Chromatography: HPLC

High-performance liquid chromatography (HPLC) is applicable to many compounds that are not suitable for GC analysis. HPLC can be used to analyze the measurement of almost all secondary metabolites, including polyphenols with oxygen in the chemical structure. In contrast to GC, HPLC is widely used for less volatile compounds, because of their polar characteristics. A solution of extracts from plants in aqueous solvents is acceptable as a sample for HPLC, which has advantages that include simple preparation and easy handling of samples.

It is important to couple a liquid chromatograph to suitable detectors, such as ultraviolet-visible (UV-VIS) and MS. UV-VIS detectors include photodiode array (PDA) detectors, which are applicable to compounds that absorb UV and VIS light at wavelengths from 190 to 830 nm. Double-bonded carbon-carbon (C=C) and carbon-heteroatom (C=O, C=N, C=S) groups are frequently observed in the chemical structures of secondary metabolites, and they can produce UV-VIS absorbance characteristics on molecules. The UV-VIS absorbance spectrum recorded by PDA is useful in determining purity and in ensuring the isolation of a compound in a single peak that has been separated via HPLC, because the spectrum reflects all structures that have UV-VIS absorbance and can determine if the single peak represents a compound that is a single substance, a compound with contamination, or a mixture of compounds.

Liquid chromatography/mass spectrometry (LC-MS) has advanced rapidly with the development of electrospray ionization (ESI) around 1990. ESI uses polar and volatile solvents to ionize compounds in solution. The ionization method for the liquid-phase sample is applicable to compounds that are polar and hydrophilic and that have a molecular weight up to approximately 100,000. In addition, fragmentation of ions rarely occurs with ESI so that peaks related to molecular ions will appear on the mass spectrum characterizing the molecular weight of the detected compound. These features of ESI-MS make it a good match for LC and samples for LC analysis, which includes plant extracts.

21.8 Reference Standard of Herbal Medicine

The demand for reference standards is growing. Every chemical analysis needs a reference standard for compounds that must be identified. Quantitative analysis in particular requires a high-quality reference standard in order to determine the content of target compounds in medicinal and aromatic plants. Chemicals with great purity (>98%), accurate chemical formulas (molecular weight, hydration), and less contamination can be used in chemical analysis as reference standards. With the bioactive compounds that are used in pharmaceuticals and herbal medicines, moreover, there is less of a difference between the production of many batches, and traceability and a stable supply are requirements for reference standards with reliable quality.

Reference standards for herbal medicines are produced either by chemical synthesis or plant extraction. With production by chemical synthesis, achieving the requirements for a reference standard is feasible, but the results are not always an improvement over plant extraction because bioactive compounds in herbal medicines sometimes have an original element to their chemical structures (chirality, enantiomer, diastereomer) and chemical compositions (ingredients in essential oils).

Since a great many herbal products have become popular, the assurance of raw plant materials for use as reference standards is a concern for human health. To reproduce the natural essence of plants and conserve original plants, cultivation remains one of the best solutions, and the technology used in PFALs can insure both quality management and a stable supply via plant cultivation.

21.9 Conclusions

Medicinal and aromatic plants are valuable as sources of natural products that contribute to human health. The extract of a plant is a rich mixture of bioactive compounds that are responsible for both positive and negative effects on mental and physical conditions or symptoms. Advances in modern chemistry and biology have revealed the structure/activity relationships of bioactive compounds, and many clinical trials have validated the safety of their therapeutic effects on brains and bodies. It is commonly understood that the desired potency of plant extracts relies on the concentrations of the active natural compounds they contain. Herbal medicines and some food additives are derived from medicinal and aromatic plants, and to prove their safety and efficacy, chemical analyses of the concentrations of active natural compounds in raw plant materials are essential. The demand for high-quality medicinal and aromatic plants that are stable and of uniform quality is increasing in industries such as pharmaceutical, food, and cosmetics that consume these plants on an ever-increasing scale of commercial production.

With conservation of the sources of plant species, cultivation of medicinal and aromatic plants is attracting global attention from these industries as well as from academia, governments, and international unions. Generally, medicinal and aromatic plants should be cultivated with special environmental requirements serving as standards for the concentrations of desired compounds since the biosynthesis and accumulation of these compounds, mostly secondary metabolites, will be greatly affected by their growth environment.

PFALs can use technology to control every environmental factor and determine the optimized environment that will best enhance the production of targeted compounds by using a plant's metabolism. The practical merits of PFALs include increased biomass yields and uniformity in the concentrations of target compounds. PFALs can provide sustainable harvesting of medicinal and aromatic plants, which will conserve valuable source-plant species for future generations.

References

Abourashed EA, Khan IA (2001) High-performance liquid chromatography determination of hydrastine and berberine in dietary supplements containing goldenseal. J Pharm Sci 90 (7):817–822

Akiyama H, Nose M, Ohtsuki N et al (2017) Evaluation of the safety and efficacy of *Glycyrrhiza uralensis* root extracts produced using artificial hydroponic and artificial hydroponic-field hybrid cultivation systems. J Nat Med 71:265–271

Attele AS, Wu JA, Yuan CS (1999) Ginseng pharmacology: multiple constituents and multiple actions. Biochem Pharmacol 58:1685–1693

Briskin DP (2000) Medicinal plants and phytomedicines. Linking plant biochemistry and physiology to human health. Plant Physiol 124:507–514

Budzinski JW, Foster BC, Vandenhoek S et al (2000) An in vitro evaluation of human cytochrome P450 3A4 inhibition by selected commercial herbal extracts and tinctures. Phytomedicine 7 (4):273–282

Canter PH, Thomas H, Ernst E (2005) Bringing medicinal plants into cultivation: opportunities and challenges for biotechnology. Trends Biotechnol 23(4):180–185

Chan TWD, But PPH, Cheng SW et al (2000) Differentiation and authentication of *Panax ginseng*, *Panax quinquefolius*, and ginseng products by using HPLC/MS. Anal Chem 72(6):1281–1287

Chen YJ, Zhao ZZ, Chen HB et al (2017) Determination of ginsenosides in Asian and American ginsengs by liquid chromatography-quadrupole/time-of-flight MS: assessing variations based on morphological characteristics. J Ginseng Res 41:10–22

Einbond LS, Negrin A, Kulakowski DM et al (2017) Traditional preparations of kava (*Piper methysticum*) inhibit the growth of human colon cancer cells in vitro. Phytomedicine 24:1–13

Gastpar M, Klimm HD (2003) Treatment of anxiety, tension and restlessness states with Kava special extract WS® 1490 in general practice: a randomized placebo-controlled double-blind multicenter trial. Phytomedicine 10:631–639

Hamid HA, Ramli ANM, Yusoff MM (2017) Indole alkaloids from plants as potential leads for antidepressant drugs: a mini review. Front Pharmacol 8:96

Hamilton AC (2008) Medicinal plants in conservation and development: case studies and lessons learnt, 4 pp, Plantlife International. www.plantlife.org.uk

Houghton PJ (2001) Old yet new-pharmaceuticals from plants. J Chem Educ 78(2):175–184

Japan Kampo Medicines Manufacturers Association [JKMMA] (2016) Report on investigation of usage of the crude drugs for Kampo preparation (4) – the usage in FY2013 and FY2014. 3–11 pp. www.nikkankyo.org/aboutus/investigation/pdf/shiyouryou-chousa04.pdf

Japan Society for Analytical Chemistry [JSAC] (2011) Shokuhin-bunseki. Maruzen Publishing Co. Ltd., Tokyo, 47–63 pp (in Japanese)

Japanese Pharmacopeia, 17th Edn. [JP17] (2016) General notices: 1 pp. official monographs, crude drugs and related drugs: 1791–2012 pp. http://www.mhlw.go.jp/stf/seisakunitsuite/bunya/0000066597.html

Li WK, Gu CG, Zhang HJ et al (2000) Use of high performance liquid chromatography-tandem mass spectrometry to distinguish *Panax ginseng* C. A. Meyer (Asian ginseng) and *Panax quinquefolius* L. (North American ginseng). Anal Chem 72(21):5417–5422

Li GN, Nikolic D, van Breemen RB (2016) Identification and chemical standardization of licorice raw materials and dietary supplements using UHPLC-MS/MS. J Agric Food Chem 64 (42):8062–8070

Lu N, Bernardo EL, Tippayadarapanich C et al (2017) Growth and accumulation of secondary metabolites in perilla as affected by photosynthetic photon flux density and electrical conductivity of the nutrient solution. Front Plant Sci 8:708

Lubbe A, Verpoorte R (2011) Cultivation of medicinal and aromatic plants for specialty industrial materials. Ind Crop Prod 34(1):785–801

Martin AC, Johnston E, Xing CG et al (2014) Measuring the chemical and cytotoxic variability of commercially available kava (*Piper methysticum* G. Forster). PLoS One 9(11):7

Silverstein RM, Webster FX, Kiemle DJ (2005) Spectrometric identification of organic compounds, 7th edn. Wiley, Hoboken, 1–8 pp

Van der Voort ME, Bailey B, Samuel DE et al (2003) Recovery of populations of goldenseal (*Hydrastis canadensis* L.) and American ginseng (*Panax quinquefolius* L.) following harvest. Am Midl Nat 149(2):282–292

Vines G (2004) Herbal harvests with a future: towards sustainable sources for medicinal plants. Plantlife International. www.plantlife.org.uk

Weber HA, Zart MK, Hodges AE et al (2003) Chemical comparison of goldenseal (*Hydrastis canadensis* L.) root powder from three commercial suppliers. J Agric Food Chem 51 (25):7352–7358

Yoshimatsu K (2012) Innovative cultivation: hydroponics of medicinal plants in the closed-type cultivation facilities. J Trad Med 29(1):30–34

Chapter 22
Detection and Utilization of Biological Rhythms in Plant Factories

Hirokazu Fukuda, Yusuke Tanigaki, and Shogo Moriyuki

Abstract Biological rhythms with a period of about 24 h, called "circadian rhythms," are generated by the expressions of clock genes. In plants, circadian rhythms increase the growth rate through the daily coordination of photosynthesis and metabolism. Therefore, the detection and utilization of circadian rhythms is required to improve the plant production in plant factories. In this chapter, recently developed technologies based on circadian rhythms are described. Section 22.1 focuses on seedling diagnosis using the circadian rhythm of chlorophyll fluorescence (CF). Section 22.2 describes high-throughput growth prediction systems based on the circadian rhythm of CF. Section 22.3 provides examples of global analysis of acquired gene expression as biological information and methods for analyzing internal time (i.e., the phases of circadian rhythms) using that data. We describe how circadian rhythms can be observed by comprehensive analyses and the methods used for such analyses. Finally, Sect. 22.4 describes a basic theory for controlling circadian rhythms by environmental stimuli. Methods that enable basic control of circadian rhythms are important, as they are applicable to a variety of research questions and industrial problems.

Keywords Circadian rhythms · Chlorophyll fluorescence · Growth prediction · Transcriptome analysis · Phase response curves · Molecular timetable method

22.1 Introduction

To adapt to diurnal environmental cycles, plants employ inherent biological rhythms—also known as circadian rhythms—to regulate various physiologic events. It is thought that circadian rhythms adapt plants to their environment by allowing

---

H. Fukuda (✉) · Y. Tanigaki · S. Moriyuki
Department of Mechanical Engineering, Graduate School of Engineering, Osaka Prefecture University, Sakai, Osaka, Japan
e-mail: fukuda@me.osakafu-u.ac.jp; yu.tanigaki@me.osakafu-u.ac.jp; sssmskztmissue@yahoo.co.jp

anticipation of daily environmental changes, coordinating cellular processes, and ensuring responses to the environment. Circadian rhythms are tied systematically to environmental cycles through indigenous responses that depend upon the phase in which the environmental stimulus was received. Hence, the ability to detect and exploit dynamic plant information (e.g., circadian rhythm phases) is important in efforts to improve plant production. In this chapter, recently developed technologies based on circadian rhythms, which across cultivation processes in plant factories, are described.

22.2 Seedling Diagnosis Using the Biological Rhythm of CF

Individual plants that grow poorly can cause significant profit losses for plant factories that utilize huge electric power for cultivation. Thus, the ability to identify and cull poorly growing (i.e., low-grade) plants at an early stage using so-called seedling diagnosis and selection technology is important for minimizing economic losses in plant factories.

Imaging of CF is generally used as a highly efficient indicator of factors that affect plant growth, such as photosynthetic capacity and degree of stress (Takayama et al. 2014). CF results from the emission of red light from chlorophyll α pigments (Krause and Weis 1991; Govindjee 1995) when residual light energy is not used for photosynthetic reactions. Accurate measurement of CF thus enables the evaluation of photosynthetic photochemical reactions and the status of heat dissipation processes, without the need for physical contact with the plant (Maxwell and Johnson 2000; Takayama and Nishina 2009). The CF imaging technique, originally developed by Omasa et al. (1987) and Daley et al. (1989), has been used to evaluate the heterogeneous distribution of photosynthetic activity over the leaf surface, thus enabling detection of photosynthetic dysfunction caused by biotic or abiotic stress factors. CF imaging techniques have also been scaled up to enable analyses of whole plants (Takayama et al. 2010), tree canopies (Nichol et al. 2012), and tomatoes cultivated in a large-scale greenhouse (Takayama et al. 2011).

CF exhibits an inherent circadian rhythm resulting from the regulation of photosynthesis-associated gene expression with an approximately 24-h rhythm. Gould et al. (2009) used CF to measure circadian rhythm in a model plant, *Arabidopsis thaliana*. Figure 22.1 shows a CF image of 600 *Lactuca sativa* L. seedlings (Moriyuki and Fukuda 2016). This experiment was carried out using *L. sativa* seeds (cv. Frillice (leaf lettuce) and Batavia (leaf lettuce), fixed lines of *L. sativa* cultivars from Snow Brand Seed Co., Ltd., Sapporo, Japan), and the CF image was obtained using a complementary metal-oxide-semiconductor (CMOS) camera immediately after the blue light-emitting diode (LED) light was turned off on day 6 after sowing. To capture the circadian rhythm of CF, the measurement was repeated 6 times every 4 h on day 6 after sowing. In addition, to investigate the correlation between the circadian rhythm of CF and plant growth, we measured the fresh weight, W_i, of 153 Frillice and 148 Batavia at 17 days after sowing.

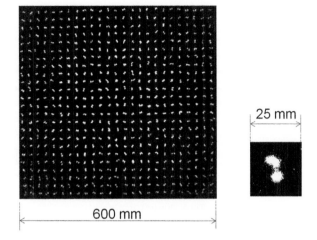

Fig. 22.1 CF image of 600 seedlings (*Lactuca sativa* L. cv. Frillice). The contrast was changed. (Moriyuki and Fukuda 2016)

Figure 22.2a, b shows the diurnal oscillations in the level of CF in both cultivars (Frillice and Batavia) from one morning to the next. All of the Frillice and Batavia exhibited a circadian rhythm and peak in CF in the evening. Plants were segregated into four groups based on fresh weight average, μ_w, and standard variation, σ_w, as follows: Group 1 ($\mu_w + \sigma_w < W_i$), Group 2 ($\mu_w < W_i \leq \mu_w + \sigma_w$), Group 3 ($\mu_w - \sigma_w < W_i \leq \mu_w$), and Group 4 ($W_i \leq \mu_w - \sigma_w$) (Fig. 22.2a, b). CF decreased with increasing weight over a certain threshold ($W_i > 7.6$ g) in Frillice; by contrast, Batavia exhibited no such tendency (Fig. 22.2c, d). In Frillice, when this plant is under a certain threshold value of CF, its growth is better. These data suggest that seedlings that use a large amount of light energy in photochemical reactions (which would decrease the CF) exhibit higher growth potential.

Figure 22.3a, b, d, e illustrates the relationship between W_i and the amplitude, A_i, and peak phase, φ_i, of circadian rhythm. No correlation was observed between these indices of circadian rhythm and W_i; that is, the correlation coefficient R was low. As shown in Fig. 22.3b, e, the center value and range of φ_i differed depending on cultivar. Hence, we also defined a baseline $\overline{\varphi_i}$ for peak phase φ_{si} to obtain phase φ'_i as a measure of environmental synchrony:

$$\varphi'_i = \overline{\varphi_i} + |\overline{\varphi_i} - \varphi_i|.$$

The $\overline{\varphi_i}$ values for Frillice and Batavia were 1.70π and 1.66π rads, respectively. Figure 22.3C and F shows the correlation between phase φ'_i and W_i, for which a weak correlation was observed in Batavia. Although correlations between each circadian index and growth were small, growth predictions could be improved by considering multiple indices of the circadian rhythm, as discussed in Sect. 22.2.

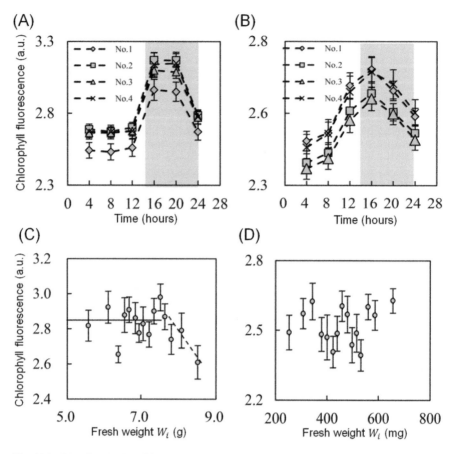

Fig. 22.2 Circadian rhythm of CF and the relationship between CF and fresh weight. Change in CF over the course of 1 day in Frillice (**a**) and Batavia (**b**), segregated into four categories. White and gray backgrounds indicate light and dark conditions, respectively. (**c, d**) Relationship between average fresh weight W_i and average CF for each 10 Frillice (**c**) and Batavia (**d**). Error bars indicate standard error (Moriyuki and Fukuda 2016)

22.3 High-Throughput Growth Prediction Systems Using Biological Rhythm

Plants that grow poorly and do not meet quality standards required for sale cause significant economic losses for plant factories (Kozai et al. 2015). Even in plants of the same variety and from the same seed lots, individual differences can lead to poor growth. Thus, identifying and culling low-grade seedlings at an early stage using so-called seedling diagnosis and selection technology is important to enhance the profitability of plant factories. This technology predicts growth using dynamic plant information from seedlings and disposes of seedlings that are predicted to grow poorly (Fukuda et al. 2011).

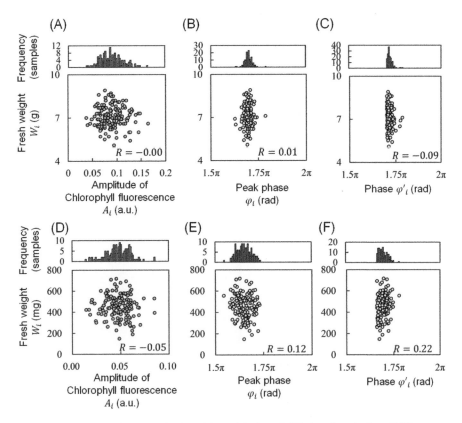

Fig. 22.3 Correlation between fresh weight and index of CF circadian rhythm. Frillice: **a–c**; Batavia: **d–f**. (**a, d**) Amplitude: A_i, (**b, e**) peak phase, φ_i, and (**c, f**) phase φ'_i. Each histogram shows the frequency (number of plant samples) for the index in horizontal axis. (Moriyuki and Fukuda 2016)

In large-scale plant factories, dynamic plant information values are statistically stable because over 1000 plants are analyzed every day. Therefore, the accuracy of growth predictions has improved as automatic data acquisition systems and databases to store dynamic plant data have been constructed. In general, multiple visual inspections of the leaf size, color, and shape of every seedling provide indices for assessing seedling growth in commercial factories.

We developed a high-throughput growth prediction methodology based on CF for *L. sativa* seedlings in a commercial large-scale plant factory at Osaka Prefecture University, which produces about 5000 *L. sativa* plants daily. The CF of each seedling was measured 6 times every 4 h at 6 days after sowing to characterize the circadian rhythm, as described in Sect. 22.1. Multiple dynamic variables, including leaf area and intensity of CF, were obtained via CF imaging. Finally, we assessed each variable as a predictor of growth and then combined the variables by machine learning to identify superior indices for seedling diagnosis.

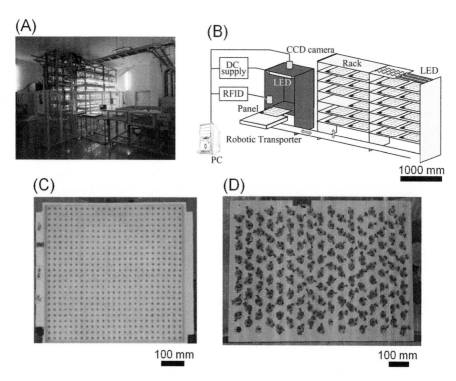

Fig. 22.4 Illustration of the high-throughput growth prediction system and individual panels. Photograph of the seedling diagnosis and selection system for *L. sativa* seedlings (**a**), diagram of the system (**b**), photograph of the greening panel in Batavia (**c**), and photograph of the raising panel in Frillice (**d**). (Moriyuki and Fukuda 2016)

22.3.1 Automatic CF Measurement System

In large-scale plant factories, seedling diagnosis and plant transplantation must be automated. Thus, we developed a seedling diagnosis system (Fig. 22.4a, b) capable of diagnosing over 7200 seedlings per day in such a factory. This seedling diagnosis system consists of a carrier robot for greening panels (Fig.22.4c), a seedling diagnosis apparatus, and a transplantation robot. The diagnosis apparatus consists of a dark box (900 mm wide, 900 mm deep, and 1200 mm high), a highly sensitive CMOS camera in the upper dark box, and blue LED panels (wavelength $\lambda = 470$ nm) in the dark box to excite the plant chlorophyll. In addition, the apparatus includes a PC-controlled CMOS camera, an LED controller, a radio frequency identifier (RFID) system, a digital input/output unit, and an automatic program for acquisition/analysis of leaf area, CF, and circadian rhythm data.

At the time of seedling diagnosis on day 6 after sowing, the greening panel carrier robot of the diagnosis system automatically carries a target panel to the dark box. As detailed in Sect. 22.1, seedlings are immediately illuminated with blue LED light for 2 s to excite chlorophyll, and the CF of each plant is then recorded using the CMOS camera. This measurement is repeated 6 times every 4 h. The RFID system is utilized to input the results of seedling diagnosis to the transplantation robot, and a digital input/output unit controls the opening and closing of the dark box shutter. Based on the results, the transplant robot automatically transplants only superior seedlings from the greening panel to the raising panel (Fig. 22.4d).

22.3.2 Growth Prediction System Based on Machine Learning

Neural networks are inherently capable of learning unknown nonlinear properties of data (Chen et al. 1990). In our system, plant growth prediction employs a neural network based on dynamic plant information obtained from measurements of leaf area and CF at 6 time points (e.g., at 4, 8, 12, 16, 20, and 24 h) and 4 circadian rhythm features. The circadian rhythm features are amplitude A_i, peak phase φ_i, average CF $<I_i>$, and the coefficient of the cosine curve approximation. The input layer of our neural network consisted of 16 types of dynamic plant information; the output layer consisted with fresh weight W_i at 11 days after measurement. In our experiment, fresh weight W_i (153 Frillice and 148 Batavia) was measured at 17 days after sowing. As training data in a backpropagation method, 70% of all plant data were used. We employed the neural network 40 times using several types of input data and estimated the average correlation coefficient R and standard error. Figure 22.5 shows the magnitude of the correlation coefficient $|R|$ between each index and W_i. White background in Fig. 22.5 indicates a single index; that is, single data points were used from leaf area at each time point, CF at each time point, amplitude A_i, peak phase φ_i, average CF $<I_i>$, and the determination coefficient. Gray background in Fig. 22.5 indicates multiple indices: leaf area (2, 3, and 6 points), CF (2, 3, and 6 points), circadian rhythm features (2, 3, and 4 kinds), and all dynamic plant indices. We defined 2 points of leaf area and CF as data acquired at 2 time points (12 and 24 h) and 3 points of leaf area as data acquired at 3 time points (8, 16, and 24 h). In addition, two types of circadian rhythm were defined by A_i and φ_i, three types were defined by A_i, φ_i, and $<I_i>$, and four types were defined by all of these parameters plus the determination coefficient. Multiple indices were better predictors of plant growth than single indices. Fukuda et al. (2011) reported improvements in plant productivity even with small values of $|R|$. Therefore, as suggested in Fig. 22.5, increasing the value of $|R|$ would lead to improved productivity.

The high-throughput growth prediction system we developed is capable of automatically identifying and selecting plants exhibiting poor growth based on biological rhythm data obtained at an early stage of growth. The system automatically measures leaf area, CF, and circadian rhythm parameters; improvements in growth prediction are integrated via machine learning.

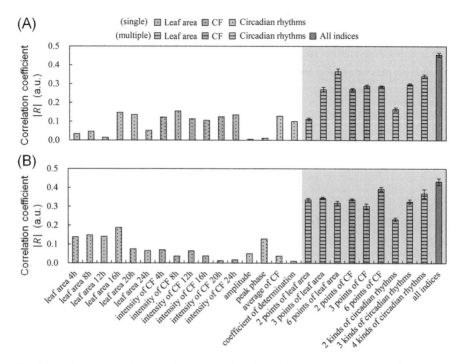

Fig. 22.5 Correlation between indices and fresh weight determined using machine learning in Frillice (**a**) and Batavia (**b**). White background indicates a single index, and gray background indicates multiple indices. Error bars indicate standard error. (Moriyuki and Fukuda 2016)

22.4 Detection of Transcriptome Biological Rhythms Using the Molecular Timetable Method

The quality of vegetables depends on plant metabolism, which is controlled by gene expression. The oscillation of diurnal gene expression results from both environmental cycles and regulation by the circadian clock. Regulation by the circadian clock ensures the stability of physiologic processes and enables adaptation to diurnal environmental cycles. Thus, it is important to measure circadian rhythms in assessing environmental regulation of vegetable quality.

Of the several methods available to measure circadian rhythm, transcriptome analysis (comprehensive gene expression analysis) is a particularly accurate approach. Transcriptome analysis involves global capture and analysis of gene transcripts and is widely used in studies of animals, plants, and insects (Scherf et al. 2000; Rifkin et al. 2003; Lister et al. 2008). In transcriptome analysis, ribonucleic acid (RNA) is extracted from the tissue of interest, and the sequences (or "reads") are determined using a sequencer (next-generation sequencing). All of the reads are then mapped to a reference sequence (the predicted transcriptome models in the National Center for Biotechnology Information (NCBI) database) using software such as Bowtie2 (Langmead and Salzberg 2012). The expression level of specific genes is then calculated from the amount of mapped reads obtained. By acquiring data over the course of time, it is possible to determine temporal patterns in gene expression. Due to decreased cost resulting from recent improvements in analytical equipment and technology, the use of transcriptome analysis in agriculture has increased (Tanigaki et al. 2017).

22.4.1 Effects of Environmental Variations on the Transcriptome

In predicting the phenome such as growth and nutritional quality using transcriptome analysis, gene annotation and sequence information (genetic information) are very important. Studies of plants for which there is a lot of genetic information available, such as tomato, can be conducted smoothly. As a representative example, Figs. 22.6 and 22.7 show the results of transcriptome analyses of gene expression in tomato leaves every 2 h for 48 h in a sunlight-type plant factory (Tanigaki et al. 2015). On the final day of sampling, it rained, and the environment changed drastically (Fig. 22.6). Nevertheless, gene expression showed stable periodicity (Fig. 22.7). Analysis of temporal patterns in gene expression revealed that 1516 genes exhibited 24-h expression patterns ($p < 0.05$; false discovery rates < 0.05). Furthermore, we found that the expression of key enzymes in major plant hormone pathways is periodic and under circadian rhythm control. Analysis of the expression pattern and effect of environmental conditions on the expression of stomata-related genes suggested that tomato plants do not open the stomata to inhibit transpiration and that production of abscisic acid is synchronous. This physiological information will help to design the environmental control in plant factories.

22.4.2 Analysis of Circadian Rhythm Using the Molecular Timetable Method

A method for measuring diurnal oscillations based on the collective behavior of genes (i.e., the molecular timetable method) was established to assay circadian rhythm (Ueda et al. 2004; Higashi et al. 2016a). In the molecular timetable method, "time-indicating genes" are selected first, and then the time of peak expression of each time-indicating gene (molecular peak time) is evaluated by fitting gene expression patterns and cosine curves. The time-indicating genes are arranged in the order of expression timeline, and then the gene expression level is presented as a scatter diagram at an arbitrary time. The body time (the phase of circadian rhythm) is estimated from the peak of the scatter diagram (Fig. 22.8).

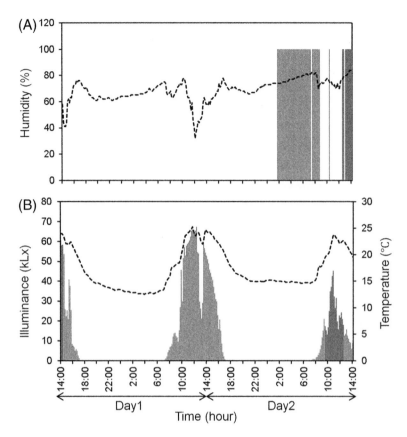

Fig. 22.6 Example of cultivation conditions in a sunlight-type plant factory at Ehime University, Japan. Tomato leaves were sampled every 2 h for 2 days (14:00–14:00 h). (**a**) Humidity (dash line) and periods of rainfall (gray areas). (**b**) Illuminance (gray areas) and temperature (dash line, second axis). (Tanigaki et al. 2015)

22.4.3 Estimation of the Body Time in Transcriptome

From the data of the transcriptome analysis of tomatoes at the sunlight-type plant factory, 143 genes were selected as time-indicating genes (Higashi et al. 2016a). As time proceeded, the peak position of sinusoidal-shaped plots shifted at regular intervals, and the time of this peak position (body time) was almost the same as the external time (sampling time) (Fig. 22.9). When the time-indicating genes were reselected only from among stress-response genes, body time could also be estimated using the molecular timetable method. Although these stress-response genes exhibited noisy profiles compared with those of the 143 standard time-indicating genes, the body time estimation was sufficiently accurate.

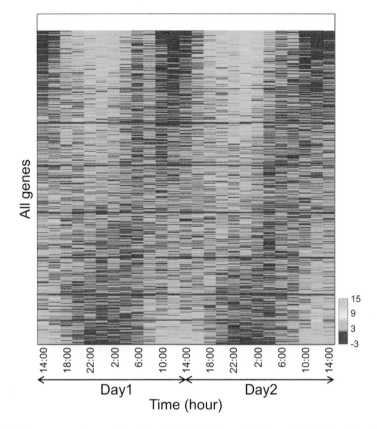

Fig. 22.7 Gene expression profile of tomato leaf in the sunlight-type plant factory described in Fig. 22.6. Gene expression is presented as log[2] values. White area at top indicates genes exhibiting low expression levels (log[2] values ≤ 0.1). (Tanigaki et al. 2015)

Similarly, from the analysis of transcriptome data for *L. sativa* (cv. Frillice) grown under two different light environments (constant light [LL] and 12-h light–12-h dark [LD] cycles) in an artificial light-type plant factory, 215 contigs (contig refers to a set of overlapping deoxyribonucleic acid (DNA) segments) were selected as time-indicating genes (Higashi et al. 2016b). These 215 contigs exhibited stable periodicity and periodic expression under both LL and LD conditions. The time of peak expression of the 215 contigs was almost the same as the external time (Fig. 22.9).

Genes or contigs exhibiting oscillating expression can be used as molecular biomarkers for determining optimal cycling of environmental conditions such as the light–dark ratio, temperature, and carbon dioxide (CO_2) concentration in plant factories. The advanced approach using plant circadian rhythm for optimization of plant cultivation conditions can also be applied to other cultivated plants for which minimal genetic information, so that this approach is available to plant factories and/or field agriculture as a specialized transcriptome analysis.

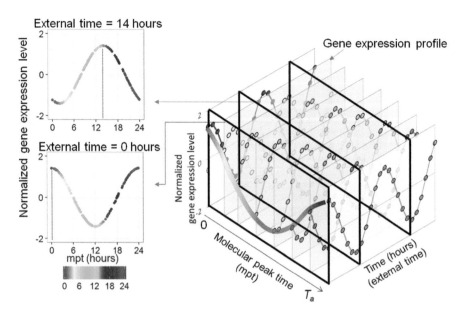

Fig. 22.8 Molecular timetable method. The sequential graphs on the right show the expression profiles of time-indicating genes arranged in order of molecular peak time (mpt). The scatter diagram on the left is a cross-sectional view obtained by cutting a graph from the right at a specific time

22.5 Control of Biological Rhythms in Plants

The circadian rhythm in plants regulates the timing of gene expression, which may be, for example, results in peak expressions of photosynthesis genes from early morning to noon, sugar transport genes from late afternoon to evening, and genes involved in fragrance production from late evening to early morning (Harmer et al. 2000). A key factor in the regulation of physiological processes by the endogenous circadian clock is circadian resonance (Dodd et al. 2005; Higashi et al. 2015), which matches the periodicity of the plant circadian rhythm to the external light-dark cycle and thereby maximizes aerial weight. For example, changing the period of the light-dark cycle from 24 h to 18, 20, and 22 h strongly suppressed the growth by approximately 50% in *L. sativa* (cv. Greenwave (leaf lettuce)) (Higashi et al. 2015). Other horticultural studies have also reported that appropriate periods of light-dark cycles provide maximum growth (Urairi et al. 2017). Therefore, the control technology of the circadian rhythms plays an important role on improvements of the productivity and plant quality.

Circadian rhythms are entrained systematically to external cycles through their indigenous responses, which depend upon the phase (body time) in which the environmental stimulus is received. Such phase-dependent responses can be described by phase response curves (PRCs), which plot shifts in the circadian rhythm phase elicited by environmental stimuli as a function of the stimulus phase

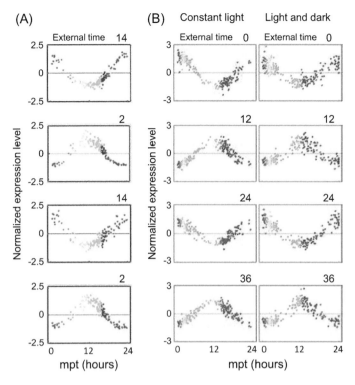

Fig. 22.9 Estimation of circadian phase using the molecular timetable method. (**a**) Tomato in a sunlight-type plant factory (Higashi et al. 2016a). (**b**) *L. sativa* (cv. Frillice) in an artificial light-type plant factory. Numbers on the top right of the panels indicate external time (sampling time). Gene expression levels are normalized. (Higashi et al. 2016b)

(Fukuda et al. 2008, 2013; Masuda et al. 2017). PRCs are essential for analyzing physiologic adaptations to the environment (e.g., recovery from jet lag in mammals and photosynthetic entrainment in plants). In addition, the plant circadian system is composed of a large number of self-sustaining cellular oscillators that synchronize with each other to produce a strong output rhythm (Ukai et al. 2012). Desynchronization of the oscillators diminishes the output circadian rhythm, whereas their resynchronization recovers the rhythm (Fukuda et al. 2013). For healthy plant growth, the activity of cellular oscillators must remain synchronous in order to sustain stable circadian rhythm as well as synchronous to environmental cycles.

Figure 22.10 shows the circadian rhythm response to the application of a 2-h pulse of darkness under constant light conditions in a model transgenic *Arabidopsis thaliana Circadian Clock Associated 1::luciferase* (*CCA1::LUC*) plants, in which a modified firefly luciferase (*LUC*) gene has been fused to the *CIRCADIAN CLOCK-ASSOCIATED1* (*CCA1*) promoter. This *CCA1::LUC* plants exhibited only very weak bioluminescence, which was proportional to the expression rate of the clock

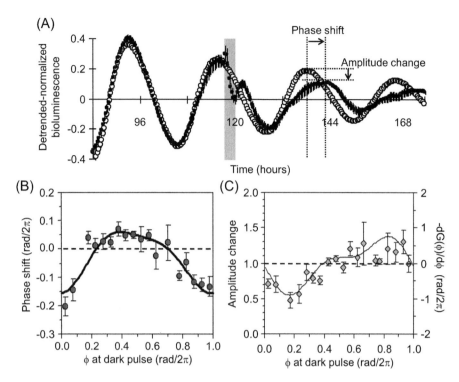

Fig. 22.10 Circadian response of *Arabidopsis thaliana CCA1::LUC* to application of a 2-h dark pulse under constant light conditions. (**a**) Bioluminescence signals were averaged over plants with and without pulse perturbation. The gray rectangle indicates the 2-h dark pulse. (**b**) Phase response curve for the dark pulse, indicating the shift in the plant circadian rhythm phase against phase ϕ, during which the dark pulse was applied. The solid line $G(\phi)$ shows the fitted curve. (**c**) Amplitude response curve for the dark pulse, indicating the amplitude change in the plants against phase ϕ of the pulse perturbation. The solid line represents $-dG(\phi)/d\phi$. (Fukuda et al. 2013)

gene *CCA1*. The dark pulse induced a phase shift and amplitude change in the circadian rhythm. In Fig. 22.10B, the PRC is drawn with respect to the circadian phase ϕ, where $\phi = 0$ and π rad correspond to subjective dawn and dusk, respectively. The PRC for the 2-h dark pulse was fitted with a sinusoidal curve $G(\phi)$ (Fukuda, et al. 2008). Figure 22.10c shows an amplitude response curve for the same plants. The dark pulse applied at $\phi = 0.2$ rad/2π (body time approximately noon) strongly suppressed the circadian oscillation (amplitude decreasing), whereas it enhanced the circadian oscillation (amplitude increasing) at $\phi = 0.85$ rad/2π (body time approximately predawn).

The change in the amplitude of circadian rhythm was caused primarily by the synchronization of the population of self-sustaining cellular circadian rhythms. The effect of the pulse perturbation on the synchronized state depends upon the negative slope of the PRC (Arai and Nakao 2008). If $-dG(\phi)/d\phi < 0$ or $-dG(\phi)/d\phi > 2$, synchronization is weakened, whereas synchronization is strengthened if $0 < -dG$

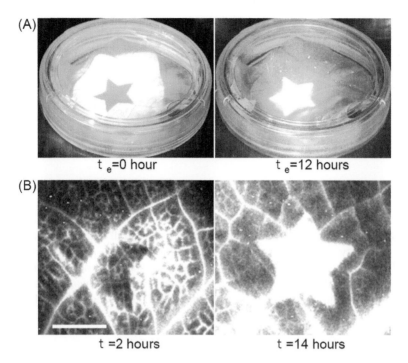

Fig. 22.11 Circadian rhythms with star-shaped initial conditions in an *AtCCA1::LUC L. sativa* leaf induced by spatiotemporal illumination using an LCD projector. (**a**) Bright and dark star pattern illumination. (**b**) Bioluminescence images; the star-shaped region was at subject dusk (left) and subject dawn (right) (scale bar: 10 mm). (Seki et al. 2015)

$(\phi)/d < 2$. The negative slope of $-dG(\phi)/d\phi$ was indeed consistent with the amplitude response curve (Fig. 22.10c).

Figure 22.11 shows entrainment of the cellular circadian rhythm in a *L. sativa* leaf brought about by spatially controlled illumination (Seki et al. 2015). The alternating bright and dark star-shaped images were applied to a *L. sativa* leaf over a 12-h period for 3 days using a liquid crystal display (LCD) projector (Fig. 22.11a). In this study, measurement of circadian rhythm was carried out using transgenic *L. sativa* (*Lactuca sativa* L. cv. Greenwave *AtCCA1::LUC*). The bioluminescence exhibited a circadian rhythm (Higashi, et al. 2014), which could be observed in almost all cells of the leaves, even under constant dark conditions, as reported by Ukai et al. (2012). Figure 22.11b shows bioluminescence images obtained under constant darkness (DD). White and black star-shaped bioluminescence patterns were observed, indicating that the phases of the star-shaped patterns were almost the inverse of each other. These results demonstrate the possibility of controlling cellular circadian rhythms via spatially controlled illumination.

The spatial control of circadian rhythms has horticultural significance. The circadian clock gene is upstream of the *FLOWERING LOCUS T* (*FT*) gene, which is involved in floral induction (Endo et al. 2014). Flowering can be controlled by regulating expression of the *FT* gene via local projection, such as changing the light quality or period, depending on the individual organs. The influence on plant growth also varies with wavelength (Lin et al. 2013). Therefore, spatially controlled illumination with optimal wavelengths and day lengths for significant organs could increase plant growth and quality.

22.6 Conclusion

Circadian rhythms are observed in all cultivation processes from germination to harvest in plant factories. Since the significance of circadian rhythms in physiological processes, circadian rhythms are useful indices to assess the growth status. Seedling diagnosis using the circadian rhythm of chlorophyll fluorescence improves the productivity and stability in a plant factory. In addition, precise detection of circadian rhythms in comprehensive gene expression data by the molecular timetable method provides the knowledge about optimization of environments. The control methodology of circadian rhythm is also applicable to precise regulation of light and temperature to maximize the plant production. Further research and development about detection and utilization of circadian rhythms would provide key technologies for enhancing the plant growth and quality in plant factories.

References

Arai K, Nakao H (2008) Phase coherence in an ensemble of uncoupled limit-cycle oscillators receiving common Poisson impulses. Phys Rev E 77(3):036218(1–17)

Chen S, Billings SA, Grant PM (1990) Non-linear system identification using neural networks. Int J Control 51(6):1191–1214

Daley PF, Raschke K, Ball JT, Berry JA (1989) Topography of photosynthetic activity of leaves obtained from video images of chlorophyll fluorescence. Plant Physiol 90(4):1233–1238

Dodd AN, Salathia N, Hall A, Ke'vei E, To'th R, Nagy F, Hibberd JM, Millar AJ, Webb AAR (2005) Plant circadian clocks increase photosynthesis, growth, survival, and competitive advantage. Science 22:630–633

Endo M, Shimizu H, Nohales MA, Araki T, Kay SA (2014) Tissue-specific clocks in *Arabidopsis* show asymmetric coupling. Nature 515:419–422

Fukuda H, Uchida Y, Nakamichi N (2008) Effect of a dark pulse under continuous red light on the *Arabidopsis thaliana* circadian rhythm. Environ Control Biol 46(2):123–128

Fukuda H, Ichino T, Kondo T, Murase H (2011) Early diagnosis of productivity through a clock gene promoter activity using a luciferase bioluminescence assay in *Arabidopsis thaliana*. Environ Control Biol 49(2):51–60

Fukuda H, Murase H, Tokuda IT (2013) Controlling circadian rhythms by dark-pulse perturbations in *Arabidopsis thaliana*. Sci Rep 3:1533(1–7)

Gould PD, Diaz P, Hogben C, Kusakina J, Salem R, Hartwell J, Hall A (2009) Delayed fluorescence as a universal tool for the measurement of circadian rhythms in higher plants. Plant J 58 (5):893–901

Govindjee (1995) Sixty-three years since Kautsky: chlorophyll α fluorescence. Aust J Plant Physiol 22:131–160

Harmer SL, Hogenesch JB, Straume M, Chang HS, Han B, Zhu T, Wang X, Kreps JA, Kay SA (2000) Orchestrated transcription of key pathways in *Arabidopsis* by the circadian clock. Science 290:2110–2113

Higashi T, Kamitamari A, Okamura N, Ukai K, Okamura K, Tezuka T, Fukuda H (2014) Characterization of circadian rhythms through a bioluminescence reporter assay in *Lactuca sativa* L. Environ Control Biol 52(1):21–27

Higashi T, Nishikawa S, Okamura N, Fukuda H (2015) Evaluation of growth under non-24 h period lighting conditions in *Lactuca sativa* L. Environ Control Biol 53(1):7–12

Higashi T, Tanigaki Y, Takayama K, Nagano AJ, Honjo MN, Fukuda H (2016a) Detection of diurnal variation of tomato transcriptome through the molecular timetable method in a sunlight-type plant factory. Front Plant Sci 7(87):1–9

Higashi T, Aoki K, Nagano AJ, Honjo MN, Fukuda H (2016b) Circadian oscillation of the lettuce transcriptome under constant light and light-dark conditions. Front Plant Sci 7:1114(1–10)

Kozai T, Niu G, Takagaki M (eds) (2015) Plant factory, 1st edition -an indoor vertical farming system for efficient quality for production. Academic, Cambridge, MA

Krause GH, Weis E (1991) Chlorophyll fluorescence and photosynthesis: the basics. Annu Rev Plant Physiol Plant Mol 42:313–349

Langmead B, Salzberg SL (2012) Fast gapped-read alignment with bowtie 2. Nat Methods 9 (4):357–359

Lin KH, Huang MY, Huang WD, Hsu MH, Yang ZW (2013) The effects of red, blue, and white light-emitting diodes on the growth, development, and edible quality of hydroponically grown lettuce (*Lactuca sativa* L var capitata). Sci Hortic 150:86–91

Lister R, O'Malley R, Tonti-Filippini J, Gregory B, Berry C, Millar A, Ecker J (2008) Highly integrated single-base resolution maps of the epigenome in *Arabidopsis*. Cell 133(3):523–536

Masuda K, Kitaoka R, Ukai K, Tokuda IT, Fukuda H (2017) Multicellularity enriches the entrainment of *Arabidopsis* circadian clock. Sci Adv 3(10):e1700808

Maxwell K, Johnson GN (2000) Chlorophyll fluorescence - a practical guide. J Exp Bot 51:659–668

Moriyuki S, Fukuda H, (2016) High-throughput growth prediction for *Lactuca sativa* L. seedlings using chlorophyll fluorescence in a plant factory with artificial lighting. Front Plant Sci 7:394 (1–8)

Nichol CJ, Pieruschka R, Takayama K, Förster B, Kolber Z, Rascher U, Grace J, Robinson SA, Pogson B, Osmond B (2012) Canopy conundrums: building on the biosphere 2 experience to scale measurements of inner and outer canopy photoprotection from the leaf to the landscape. Funct Plant Biol 39(1):1–24

Omasa K, Shimazaki K, Aiga I, Larcher W, Onoe M (1987) Image analysis of chlorophyll fluorescence transients for diagnosing the photosynthetic system of attached leaves. Plant Physiol 84(3):748–752

Rifkin SA, Kim J, White KP (2003) Evolution of gene expression in the Drosophila melanogaster subgroup. Nat Genet 33(2):138–144

Scherf U, Ross D, Waltham M, Smith L, Lee J, Tanabe L, Kohn K, Reinhold W, Myers T, Andrews D, Scudiero D, Eisen M, Sausville E, Pommier Y, Botstein D, Brown P, Weinstein J (2000) A gene expression database for the molecular pharmacology of cancer. Nat Genet 24 (3):236–244

Seki N, Ukai K, Higashi T, Fukuda H (2015) Entrainment of cellular circadian rhythms in *Lactuca sativa* L. leaf by spatially controlled illuminations. J Biosens Bioelectron 6(4):186(1–186(5

Takayama K, Nishina H (2009) Chlorophyll fluorescence imaging of the chlorophyll fluorescence induction phenomenon for plant health monitoring. Environ Control Biol 47(2):101–109

Takayama K, Sakai Y, Oizumi T, Nishina H (2010) Assessment of photosynthetic dysfunction in a whole tomato plant with chlorophyll fluorescence induction imaging. Environ Control Biol 48 (4):151–159

Takayama K, Nishina H, Arima S, Hatou K, Ueka Y, Miyoshi Y (2011) Early detection of drought stress in tomato plants with chlorophyll fluorescence imaging -practical application of the speaking plant approach in a greenhouse. Preprints of the 18th IFAC World Congress, pp 1785–1790

Takayama K, Hirota R, Takahashi N, Nishina H, Arima S, Yamamoto K, Sakai Y, Okada H (2014) Development of chlorophyll fluorescence imaging robot for practical use in commercial greenhouse. Acta Hort (1037):671–676

Tanigaki Y, Higashi T, Takayama K, Nagano AJ, Honjo MN, Fukuda H (2015) Transcriptome analysis of plant hormone-related tomato (*Solanum lycopersicum*) genes in a sunlight-type plant factory. PLoS One 10(12):e0143412

Tanigaki Y, Higashi T, Takayama K, Nagano AJ, Honjo MN, Fukuda H (2017) Transcriptome analysis of a cultivar of green perilla (*Perilla frutescens*) using genetic similarity with other plants via public databases. Environ Control Biol 55(2):77–83

Ueda HR, Chen W, Minami Y, Honma S, Honma K, Iino M, Hashimoto S (2004) Molecular-timetable methods for detection of body time and rhythm disorders from single-time-point genome-wide expression profiles. Proc Natl Acad Sci USA 101(31):11227–11232

Ukai K, Inai K, Nakamichi N, Ashida H, Yokota A, Hendrawan Y, Murase H, Fukuda H (2012) Traveling waves of circadian gene expression in lettuce. Environ Control Biol 50(3):237–246

Urairi C, Shimizu H, Nakashima H, Miyasaka J, Ohodoi K (2017) Optimization of light-dark cycles of *Lactuca sativa* L. in plant factory. Environ Control Biol 55(2):85–91

Chapter 23
Automated Characterization of Plant Growth and Flowering Dynamics Using RGB Images

Wei Guo

Abstract Monitoring and measuring the phenotypic traits of plants to understand their growth using computer vision techniques are becoming increasingly important in agriculture. These techniques are expected to replace destructive and labor-intensive traditional plant investigation methods. In this chapter, we first introduce several innovative computer vision-based techniques developed to measure the growth dynamics of plants under outdoor conditions, with the aim of understanding plant formation under dynamic interaction between genotype and environment. Then, the potential possibilities of adapting such techniques to plant factory industries to maximize productivity are discussed.

Keywords Plant phenotyping · Image processing · Machine learning

23.1 Introduction

As an efficient and nondestructive measurement approach, computer vision-based techniques are becoming readily available for both indoor and outdoor agronomic applications, especially as a powerful tool for high-throughput plant phenotyping in breeding industries (Houle et al. 2010; Furbank and Tester 2011; Minervini et al. 2015). For example, in order to help researchers and breeders understand the performance of crop genetic resources, image-based methods for extracting and estimating phenotypic characteristics such as canopy coverage, leaf area index, and plant height (Guo et al. 2013; Yeh et al. 2014; Jiang et al. 2015; Liu et al. 2015) and specific plant organs such as leaf, fruit, flower, and grain (Yoshioka et al. 2004; Remmler and Rolland-Lagan 2012; Yamamoto et al. 2014, 2015; Guo et al. 2015; Lu et al. 2017) have been proposed. Many researchers have indicated that with

---

W. Guo (✉)
International Field Phenomics Research Laboratory, Institute for Sustainable Agro-ecosystem Services (ISAS), Graduate School of Agricultural and Life Sciences, The University of Tokyo, Nishi-Tokyo, Tokyo, Japan
e-mail: guowei@isas.a.u-tokyo.ac.jp

computer vision-based solutions, phenotyping of crops under outdoor conditions is more challenging than under indoor conditions. This is because even for the same region of interest (ROI)—i.e., an individual crop or an organ of a crop—under outdoor conditions, the images acquired by digital cameras are affected by a wide variety of environmental parameters, such as sunlight, wind, and rainfall. Consequently, the development of true color (RGB, red green blue) image processing algorithms for outdoor conditions requires highly specialized knowledge in both agronomic and engineering fields. We believe an algorithm developed for outdoor environmental crop analysis can be easily adapted for indoor use such as in a plant factory environment.

In this chapter, we first introduce several innovative image analysis techniques and applications for monitoring and quantifying crop dynamics under natural field conditions using only RGB images. Then, we discuss the potential possibilities of adapting those techniques to future plant factory industries.

23.2 Computer Vision Techniques Under Outdoor Conditions

To develop a successful computer vision solution for plant research, four fundamental steps need to be considered: image preprocessing, image segmentation, ROI detection, and phenotypic traits extraction, as shown in Fig. 23.1. Compared with images taken under indoor light conditions, for images take under outdoor light conditions, it is very difficult to have good "image segmentation" and "ROI detection" performance because factors such as lighting variations, shadows, and backgrounds are not controlled.

23.2.1 Image Segmentation

Image segmentation is a fundamental step in virtually all computer vision-based solutions. In this step, an image is separated into regions that are homogeneous with respect to some characteristics, such as color or texture. The performance of the

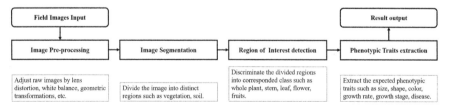

Fig. 23.1 Workflow of computer vision-based solution for plant research

Fig. 23.2 Example of the performance of the "DTSM" vegetation segmentation method under complicated field conditions

segmentation affects future processes such as detection and identification of the ROI. In plant-related research under field conditions, segmentation of regions containing vegetation is also an important research target, and several methods have been proposed. The proposed methods include traditional color space-based segmentation (Philipp and Rath 2002; Panneton and Brouillard 2009; Liu et al. 2012) and new color space-based segmentation (Woebbecke et al. 1995; Meyer and Neto 2008). However, most of the proposed methods require threshold processing of each image, which is not efficient for large numbers of time series images. A powerful new machine learning-based approach for segmentation of vegetation from the background in images taken outdoor that utilizes a series of images taken automatically regardless of light conditions has been proposed. The proposed method, called the "Illumination invariant Decision tree based vegetation segmentation model (DTSM)" (Guo et al. 2013), exhibited highly robust segmentation performance under varying outdoor light conditions for wheat. Further, its segmentation accuracy, particularly under sunny condition, is much better than that of previously proposed segmentation methods. Figure 23.2 shows an example of the performance of the "DTSM" vegetation segmentation method under complicated field conditions.

23.2.2 Characterizing Plant Growth by Canopy Coverage Estimation

"Plant canopy coverage ratio" is a well-known parameter that is indicative of plant growth. It is usually defined as the percentage of the orthogonal projection area relative to the area of crop foliage in the horizontal plane. This ratio is reported to be highly correlated with leaf area index (LAI), canopy light interception, nitrogen content, and crop yield (Fukushima et al. 2003; Campillo et al. 2008; Takahashi et al. 2012). As the proposed DTSM method provides efficient extraction of crop vegetation regions from images taken under various field conditions, the canopy coverage ratio within a given photograph can be calculated easily by dividing the total number

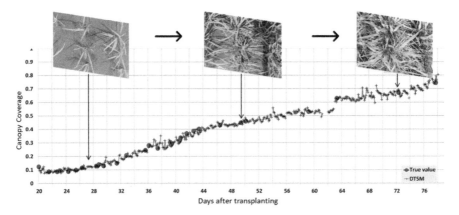

Fig. 23.3 Example of canopy coverage evaluation using EasyPCC

of pixels in the vegetation segments of the image by the total number of pixels in the entire image. Based on the algorithm of "DTSM," a software application for measurement of the canopy coverage ratio applicable to time series images taken under various light conditions, called "Easy Plant Coverage Calculation tool (EasyPCC)" (Fig. 23.3), has also been released for research use (Guo et al. 2017). To evaluate its ability to estimate plant coverage ratio, "EasyPCC" was applied to image datasets of paddy rice acquired daily from 20 days after transplanting to one week before heading from about 08:00 to 16:00 local time in a field at the Institute for Sustainable Agro-ecosystem Services, University of Tokyo. Figure 23.3 shows that the coverage ratio estimation by the tool had an accuracy of 99%. By using an approach such as this, the growth curve of crops could be easily produced.

23.2.3 Characterizing the Flowering Dynamics of Paddy Rice

Flowering is one of the most important phenotypic characteristics of paddy rice as it defines the critical growing stage of rice. Diurnal timing within a day is very important as heat affects pollen fertility and pollination efficiency; high heat results in decreasing yield and degradation of its quality. Although methods such as the color-based method for lesquerella and the spectral reflectance-based method for winter wheat (Scotford and Miller 2004; Thorp and Dierig 2011) have been proposed to identify flowers under natural conditions, these are sufficient for paddy rice flowering identification. Consequently, we proposed an algorithm that combines the local feature descriptor scale-invariant feature transform (SIFT), the image representation method bag of visual words (BoVWs), and the machine learning model support vector machine (SVM) to detect the flowering panicles of paddy rice in normal RGB images taken under natural field conditions. The proposed method is based on generic object recognition technology, which is still challenging in

Fig. 23.4 Example of flowering detection and counting of paddy rice; red dots indicate the detected flowering panicles (planting density: 12 plants/m^2)

machine vision. The method has an accuracy of approximately 80% and efficiently estimates the flowering stage and flowering timing of field planted rice (Guo et al. 2015). Further, it has been validated on a new dataset collected from a field in which a popular Japanese paddy variety called *Koshihikari* was planted with three different plant densities, as shown in Figs. 23.4 and 23.5.

The performance of the proposed method was evaluated by monitoring the diurnal/daily flowering pattern and the flowering extent of paddy rice during the flowering period (Figs. 23.4 and 23.5).

23.3 Potential Possible Usage of Image Analysis Techniques by Future Plant Factories

23.3.1 Characterizing the Growth Dynamics of Leafy Vegetables

In plant factories, especially those that provide a fully controlled growth environment that includes artificial lighting, temperature, and humidity, high quality and stable images can be acquired easily. This facilitates very detailed measurement of crops such as leafy vegetables. For example, Fig. 23.6 shows detailed measurement of leafy vegetables taken in the initial growth stage. Each individual plant was measured with several parameters, such as area, canopy coverage, and other morphological features; the growth dynamics of each seedling could be observed easily.

23.3.2 Detecting and Counting Fruits

Crop load, a quantitative parameter that is generally defined as the number of fruits per plant, is a very important phenotypic trait of fruits and vegetables as it directly

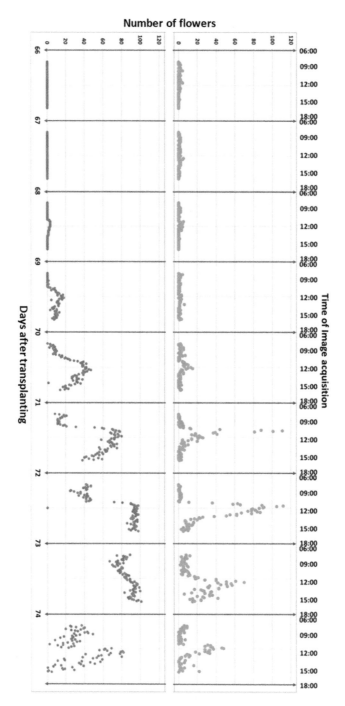

Fig. 23.5 Comparison of manually and automatically determined numbers of flowers. Upper image: number of blocks adjudged to contain the flowering parts of panicles. Lower mage: number of visually counted flowering panicles. The images were acquired every 5 min from 08:00 to 16:00 during the flowering period between days 66 and 75 after transplanting

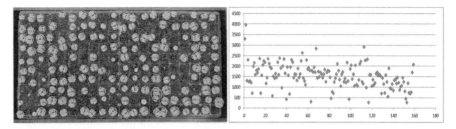

Fig. 23.6 Example of detailed measurement taken of the growth characteristics of crop seedlings

Fig. 23.7 Example of tomato detection and counting (note: green, red, and overlapped tomatoes are all counted)

affects the quality and quantity of the products. Both breeders and farmers are seeking ways to understand crop load levels automatically in order to reduce labor costs associated with cultivation management and harvesting and to optimize the amount of materials required such as fertilizers and agricultural chemicals. The flowering detection method discussed above can be used to detect and count mature and immature fruits grown in solar-type plant factories. This would make long-term prediction of fluctuations in yield possible (Fig. 23.7).

23.4 Conclusion and Discussion

In this chapter, we introduced several computer vision-based solutions for characterizing the growth dynamics of plants under natural outdoor field conditions. As only RGB images are needed, using these solutions, the growth of plants can be monitored at a very low cost, bringing automatic and efficient plant management to the agronomic industry. We believe the solutions developed for outdoor environments can be easily adapted to plant factory environments. In particular, for plant factories with artificial lighting, in which light, air, and temperature are fully controlled (Kozai and Niu 2015), the growth situation could be automatically monitored and measured via images in real time. Nevertheless, the information that can be obtained from RGB images is limited, invisible phenotypic traits such as photosynthesis, chlorophyll, stresses, and early diseases cannot be measured by the solutions introduced here. However, by integrating specific spectral cameras, such as multiple/hyperspectral, fluorescence, and thermal cameras, especially under controlled lighting conditions, it is already possible to monitor those invisible but very important growth dynamics of plants. In future plant factories, multi-camera systems are expected to enable the collection of both visible and invisible plant growth information in real time. This will facilitate maintenance of the resources necessary for plant growth, such as light energy, water, CO_2, and inorganic fertilizer (Kozai 2013), at the most appropriate values, resulting in maximum production with minimum energy input.

Acknowledgments Part of the approaches introduced in this chapter are partially funded by the CREST Program "Knowledge Discovery by Constructing AgriBigData" and the SICORP Program "Data Science-based Farming Support System for Sustainable Crop Production under Climatic Change," Japan Science and Technology Agency.

References

Campillo C, Prieto M, Daza C (2008) Using digital images to characterize canopy coverage and light interception in a processing tomato crop. Hortscience 43:1780–1786
Fukushima A, Kusuda O, Furuhata M (2003) Relationship of vegetation cover ratio to growth and yield in wheat. Rep Kyushu Branch Crop Sci Soc Japan:33–35
Furbank RT, Tester M (2011) Phenomics—technologies to relieve the phenotyping bottleneck. Trends Plant Sci 16:635–644. https://doi.org/10.1016/j.tplants.2011.09.005
Guo W, Rage UK, Ninomiya S (2013) Illumination invariant segmentation of vegetation for time series wheat images based on decision tree model. Comput Electron Agric 96:58–66. https://doi.org/10.1016/j.compag.2013.04.010
Guo W, Fukatsu T, Ninomiya S (2015) Automated characterization of flowering dynamics in rice using field-acquired time-series RGB images. Plant Methods 11:7. https://doi.org/10.1186/s13007-015-0047-9
Guo W, Zheng B, Duan T et al (2017) EasyPCC: benchmark datasets and tools for high-throughput measurement of the plant canopy coverage ratio under field conditions. Sensors (Switzerland) 17:1–13. https://doi.org/10.3390/s17040798
Houle D, Govindaraju DR, Omholt S (2010) Phenomics: the next challenge. Nat Rev Genet 11:855–866. https://doi.org/10.1038/nrg2897
Jiang N, Yang W, Duan L et al (2015) A nondestructive method for estimating the total green leaf area of individual rice plants using multi-angle color images. J Innov Opt Health Sci 8:1550002. https://doi.org/10.1142/S1793545815500029
Kozai T (2013) Resource use efficiency of closed plant production system with artificial light: concept, estimation and application to plant factory. Proc Jpn Acad Ser B Phys Biol Sci 89:447–461. https://doi.org/10.2183/pjab.89.447
Kozai T, Niu G (2015) Plant factory as a resource-efficient closed plant production system. Elsevier Inc
Liu Y, Mu X, Wang H, Yan G (2012) A novel method for extracting green fractional vegetation cover from digital images. J Veg Sci 23:406–418. https://doi.org/10.1111/j.1654-1103.2011.01373.x

Liu T, Wu W, Chen W et al (2015) Automated image-processing for counting seedlings in a wheat field. Precis Agric 17:392. https://doi.org/10.1007/s11119-015-9425-6

Lu H, Cao Z, Xiao Y et al (2017) Towards fine-grained maize tassel flowering status recognition: dataset, theory and practice. Appl Soft Comput 56:34–45. https://doi.org/10.1016/j.asoc.2017.02.026

Meyer GE, Neto JC (2008) Verification of color vegetation indices for automated crop imaging applications. Comput Electron Agric 63:282–293

Minervini M, Scharr H, Tsaftaris SA (2015) Image analysis: the new bottleneck in plant phenotyping [applications corner]. IEEE Signal Process Mag 32:126–131. https://doi.org/10.1109/MSP.2015.2405111

Panneton B, Brouillard M (2009) Colour representation methods for segmentation of vegetation in photographs. Biosyst Eng 102:365–378. https://doi.org/10.1016/j.biosystemseng.2009.01.003

Philipp I, Rath T (2002) Improving plant discrimination in image processing by use of different colour space transformations. Comput Electron Agric 35:1–15. https://doi.org/10.1016/S0168-1699(02)00050-9

Remmler L, Rolland-Lagan AG (2012) Computational method for quantifying growth patterns at the adaxial leaf surface in three dimensions. Plant Physiol 159:27–39. https://doi.org/10.1104/pp.112.194662

Scotford IM, Miller PCH (2004) Estimating tiller density and leaf area index of winter wheat using spectral reflectance and ultrasonic sensing techniques. Biosyst Eng 89:395–408. https://doi.org/10.1016/j.biosystemseng.2004.08.019

Takahashi K, Rikimaru A, Sakata K, Endou S (2012) A study of the characteristic of the observation angle on the terrestrial image measurement of paddy vegetation cover. Japan Soc Photogramm Remote Sens 50:367–371 (In Japanese)

Thorp KR, Dierig DA (2011) Color image segmentation approach to monitor flowering in lesquerella. Ind Crop Prod 34:1150–1159. https://doi.org/10.1016/j.indcrop.2011.04.002

Woebbecke DM, Meyer GE, Von BK, Mortensen DA (1995) Color indices for weed identification under various soil, residue, and lighting conditions. Trans ASAE 38:259–269. https://doi.org/10.13031/2013.27838

Yamamoto K, Guo W, Yoshioka Y, Ninomiya S (2014) On plant detection of intact tomato fruits using image analysis and machine learning methods. Sensors (Basel) 14:12191–12206. https://doi.org/10.3390/s140712191

Yamamoto K, Ninomiya S, Kimura Y et al (2015) Strawberry cultivar identification and quality evaluation on the basis of multiple fruit appearance features. Comput Electron Agric 110:233–240. https://doi.org/10.1016/j.compag.2014.11.018

Yeh YHF, Lai TC, Liu TY et al (2014) An automated growth measurement system for leafy vegetables. Biosyst Eng 117:43–50. https://doi.org/10.1016/j.biosystemseng.2013.08.011

Yoshioka Y, Iwata H, Ohsawa R, Ninomiya S (2004) Quantitative evaluation of flower colour pattern by image analysis and principal component analysis of Primula sieboldii E. Morren. Euphytica 139:179–186. https://doi.org/10.1007/s10681-004-3031-4

Chapter 24
Toward Nutrient Solution Composition Control in Hydroponic System

Toyoki Kozai, Satoru Tsukagoshi, and Shunsuke Sakaguchi

Abstract The concept and method for estimating the uptake rate of water and ions by hydroponically grown plants based on the water and ion balance in the culture beds are proposed. The method is also used to estimate the transpiration rate and the increase in fresh weight of plants. The concept of estimating the ion concentration of NH_4^+, Mg^{2+}, Cl^-, PO_4^{3-} ($H_2PO_4^-$), and SO_4^{2-} using the measured data on NO_3^-, K^+, Ca^{2+}, Na^+, EC(electric conductivity), and pH is described. EC of the nutrient solution is related to the total sum of the Eq (equivalent) concentration of all types of ions.

Keywords EC · Equivalent · Ion-selective sensor · Ion balance · Respiration rate · Valence · Water balance

24.1 Introduction

From the time that large-scale hydroponics was first developed for commercial production of vegetables in the 1940s, the main control variables of the nutrient solution have been EC, pH (indicating acidity or alkalinity of the solution), temperature, and the flow rate of nutrient solution into the culture beds. EC is an indicator of the total sum of the ion equivalent concentration (Eq kg^{-1}) (see Chap. 12 for details) of the nutrient solution.

T. Kozai (✉)
Japan Plant Factory Association (NPO), Kashiwa, Chiba, Japan
e-mail: kozai@faculty.chiba-u.jp

S. Tsukagoshi
Center for Environment, Health and Field Sciences, Chiba University, Kashiwa, Chiba, Japan
e-mail: tsukag@faculty.chiba-u.jp

S. Sakaguchi
PlantX Corp, Kashiwa, Chiba, Japan
e-mail: sakaguchi@plantx.co.jp

In this chapter, a nutrient solution control algorithm for the next-generation hydroponics is discussed, and a basic control procedure is proposed based on the water and ion balance in the hydroponic culture beds. Here, the nutrient film technique (NFT) system and deep flow technique (DFT) system (Son et al. 2015) are assumed as the hydroponic systems, because they are relatively simple in structure and are the most popular ones used in plant factories with artificial lighting (PFAL) in many countries. In these types, evaporation from the culture beds is often negligibly small. As the next step, the concept and methodology discussed below can be applied, after some modification, in other types of hydroponic systems such as the drip (trickle) irrigation system and the ebb and flow system.

24.2 Current Limitation of Nutrient Solution Control in Hydroponics

One of the advantages of hydroponics is the ability to control the supply rate of the raw water (groundwater, rainwater, or city water), stock nutrient solution with a predetermined nutrient ion composition and pH control agents to the nutrient solution tank. In many hydroponic systems, it is also possible to control or monitor the dissolved oxygen concentration (DO) and the temperature of the nutrient solution. The intention of controlling these factors is to control the rate of water and nutrient ion uptake, transpiration, photosynthesis, and consequently the plant growth and quality of produce. The supply rate of raw water, stock solution, and pH control agents is controlled to maintain the set points of the EC, pH, and level of nutrient solution in the nutrient solution tank.

With this control method, the concentration of each type of ion in the nutrient solution tank fluctuates somewhat with time, and it is difficult to stabilize the nutrient composition and its strength, even when the EC and pH are well controlled at their set points. The main reasons for the fluctuation in ion concentration even under constant EC and pH are given below:

1. The formula (nutrient element composition and strength) is not suitable for the plants under the particular environment.
2. The formula is not adjusted to the ion composition and strength of the raw water being used.
3. Cl^-, Na^+, SO_4^{2-}, and/or other ions tend to accumulate in the nutrient solution, which affects the EC and pH, due to an imbalance between the ion supply and uptake rates.
4. The chemical composition of the pH control agents (e.g., H_3PO_4, HNO_3, and H_2SO_4 for lowering pH and KOH, K_2CO_3, and K_2SiO_3 for raising pH) affects the composition of the nutrient solution to some extent, but its effect is not considered in the EC control system.

5. Ion uptake characteristics of plants depend on the growth stage, plant species, plant biomass density (kg m^{-2}), etc., but these characteristics are not considered in the formula of the nutrient solution.
6. The ratio of a particular nutrient ion uptake rate to the water uptake rate (often called n/w (nutrient/water) ratio) is different from the ratio of other nutrient ions, which results in a change in the composition of the nutrient solution.

Other factors affecting the properties of the nutrient solution include (1) propagation and photoautotrophic growth of algae, which absorb nutrient elements in the solution, (2) presence of dead algae and plant roots in the nutrient solution, (3) propagation and heterotrophic growth of microorganisms through decomposition of dead algae and plant roots, (4) efflux of organic acids and/or volatile (allelopathic) gases produced from the plants roots, and (5) the presence and propagation of pathogenic microorganisms in the culture beds. Currently, little is known about such nutrient solution properties affected by ecological processes in the culture beds.

Thus, it is necessary to develop a new method and system to overcome the issues mentioned above. As a standard approach, extensive research has been carried out to develop ion-selective sensors for all or most types of nutrient ions (e.g., Son and Takakura 1987; Son and Okuya 1991; Shin and Son 2016; Jung et al. 2016; Chen 2017; Son et al. 2015).

The method described below takes a different, unique approach in that the ion concentration of NH_4^+, Mg^{2+}, Cl^-, PO_4^{3-}, and SO_4^{2-} is estimated based on the measured data on NO_3^-, K^+, Ca^{2+}, Na^+, EC, and pH, using inexpensive ion concentration sensors.

24.3 Ion-Specific Nutrient Solution Control in Hydroponics

24.3.1 Water and Ion Balance Characteristics

The flow rate of nutrient solution into and out from the culture beds through piping can be measured relatively easily (Fig. 24.1) using a flowmeter. The rate of water uptake and transpiration of plants in the culture beds can be estimated based on the water balance in the culture beds.

Then, the rate of increase in fresh and dry weight and the rate of ion uptake of the plants can be estimated as described in the following sections. This estimation is possible because the hydroponic culture beds are isolated from the floor soil. This advantage should be fully utilized in the next-generation plant factory with artificial lighting (PFAL) to estimate the flow of water and ions in the PFAL, to improve the yield and quality of produce and the cost performance of hydroponics. Adoption of this method can realize steady improvements in yield and quality of produce, especially when the plant growth data as well as the above-described data are measured continuously to find the relationships between the environmental data and plant growth data.

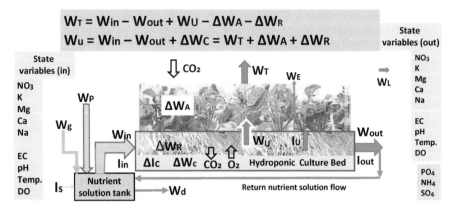

Fig. 24.1 Estimating the rate of water uptake (WU), transpiration (WT), nutrient element uptake (IU), and plant growth ($\Delta WA + \Delta WR$) in hydroponics. Evaporation from culture beds is not considered in the above diagram. *DO* dissolved oxygen concentration. For meanings of symbols, see Table 13.1

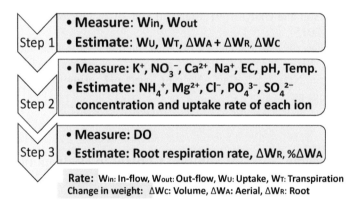

Fig. 24.2 Estimating the major variables in the hydroponic culture beds. For meanings of symbols, see Table 13.1

The algorithm for nutrient solution control consists of three basic steps as shown in Fig. 24.2. They are (1) water balance, (2) ion balance, and (3) parameter estimation. For the meanings of the symbols and the units of the variables in the figure, see Table 24.1. For further details, see Kozai et al. (2015). The basic procedures described above may seem to be simple, but each basic procedure consists of many sub-procedures. All the sub-procedures will be programmed and sold as an inexpensive and handy commercial product with or without the related sensors in the near future.

Table 24.1 Symbols, meanings, and units of variables shown in Fig. 24.1

No.	Symbol	Meaning	Unit
1	Δt	Time interval of estimation	h
2	ΔW_A	Change rate in weight of aerial parts of plants	kg h^{-1}
3	ΔW_C	Change rate in weight of nutrient solution in the culture beds	kg h^{-1}
4	ΔW_D	Change rate in dry weight of plants in the culture beds	kg h^{-1}
5	ΔW_R	Change rate in weight of root parts of plants	kg h^{-1}
6	ΔX_{in}	Change rate in water vapor concentration of air inside the culture room	kg m^{-3} h^{-1}
7	C_{in}	CO_2 concentration in the culture room	µmol mol^{-1}
8	C_{out}	CO_2 concentration outside the culture room	µmol mol^{-1}
9	DO	Dissolved oxygen concentration	mg kg^{-1}
10	EC	Electric conductivity	dS m^{-1}
11	I_{in}	Flow-in rate of ion i ($i = 1, \ldots, k$) at the inlet of the culture beds	µmol h^{-1}
12	I_{out}	Flow-out rate of ion i ($i = 1, \ldots, k$) at the outlet of the culture beds	µmol h^{-1}
13	I_U	Uptake rate of ion i ($i = 1, \ldots, k$) by plants in the culture beds	µmol h^{-1}
14	k_C	Conversion factor (1.964) from volume to mass	kg m^{-3}
15	N	Number of air exchanges in the culture room	h^{-1}
16	pH	Potential of hydrogen (acidity/basicity)	-
17	R_R	Root respiration (O_2 absorption) rate in the culture beds	kg h^{-1}
18	Temp.	Temperature of nutrient solution at the inlet or outlet	°C
19	V_R	Volume of air in the culture room	m^3
20	W_E	Evaporation rate from the culture beds	kg h^{-1}
21	W_g	Supply rate of raw water into the nutrient solution tank	kg h^{-1}
22	W_L	Loss rate of water vapor through air gaps to the outside	kg h^{-1}
23	W_{in}	Flow-in rate of nutrient solution at the inlet of the culture beds	kg h^{-1}
24	W_{out}	Flow-out rate of nutrient solution at the outlet of the culture beds	kg h^{-1}
25	W_P	Flow-in rate of water condensed and collected at the cooling panels of air conditioners and returned to the nutrient solution tank	kg h^{-1}
26	W_C	CO_2 supply rate to keep its concentration at the set point	kg h^{-1}
27	W_{CL}	Loss rate of CO_2 to the outside due to air exchanges	kg h^{-1}
28	W_{WL}	Loss rate of water vapor to the outside due to air exchanges	kg h^{-1}
29	W_T	Transpiration rate of plants	kg h^{-1}
30	W_U	Water uptake rate of plants	kg h^{-1}
31	X_{in}	Water vapor concentration of air inside the culture room	kg m^{-3}
32	X_{out}	Water vapor concentration of air outside the culture room	kg m^{-3}

24.3.2 Water Balance

Water balance in the culture beds can be analyzed using the procedures given below. These procedures can be conducted once every hour or so:

1. Flow-in rate of nutrient solution at the inlet of the culture beds, W_{in}, and flow-out (drained) rate at the outlet of the culture beds, W_{out}, as well as the supply rate of raw water into the nutrient solution tank, W_g, are measured separately using flowmeters.
2. Water uptake rate of plants, W_u, in each culture bed and/or group of culture beds is estimated by $W_u = (W_{in} - W_{out} + \Delta W_C)$. See Table 24.1 for the meanings of variables.
3. Transpiration rate, W_T, is estimated as the sum of "water vapor condensation rate at the cooling panels of the air conditioners" and "water vapor loss rate to the outside due to air exchange," W_{WL}. $W_T = W_P - W_{WL}$.
4. W_{WL} is estimated by the equation $W_{WL} = k \times N \times V_R \times (X_{in} - X_{out})$ where k is the conversion factor, N is the number of air exchanges per hour in the culture room, V_R is the volume of air in the culture room (m³) and $(X_{in} - X_{out})$ is the difference in water vapor concentration in the air between inside and outside the culture room.
5. Water weight increase (fresh weight minus dry weight) in plants, $(W_U - W_T)$, in each culture bed or group of culture beds is estimated.
6. Weight increase of carbohydrates such as starch $(C_6H_{10}O_5)n$ in plants is estimated using the data on the net photosynthetic rate of plants calculated based on the CO_2 balance in the culture room (Kozai et al. 2015). Weight increase of carbohydrates is nearly equal to (or slightly lower than) its dry weight increase.
7. Net photosynthetic rate of plants in the culture room, P_n, is estimated by $P_n = (W_C - W_{CL} + \Delta C_{in} \times V_R)$ where $W_{CL} = k_C \times V_R \times N \times (C_{in} - C_{out})/10^6$.
8. Total fresh (water and carbohydrates) weight increase in plants is estimated as the sum of water and carbohydrate (or dry) weight increases. In the case of leafy vegetables, the percent dry matter weight of the aerial part is 4–5% of its total fresh weight and that of the root part is 5–6%.
9. Since the total fresh weight of the aerial and root parts is measured separately after harvesting using an electronic weighing balance, the estimated data on fresh weight can be compared with the measured data. In actual plant production operation, damaged and/or yellowed leaves are removed after physically separating the roots from the plants and before packing the marketable part in bags or boxes. Then, it is possible to calculate the percentage by weight of the aerial part and that of the marketable part to the total fresh weight.
10. Change in weight (or volume) of nutrient solution, $\Delta W_C = (W_{in} - W_{out} - W_U + \Delta W_C)$, in each culture bed or group of culture beds is estimated. ΔW_C can also be calculated as a product of the change in nutrient solution level (m) and the surface area of the culture beds (m²).

11. Supply rate of raw water into the nutrient solution tank, W_g, is measured using a flowmeter.
12. Water use efficiency (WUE) can be estimated by the following equation (Kozai et al. 2015):
    $$\text{WUE} = (W_P + \Delta W_A + \Delta W_R)/W_g = (W_g - W_L)/W_g.$$

Water used for cleaning the floor and culture panels is not considered in the above equation.

The values of the rate variables such as transpiration, W_T, and water uptake, W_U, are statistically correlated with the environmental variables and state variables for the plants such as fresh weight, ΔW_A and ΔW_R.

24.3.3 Ion Balance

Ion balance in the culture beds can be analyzed once every hour or so using the procedures described below:

1. The supply rate of stock solution to be added to raw water for preparing the nutrient solution is measured. The supply rate of pH control agents to be added to the nutrient solution tank is also measured. These rates are converted to the supply rate of ion i ($i = 1, \ldots, k$) in units of µmol h^{-1} and of water in units of kg h^{-1}.
2. The ion concentration of K^+, NO_3^-, Ca^{2+} and Na^+ at the inlet and outlet of the culture beds is measured every hour or manually at least once a day. These ion concentrations can be measured relatively easily using an ion-selective electrode or specific ion electrode sensor.
3. EC, pH, and temperature of the nutrient solution in the nutrient solution tank (or at the inlet) and at the outlet of the culture beds are measured using the respective sensors.
4. The ion concentration of NH_4^+, Mg^{2+}, Cl^-, PO_4^{3-}, and SO_4^{2-} at the outlet is estimated every hour or continuously (at least once a day manually), using the data on NO_3^-, K^+, Ca^{2+}, Na^+, EC, pH, and temperature by either of the following methods: (1) multivariate statistical analysis for a multiple regression line with contributing factors and (2) simultaneous equations for the ion balance of the nutrient solution, considering the activity coefficient of each ion (Son and Takakura 1987). This estimation is currently necessary, because it is difficult to measure the concentration of NH_4^+, Mg^{2+}, Cl^-, PO_4^{3-}, and SO_4^{2-} ions using relatively inexpensive and handy ion-selective sensors.
5. Uptake rate of ion i ($i = 1, \ldots, k$), I_u, is estimated based on the water and ion balance equations: $I_u = W_{in} \times I_{out} - W_{out} \times I_{out} + \Delta I_c \times \Delta V$. The values of the rate variables for nutrient uptake are statistically correlated with the environmental variables and state variables for the plants.

6. Ion use efficiency (IUE) for each nutrient element, I, is calculated by $IUE_I = I_u/I_s$ where I_u is the uptake rate of ion i by plants and I_s is the supply rate of ion i to the culture beds.
7. The total mEq kg^{-1} (or L^{-1}) and mEq kg^{-1} of each ion is calculated, and the stock solution composition is adjusted to maintain the set points of mEq kg^{-1} for the total ions and each ion.
8. The change in pH is estimated, if possible, using the data on the H^+ ion balance in the culture beds. The estimated pH is compared with the measured pH. The pH is decreased when a cation such as NH_4 is absorbed by the plants and is increased when an anion such as H^+ is absorbed by the plants.

When using inexpensive ion concentration meters, it is necessary to calibrate the meters every month using several nutrient solutions with known composition and concentrations. Then, the coefficient values of the calibration curves can be stored in an Excel file so that the calibrated concentrations are automatically calculated in the file by inputting the measured values. This calibration can be conducted with unused nutrient solutions, so that the correct value of each ion concentration is known a priori.

24.4 Respiration Rate of Roots and Fresh Root Weight

1. Dissolved oxygen concentration at the inlet, DO_{in}, and outlet, DO_{out}, of the culture beds is measured.
2. In hydroponics with culture panels floating on nutrient solution in the culture beds, the oxygen absorption rate or respiration (CO_2 emission) rate of plant roots in the culture beds, R_R, is estimated by $R_R = (DO_{in} \times W_{in} - DO_{out} \times W_{out} + \Delta W_C \times (DO_{in} + DO_{out})/2$. In the case of an air gap layer between the culture panels and nutrient solution surface, it is necessary to consider the dissolution of O_2 in the air into the nutrient solution.
3. Percent dissolved oxygen concentration over saturated oxygen concentration at the inlet and outlet is calculated. Note that the saturated DO decreases exponentially with increasing temperature of the nutrient solution.
4. Using the above calculated root respiration rate, R_R, root dry or fresh weight is estimated assuming that the root respiration rate is nearly proportional to the root dry/fresh weight and exponentially increases with increasing root temperature. Dry matter percent of roots is 1–2% higher than that (4–5%) of aerial parts.
5. Fresh weight increase of aerial parts of plants per unit time interval of Δt, ΔW_A, is estimated by $\Delta W_A = (W_T - W_U) - \Delta W_R$.
6. When the culture beds are not thermally well-insulated and/or the nutrient solution is heated by the heat generated from lamps installed above or below the culture beds, the heat energy balance of the culture beds is analyzed using the data on the flow rate of the nutrient solution and the temperature difference between the inlet and outlet of the culture beds.

24.5 Conclusions

Today's hydroponics is considered to be a well-developed cultivation system with a high degree of controllability. However, in the next-generation hydroponics, the concentration of each type of ion needs to be measured separately. The control software for hydroponics also needs to be substantially improved. Furthermore, a summary of the data collected and the calculated indices showing the system performance need to be visualized for continuously enhancing system performance. These improvements can be realized relatively easily for hydroponics in PFALs with a well-controlled environment.

The hardware for hydroponics also needs to be improved, although the details are not discussed in this chapter. For example, the total volume of nutrient solution in hydroponics should be minimized to improve its controllability and to reduce the cost for hardware. The flow rate of nutrient solution in the culture beds should be controlled depending on the volume/weight of roots in the culture beds. Manpower for keeping the culture beds, nutrient tank, and piping clean should be minimized. The replacement ratio of old nutrient solution with new solution should also be minimized.

Acknowledgment We thank all the members of the Committee on Hydroponic Solution Control (chairperson, Yutaka Shinohara). Some important ideas described in this chapter were obtained during discussions at the committee meetings, especially the idea of estimating the ion concentration of NH_4^+, Mg^{2+}, Cl^-, PO_4^{3-}, and SO_4^{2-} using the measured data on NO_3^-, K^+, Ca^{2+}, Na^+, EC, and pH.

References

Chen L-C (2017) Next-generation ion-sensing technology for nutrient element monitoring in hydroponics. In: Proceeding for the international forum for advanced protected horticulture, Taipei, Taiwan, pp 53–62

Jung DH, Kim HJ, Choi GL, Ahn TI, Son JE, Sudduth KA (2016) Automated lettuce nutrient solution management using an array of ion-selective electrodes. Trans ASABE 58(5):1309–1319

Kozai T, Niu G, Takagaki M (eds) (2015) Plant factory: an indoor vertical farming system for efficient quality food production. Academic, Amsterdam, p 405

Shin JH, Son JE (2016) Changes in electrical conductivity and moisture content of substrate and their subsequent effects on transpiration rate, water use efficiency, and plant growth in the soilless culture of paprika (*Capsicum annuum* L.). Hort Environ Biotechnol 56(2):178–185

Son JE, Okuya T (1991) Prediction of electrical conductivity of nutrient solution. J Agr Met 47(3):159–163 (in Japanese with English summary captions)

Son JE, Takakura T (1987) A study on automatic control of nutrient solution in hydroponics. J Agr Met 43(2):147–151 (in Japanese with English summary captions)

Son JE, Kim HJ, Ahn T (2015) Chapter 17: hydroponic systems. In: Kozai T, Niu G, Takagaki M (eds) Plant factory: an indoor vertical farming system for efficient quality food production. Academic, Amsterdam, pp 213–222

Chapter 25
Phenotyping- and AI-Based Environmental Control and Breeding for PFAL

Eri Hayashi and Toyoki Kozai

Abstract Collaborative national research project on phenotyping- and artificial intelligence (AI)-based smart plant factories with artificial lighting (PFALs) has been launched since September 2017 in Kashiwanoha, Japan. Research objective and topics of this ongoing project are described. After discussing the development of research achievement, phenotyping- and AI-based environmental control and breeding, and commercialization of research achievement, future prospect of sustainable society with phenotyping- and AI-based PFALs is introduced.

Keywords Phenotyping · Artificial intelligence · Environmental control · Breeding · Cultivation system module (CSM) · Plant factory · Open data

25.1 Introduction: Project Background

In September 2017, Japan Plant Factory Association (JPFA) launched a collaborative national research project on phenotyping- and artificial intelligence (AI)-based environmental control and breeding for plant factories with artificial lighting (PFALs), in collaboration with the National Institute of Advanced Industrial Science and Technology (AIST), Kajima Corporation, and Chiba University.

This is a part of national AI and robotics project, Strategic Advancement of Multi-Purpose Ultra-Human Robot and Artificial Intelligence Technologies (SamuRAI) project commissioned by the New Energy and Industrial Technology Development Organization (NEDO), Japan. This NEDO research program aims to accelerate societal implementation of AI along with continuing technology development of AI by merging other fields' technologies, such as robotics, materials, and devices, to address social challenges. It is a pilot research and development project, which is to be implemented until March 2019 prior to the launch of full-scaled projects, in

E. Hayashi (✉) · T. Kozai
Japan Plant Factory Association (NPO), Kashiwa, Chiba, Japan
e-mail: ehayashi@npoplantfactory.org; kozai@faculty.chiba-u.jp

Fig. 25.1 Japan Plant Factory Association, part of Kashiwanoha campus of Chiba University and national AI (artificial intelligence) research center of AIST to be founded. AIST stands for the National Institute of Advanced Industrial Science and Technology

coordination with the national AI research center of AIST to be founded in coming years in Kashiwanoha, a smart city in Chiba prefecture, Japan.

Building on the years of research history, expertise, and environmental datasets generated at the Kashiwanoha campus of Chiba University and JPFA, this collaborative phenotyping- and AI- based smart PFALs research is conducted mainly in the PFAL facility located in Kashiwanoha campus (Fig. 25.1).

25.2 Research Objective

While much of the current and past plant phenomics have been focusing on field phenotyping on a large scale, the concept of this phenotyping- and AI-based smart PFALs project is to target PFAL as phenotyping laboratory, concurrently with commercial plant production facility. As discussed in other chapters earlier, PFAL has a capability of controlling any environmental factors required for plant production and of maximizing resource outputs while minimizing inputs. Furthermore, since PFAL is equipped with a thermally insulated airtight cultivation room, resource use efficiency (RUE) can be measured with minimum disturbance by climate outside, i.e., much lower noise under controlled environments for phenotyping. Big data from daily mass production of crops can be obtained in PFAL.

Although PFAL can control any environmental aspects more than any other plant production methods, there is still much space to be improved to fulfill its potential. One of the current challenges is to control environment more precisely and timely in accordance to the circumstances of dynamic behavior of plants (phenotypic trait dynamics), resource inputs/outputs, or market needs. Moreover, automation of acquiring phenotyping data besides environmental data during the whole cultivation processes is much needed.

It remains uncertain how many plant recipes or production methodologies need to be established. This is because phenotypic traits can vary depending on environment, genotype, and plant management. In order to achieve successful phenotyping-based environmental control, it stands to reason that AI plays a prominent role particularly to identify optimal plant recipes for on-demand production. Factors that need to be considered include, but not limited to, numerous crops or cultivars, set points of each environmental factor, plant phenotypic traits, market-driven plant design goals (e.g., nutrition, flavor, and shape), timing of harvesting in accordance to market demands, plant applications (e.g., food, nutraceutical, pharmaceutical, medical, etc.), production goals, and process management.

With the view toward spurring this ongoing research activities, the phenotyping- and AI-based smart PFALs project takes full advantage of utilizing mechanistic models which have already been developed here in Chiba University, Kashiwanoha campus, to analyze rate variables with unit of time of environmental and phenotypic data such as net photosynthetic rate, transpiration rate, water uptake rate, etc., besides state variables without unit of time such as temperature and CO_2 concentration (Kozai et al. 2015). These mechanistic models include mass and energy balance models and plant growth and development models. In addition to both mechanistic, multivariate statistical and behavior (or surrogate) models, artificial intelligence models with the use of neural network or deep learning should be effectively utilized particularly for image analysis and other analysis that deal with phenotypic traits including flavors, textures, and other factors, which have not been examined with existing mechanistic models up to date (Kozai 2018) (Fig. 25.2).

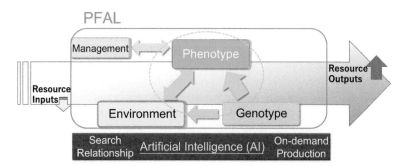

Fig. 25.2 Phenotyping- and AI-based environmental control: big data analysis of phenotype, environment, genotype, resource inputs/outputs, and management

25.3 Major Research Topics on Phenotyping- and AI-Based Smart PFALs

This ongoing project, phenotyping- and AI- based smart PFALs, is composed of two major and one sub-major subjects: (1) to research on AI-based phenotyping in PFAL and (2) to develop cultivation system module (CSM) for smart PFALs, and (3) breeding of leafy lettuce and other vegetables suited to PFAL.

With regard to (1) research on AI-based phenotyping, research and development of (a) plant phenotyping unit and (b) analysis technology of AI-based phenotyping are to be implemented. The major challenge is to develop non-invasive, automated, and continuous phenotyping methods that accommodate the extreme space constraint during the whole cultivation process.

As for the (2) and (3) research on CSM for smart PFALs, (c) CSM and (d) phenotyping- and AI-based environmental control technology are to be developed. To overcome the existing challenges in reproducibility of research outcome from laboratory to various commercial production sites, the smart PFAL with CSMs will achieve scalability, controllability, and adaptability. Throughout the research project, methodologies to analyze the relationship between phenotypic and environmental data are to be developed and to be implemented in smart PFALs (Table 25.1).

25.4 Development of Research Achievement

25.4.1 Phenotyping- and AI-Based Environmental Control and Breeding

Phenotyping- and AI-based smart PFALs can be a productive tool for plant breeding for both PFAL and non-PFAL (greenhouse and open field). With the controllable environment, breeding process will potentially become much more efficient and will have a large increase in the speed of breeding. Furthermore, it might allow the breeding industry to become diversified in terms of breeders and cultivars. There is a decent possibility we will experience the shifts from centralized breeding/seed industry to decentralized breeding/seed society where any farmers and any PFAL systems could take part in the breeding process. It should be noted that, with the

Table 25.1 Project research outline

(1) Research on AI-based phenotyping in PFAL
(a) To develop plant phenotyping unit
(b) To develop analysis technology of AI-based phenotyping
(2) Development of smart PFAL with CSMs
(c) To develop CSM for smart PFALs
(d) To develop phenotyping- and AI-based environmental control technology

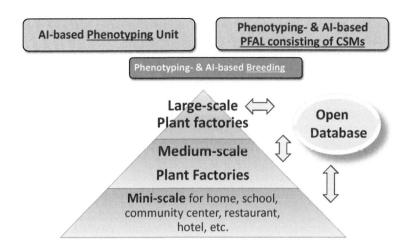

Fig. 25.3 Commercialization of research achievement: AI-based-phenotyping units (hardware and software), phenotyping- and AI-based PFAL consisting of CSMs, new cultivars developed by breeding in phenotyping- and AI-based PFAL, open database

analysis of the relationship among phenome, genome, environment, resource inputs/outputs, and management, phenotyping- and AI-based environmental control would generate big data out of the PFAL plant production facility for developing diverse crop varieties, possibly custom-made cultivars for each farmer.

25.4.2 Commercialization of Research Achievement

This research project progresses with a view of commercialization of research outcome. This commercialization will include AI-based-phenotyping units (hardware and software) and phenotyping- and AI-based PFAL consisting of CSMs. Moreover, new cultivars developed by breeding in phenotyping- and AI-based PFAL will likely be released in the near future. For instance, this breeding might include "tip burn-resistant, high speed growing lettuce."

As the research data accumulate through the project, open database will be implemented in order to connect each production system in multiple sites and to share associated data including numerous plant recipes. This will facilitate not only multi-faced academic researches and PFAL industrial developments but also breeding development, sustainability of plant production, people's living, etc. It is certain that the open data will make the industry develop in sustainable ways both technically and economically (Kozai et al. 2016). Open database will create a smart PFAL ecosystem, in conjunction with multiple plant data service businesses for data processing, PFAL application software industry users or consumers, entertainment, education, etc (Fig. 25.3).

25.5 Conclusion: Future Prospect of Sustainable Society with Phenotyping- and AI-Based Smart PFALs

We live in an era of seeking creative solutions to accelerating major global challenges on food, energy, and resource issues. Instead of being bombarded with the damages caused by worldwide climate change, we would rather need to establish efficient ways and tools as much as possible to positively cope with those issues. Rapidly increasing global population, decreasing arable land area, shortage of water, aging populations, and shortage of farmers urge us to establish promising methods to produce plants as much efficiently and stably as possible.

Through the phenotyping- and AI-based smart PFALs development, we are headed toward highly efficient and stable production of high-quality plants, development of multiple plant recipes, high-speed and diversified breeding, establishment of sustainable plant production systems and techniques, and on-demand plant production, among other possibilities.

In the years ahead, after achieving the development of on-demand production of numerous crops based on the PFAL big data analysis of phenome, genome, environment, resource inputs/outputs, and management, it is predicted that all consumers, restaurants, retailers, and PFAL farmers will be connected with each other by AI-based agricultural data science in agri-smart city Kashiwanoha and, moreover, throughout the world (Figs. 25.4 and 25.5).

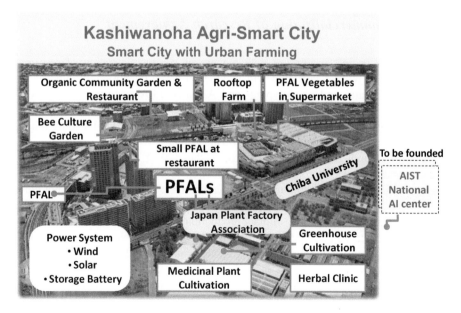

Fig. 25.4 Kashiwanoha, agri-smart city

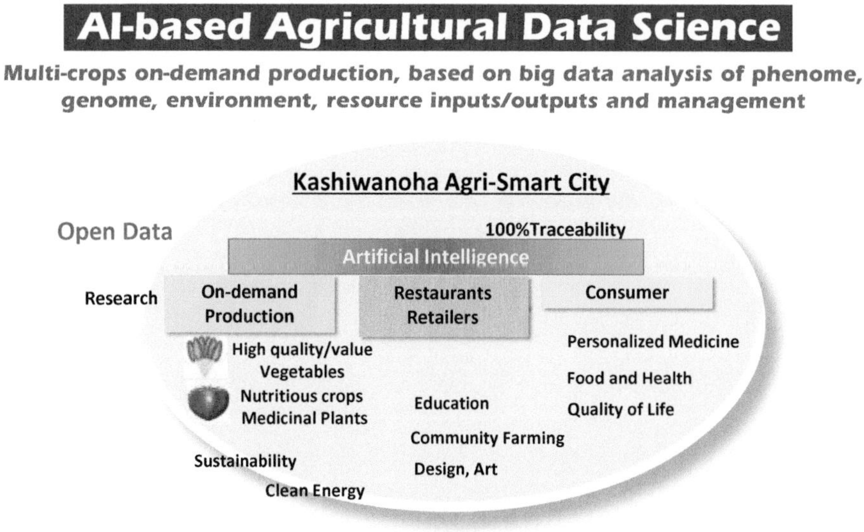

Fig. 25.5 AI-based agricultural data science: multi-crops on-demand production, based on big data analysis of phenome, genome, environment, resource inputs/outputs, and management

Acknowledgments This chapter is based on results obtained from a project, artificial intelligence (AI)-based plant phenomics in plant factories with artificial lighting (PFALs), commissioned by the New Energy and Industrial Technology Development Organization (NEDO). The authors would like to thank Toru Maruo and Na Lu of Chiba University; Takeshi Nagami, Tamio Tanikawa, Yoshihiro Nakabo, Takeshi Masuda, Kiyoshi Fujiwara, and Takumi Kobayashi of the National Institute of Advanced Industrial Science and Technology (AIST); Masashi Fukui, Hiroki Sawada, Mariko Hayakumo, and Ryoji Sheena of Kajima Corporation; and Toshiji Ichinosawa, Toshitaka Yamaguchi, and Nozomi Hiramatsu of Japan Plant Factory Association (JPFA). The authors also express their gratitude to many other project members and advisors for their technical and administrative support.

References

Kozai T (2018) Benefits, problems and challenges of plant factories with artificial lighting (PFALs). Acta Hort. (GreenSys 2017, Beijing, China) (in press)

Kozai T, Niu G, Takagaki M (eds) (2015) Plant factory: an indoor vertical farming system for efficient quality food production. Academic, p 405

Kozai T, Fujiwara K, Runkle E (eds) (2016) LED lighting for urban agriculture. Springer. Singapore, p 454

Chapter 26
Plant Cohort Research and Its Application

Toyoki Kozai, Na Lu, Rikuo Hasegawa, Osamu Nunomura,
Tomomi Nozaki, Yumiko Amagai, and Eri Hayashi

Abstract The concept and methods of plant cohort research for production in plant factories with artificial lighting (PFALs) are discussed. In plant cohort research, the time courses of plant traits (i.e., phenomes) are measured for plant individuals together with the data on environments, human and machine interventions, and resource inputs and outputs. The components of the plant cohort research system for a PFAL include the cultivation system module (CSM), phenotyping unit, database or data warehouse, and application software for the collection, analysis, and visualization of the processed data. Possible applications of the plant cohort research system are discussed.

Keywords Breeding · Cohort research · Democratization · Internet of Things · Longitudinal research · Phenotyping · Socialization · Transparentization · Variance

26.1 Introduction

The concept and methods of plant cohort research for plant production are discussed in this chapter aiming at improving the productivity and cost performance of plant factories with artificial lighting (PFALs). The research is conducted by using a key component of the PFAL, that is, the cultivation system module (CSM), with a phenotyping unit consisting of various types of cameras and artificial intelligence (AI)-based image processing devices as described in Chap. 5. The system scheme is illustrated in Fig. 26.1. The phenotyping unit is used to reveal the plant trait

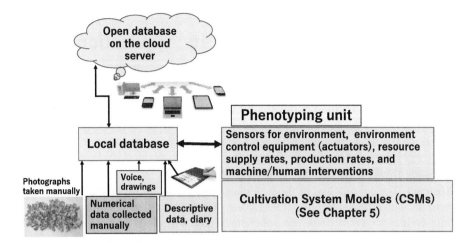

Fig. 26.1 Scheme of CSM leader with a phenotyping unit connected to an open database on the cloud server used for production and selection for breeding of plants

dynamics in a noninvasive manner. Plant cohort research for PFALs at Japan Plant Factory Assiciation in collaboration with Chiba University started in September 2017 as part of a research project on phenotyping- and artificial intelligence (AI)-based environmental control and breeding for the PFAL (see Chap. 25). A special feature of the plant cohort research is that the life cycle phenome history of an individual plant and its related data are continuously measured and analyzed for all the plants in the PFAL.

26.2 Transparentization, Democratization, and Socialization of the PFAL

Big data analysis of the plant trait dynamics in relation to the environment, resource inputs/outputs, and human/machine interventions will make the dynamic behavior of the plants and the CSM in the cultivation room "transparent" or a "white box" rather than a "gray or black box." Note that the traceability of the production process in the PFAL with CSMs is nearly 100%.

The data obtained from the transparent CSM is stored in an open database and is planned to be shared with people around the world in the near future. In this way, the accumulation and distribution of data, knowledge, and wisdom on food production using the CSM with a minimum use of resources and minimum emission of environmental pollutants will be accelerated. The PFAL consisting of CSMs will thus help solve the food-resource-environment trilemma on a local as well as global scale. In the long term, democratic use and socialization of the PFAL will be accelerated (Harper and Siller 2015; Floridi 2014) (Fig. 26.2).

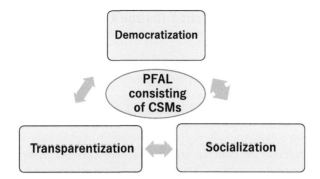

Fig. 26.2 The PFAL consisting of CSMs with a phenotyping unit enables the transparentization, democratization, and socialization of PFALs for many purposes such as education and lifelong self-learning and for solving the local as well as global food-resource-environment trilemma

26.3 Cohort or Longitudinal Research in Social Sciences

A cohort is a group of human individuals having a statistical factor such as age in common, for example, in a demographic study. Cohort research or longitudinal research is a field of observational research in which the investigator determines the exposure status of subjects and then follows them for subsequent outcomes, and the incidence of outcomes is observed over long periods of time, often many years or decades, in the exposed group and is compared with that in a nonexposed group (Shen et al. 2017).

Recent epidemiologic strategies have encouraged the exchange of cohort research information and cooperation to improve the cognition of disease etiology (a branch of medical science concerned with the causes and origins of diseases) with the use of "omics (genomics, proteomics, metabolomics, transcriptomics, etc.)" data. Omics aims at the collective characterization and quantification of pools of biological molecules that translate into the structure, function, and dynamics of an organism or organisms.

Cohort research can bridge the gap between micro and macro research in the fields of medicine, nursing, psychology, actuarial science, business analytics, ecology, and so on. For example, large-scale cohort research using a prospective multiple design and long follow-ups has explored some of the challenges in preventive medicine (Shen et al. 2017).

26.4 Plant Cohort Research: Lifetime Care of Plant Individuals

26.4.1 Outline

Using recent advanced technologies related to phenotyping and big data mining, cohort research can be conducted for plant production and breeding in the PFAL with CSMs. Plant cohort means a relatively large group of plant individuals having a statistical factor (such as cultivar, planting density, and cultivation system) in common with the plants in a PFAL for commercial plant production and/or breeding. The plant cohort research can be conducted for multiple purposes in a PFAL. For example, plant cohort research aiming to improve yield and quality by environmental control and breeding can be conducted concurrently using the same set of big data.

26.4.2 Plant Growth Stages

In a large-scale PFAL for commercial production, around 10,000–50,000 seeds are sown daily throughout the year, and they are grown under a controlled environment for 1–2 months until harvest (Fig. 26.3). The vegetative growth process of leafy greens can be divided into Stage 1, Stage 2, and Stage 3 (Table 26.1).

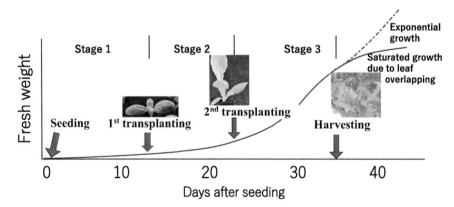

Fig. 26.3 Scheme of plant growth stages for leafy greens and human or machine interventions during the cultivation

Table 26.1 Typical plant growth stages and human or machine interventions in the PFAL growing leafy greens

	Day	Ecophysiological status	Human or machine intervention
Stage 1	0		Seeding
	1–2	Seeds germinate (under dark)	
	2–3		Light is provided for greening
	7	Cotyledons unfold under light and photosynthetic growth starts	Environmental control for promoting photosynthesis starts
	10	A few true leaves emerge	
Stage 2	14		First transplanting for spacing
	16–24	True leaves emerge successively and grow	
		Mutual shading of leaves of neighboring plants begins	
Stage 3	25		Second transplanting for re-spacing
	26	Rapid exponential growth begins	
	27	Mutual shading of leaves begins	
	30	Heavy mutual shading of leaves of neighboring plants begins	Environmental control for improving the quality of produce starts
	32	Exponential growth terminates due to heavy mutual shading of leaves	
	35		Harvesting

26.4.3 Plant Traits and Resource Input/Output

Representative plant traits are listed in Table 26.2 (some traits cannot currently be measured noninvasively). In the plant cohort research, the traits of individual plants are measured from seed sowing to harvesting by using the phenotyping unit consisting mainly of various kinds of cameras, and the data is recorded together with the time courses of the environments and human/machine intervention. Note that the visible light, thermal, microscopy, fish-eye and chlorophyll fluorescence cameras, memories, and microprocessors will continue to become less expensive and smaller.

This big dataset is analyzed for the purpose of improving the plant production process in the PFAL and the selection of plants with extreme traits for breeding. The 2D distributions of environmental factors such as photosynthetic photon flux density (PPFD) and air temperature over the tray for seedling production are also measured periodically. When the plants are moved for transplanting, for example, their 3D position (x-y-z coordinates) in the cultivation room is recorded for all plant individuals.

Representative resource input and output variables to be measured or estimated in the PFAL are listed in Table 26.3. In the past, only the plant phenome-environment relationships were mainly analyzed. However, in the commercial-use PFAL, the

Table 26.2 Representative plant traits measured or estimated at certain time and spatial intervals using a phenotyping unit in plant cohort research

State variables (without units of time)	
Physical	Three-dimensional plant canopy structure with height, leaf area, number of leaves, leaf shape
	Fresh and dry weights of aerial, root zone, yellow/wilted/damaged parts
	Color, spectral reflectivity/transmissivity, physiological disorders such as tipburn
Chemical	Concentrations of chlorophylls a/b, nitrogen, sugar, anthocyanin, phenols
	Concentrations of vitamins (ascorbic acids), antioxidant substances
	Antioxidant activity, stomatal conductance, chlorophyll fluorescence
	Taste (sweetness, bitterness, softness), texture, crispiness, flavor (volatile gases)
Rate variables (with units of time)	
Dynamic	Germination rate, leaf area growth rate, leaf emergence rate
	Net photosynthetic rate, dark respiration rate, water uptake rate, transpiration rate, dissolved O_2 uptake rate
	Nutrient element uptake rate (N, P, K, Ca, Mg, etc.)
	Periodic and nonperiodic movements of plants due to change in water potential in leaves or external force by air currents around the plants

Table 26.3 Representative resource input and output variables to be measured or estimated in the PFAL for plant cohort research

	Resource supply rate	Unit
1	Electric energy (electricity) supply rate (at time t)	kW
2	Electric energy supply rate (for 1 day)	kWh d^{-1}
3	Photosynthetic photon supply rate	mol d^{-1}
4	Water supply rate (for irrigation and floor/cultivation panel washing)	kg h^{-1}
5	CO_2 supply rate (CO_2 is absorbed by plants for photosynthesis)	kg h^{-1}
6	Seed supply rate	d^{-1}
7	Fertilizer supply rate (N, P, K, Ca, Mg, etc.)	mol d^{-1}
8	Labor (man-hours)	h d^{-1}
9	Supply rates of disposable substrate, gloves, masks, caps, cloth, etc.	d^{-1}
10	Supply rates of plastic bags and boxes for packing/packaging	d^{-1}
	Resource output rate	
1	Marketable plant (produce) production rate	kg d^{-1}
2	Plant residue production rate (unmarketable part of plants)	kg d^{-1}
3	CO_2 production rate by dark respiration	mol d^{-1}
4	CO_2 release rate to the outside by air infiltration through the walls	mol d^{-1}
5	O_2 production rate by photosynthesis	mol d^{-1}
6	Water vapor recovery rate at cooling panels of air conditioners	kg h^{-1}
7	Wastewater production rate (water for cultivation panel/floor washing)	kg d^{-1}
8	Production rate of used disposable gloves, masks, caps, cloth, etc.	d^{-1}
9	Heat energy generated by lamps and equipment	kW or kWh d^{-1}

Table 26.4 Representative environmental factors to be measured or estimated at certain time and spatial intervals in plant cohort research

	Aerial environmental factors
1	State variable: air temperature, leaf temperature
2	State variable: CO_2 concentration, VPD
3	State variable: light spectral distribution, lighting direction, lighting cycle
4	Rate variable: air current speed, air flow rate
5	Rate variable: photosynthetic photon flux density (PPFD), FR/UV flux density
	Nutrient solution factors
6	State variable: pH, EC, temperature, dissolved O_2 concentration, depth of nutrient solution in the cultivation beds, algae, microorganisms, dead roots
7	State variable: concentrations of NO_3-, PO_4^{3-}, K^+, Ca^{2+}, Mg^{2+}, Na^+, organic acids
8	Rate variable: nutrient solution flow rate

Notes: *EC* electric conductivity, *FR* far-red, *VPD* vapor pressure deficit, *UV* ultraviolet

analyses of relationships between the plant phenome (traits), environment, genome (genotype) and resource input/output are essential for understanding the economic viability, environmental friendliness, and life cycle assessment of the PFAL.

26.5 Causes of the Variation in Plant Traits and Separation of the Causes

26.5.1 Causes of the Variation in Plant Growth Rate

The variations in the traits of plants sown at the same time and place are attributable to four factors: (1) environment (Table 26.4), (2) genome (specific to species and cultivar), (3) human intervention, (4) machine intervention, and (5) unavoidable marginal error. Human and/or machine intervention, which can cause a positive or negative change in the plant traits, includes seed priming, seed coating, seed selection, seeding, transplanting or spacing, harvesting, cleaning/washing, and maintenance. A physiological disorder of a plant is not considered to be caused by a disease but instead is considered as a plant trait.

Variations in plant traits that cannot be explained by environmental variations or human and/or machine interventions are possibly due to the genetic variations of seeds, which should be random in terms of the position of the plants over the trays. Plants with an extreme genetic variation can be used for breeding purposes.

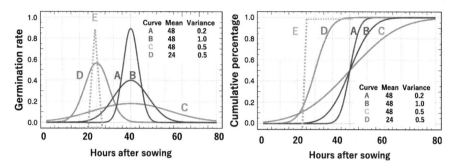

Fig. 26.4 Imaginary time courses of germination rates and cumulative germination percentages with different variances. The objective of seed germination in the PFAL is ideally to achieve Curve E, although this never occurs

26.5.2 Initial Plant Growth Rate After Seed Germination at Stage 1

A high, uniform germination rate is a necessary condition for uniform seedlings of high quality, and such seedlings are a necessary condition for obtaining a high yield and quality of produce. The ecophysiological status of germinated seeds affects the initial plant growth rate or initial fresh weight/leaf area increase rate and thus the subsequent plant growth.

For example, some lettuce plant seeds germinate within 24 h after sowing, but some others germinate 40–50 h after sowing. The first step to achieving uniform harvests is to find the causes of the variation in seed germination rate, and the second step is to remove the causes.

26.5.3 Germination Rate and Cumulative Germination Percentage of Seeds

General time courses of the germination rate and cumulative germination percentage of seeds after sowing are shown in Fig. 26.4. Curve E is ideal for the PFAL, although it never occurs. An example of actual variations in the growth of germinated seeds in a plugtray or a plastic tray ("tray" hereafter) (30 × 60 × ~3 cm) 3, 5, 7, and 10 days after sowing is shown in Fig. 26.5. In Fig. 26.5, germinated and ungerminated seeds can be distinguished by their photo images at Day 3. As for the germinated seed, radicle and hypocotyl emerged from the seed skin can also be distinguished. The average and variance of the actual germination rate are influenced by many factors, as shown in Table 26.5.

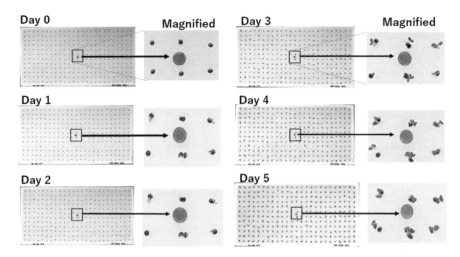

Fig. 26.5 Photographs of coated lettuce seeds taken from 50 cm above the plugtray with 300 germinating seeds. Time courses of cumulative germination percentage and germination rate can be estimated by image processing

Table 26.5 Factors that can affect the germination rate and cumulative percentage of seeds sown in the PFAL

No.	Factors
1	Cultivar, product number, lot number, year/date purchased
2	ID number of person who sowed the seeds
3	Seed type: coated or naked, pretreated, etc.
4	Year, month, day, and time of seeding
5	Seed radicle orientation on the substrate at seeding: at random or the same orientation
6	Automatic or manual seeding and tray transportation
7	Seed size, shape, color, weight, specific gravity, and injury
8	Tray cover: plastic film, sheet, paper, or nothing
9	Water quality or strength and composition of nutrient solution
10	Slightly pressurized with a sheet or no pressure on seeds
11	Automatic or manual watering and tray covering
12	Product code, manufacturer of substrate (supports)
13	Surface dryness/wetness of substrate with seeds
14	Germination under dark or light condition (PPFD and photoperiod)
15	Temperature of seeds, substrate or nutrient solution
16	Population density of algae/fungi on/in the substrate

PPFD photosynthetic photon flux density

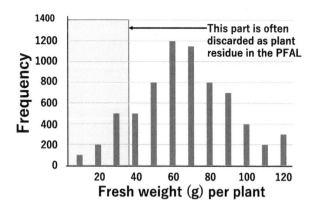

Fig. 26.6 Scheme for histogram showing the fresh weight (g) per plant grown in the PFAL. The variation in fresh weight is mainly due to environmental and genetic variations and human/machine interventions

26.5.4 Phenotyping the Germinated Seeds in Trays

In a PFAL producing 10,000 lettuce heads daily, a little more than 10,000 seeds are sown daily. The seeds are usually sown in a tray with water-soaked substrate such as foamed urethane (sponge) consisting of 300 (= 25 × 12) cubes (Fig. 26.5). Then, 35 (= 10,500/300) trays are used daily for seed sowing. Figure 26.6 shows an example of histogram of fresh weight per plant at harvest.

By photographing all the trays with seeds/plants using a phenotyping unit (cameras) four times a day for 14 days (at Stage 1: from sowing to the first transplanting), 1960 (35 × 4 × 14) digital photo images are obtained, each with about 300 seeds, germinated seeds, or seedlings with cotyledons. The simplest method for estimating the seed germination rate or plant growth rate is by subtracting the visible image taken as at time t from that taken at time (t − 1).

By using the 1960 digital photographic images, the time courses of 2D horizontal distribution maps of the germination rate and cumulative percentage with variance over the tray can be obtained. By taking photos of all the trays in seedling production stage (14 days) four times a day, 350 days a year, 686,000 (= 1960 × 350) photo images are obtained in 1 year.

Then, the horizontal distribution maps can be related to the factors listed in Table 26.5, and a method for predicting and improving the germination rate and cumulative germination percentage and/or reducing their variance can be found by using a methodology for image processing with or without the use of artificial intelligence (AI) and/or multivariate statistical analysis.

The photographic digital images mentioned above do not necessarily mean "visible camera images." The cameras include a thermal camera for measuring the surface temperature of the leaves, a chlorophyll fluorescence camera, a nitrogen content camera, a chlorophyll content camera, a spectral leaf reflectance camera, an X-ray camera, and so on (Fig. 5.19, Chap. 5). The analysis of the images taken by different types of cameras may be useful in finding a characteristic feature of a trait. For example, a comparison of visible camera images with the images taken by chlorophyll fluorescence, nitrogen content, and chlorophyll content cameras may help find an invisible physiological symptom and its causes.

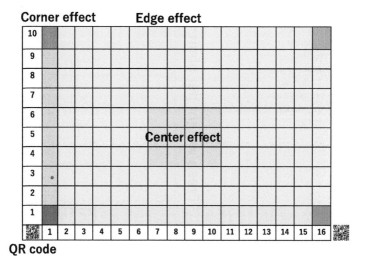

Fig. 26.7 Two-dimensional variation in the fresh weight per plant over a plastic tray can be attributed to the 2D variation in microenvironmental factors such as PPFD and air current speed over and within the plant canopy

26.5.5 Fresh Weight per Plant at Stages 2 and 3 as Affected by Environmental Factors

At first glance, the growth of the plants on the tray seems to be uniform. However, there is a relatively large variation in the fresh weight per plant, as shown in the example in Fig. 26.6. In this case, a substantial percentage of the produce is discarded as plant residue in a commercial-use PFAL. Similar variations are observed for other plant traits such as plant height, morphology, and concentrations of secondary metabolites.

Two-dimensional variation in the fresh weight per plant over a tray can be partly attributed to 2D variations in microenvironmental factors such as photosynthetic photon flux density (PPFD) and air current speed over and within the plant canopy in the tray. The fresh weight is sometimes greater at the central part (center effect) and lower at the corner and edges (corner and edge effects) (Fig. 26.7).

Then, by comparing the 2D variation in the fresh weight with that in the PPFD and air current speed, regression curves showing the effect of PPFD and air current speed on the fresh weight can be obtained (Fig. 26.8). Then, the effect of PPFD and air current speed on the fresh weight per plant can be estimated, and the 2D distribution of PPFD and/or air current speed can be improved to reduce the 2D variation in fresh weight. Once the causes of the variation in the plant trait are revealed, the method for solving the issue can be found (Fig. 26.9). In the above, "fresh weight" is just an example of the plant traits; the same or a similar method can be applied for any other plant trait. For example, the cause(s) of a well-known physiological disorder, tipburn, can be found, and its occurrence can be reduced.

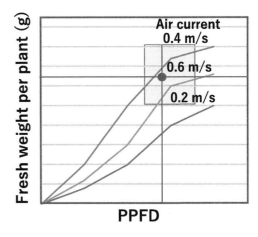

Fig. 26.8 Scheme of imaginary regression lines showing the fresh weight per plant as affected by PPFD and air current speed. The red solid circle denotes the fresh weight per plant at the center of the tray under the given PPFD and air current speed

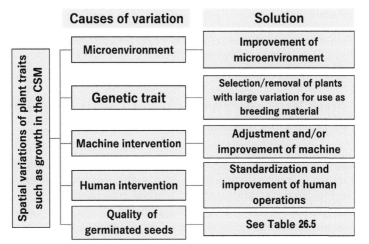

Fig. 26.9 Interrelated causes of spatial variation in plant growth in the CSM. Differentiation of the four causes is conducted by using CSM-L. Plants with ecophysiological variation are selected for further examination

Similarly, 2D variations in plant traits caused by human and/or machine interventions can be found and reduced.

26.5.6 Phenotyping at Stage 3

Plants grow fast at Stage 3 with an increase in plant height, number of leaves, and total leaf area causing the overlapping of leaves with those of neighboring plants. To obtain a 3D structure of such a plant canopy, in addition to small visible light cameras with wide angle lens, a fish-eye camera (5 × 5 mm) and/or a small

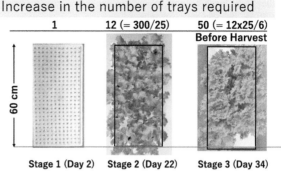

Fig. 26.10 Increase in the number of trays required for spacing by the first (Day 14) and second transplanting (Day 24) for a case in which the number of cells per tray at Stages 1, 2, and 3 was, respectively, 300, 25, and 6. The number of days required for Stages 1, 2, and 3 was, respectively, 14, 10, and 10

compound-eye camera with or without a laser emission unit might be useful because the distance between the plant canopy surface and the cameras is only 5–30 cm when the cameras are installed at the ceiling (with the lamps) of the cultivation space. A small camera can be installed on the side walls and the tray surface to take pictures of the plant canopy from the sides and underneath. A small lens attached at one end of a fiberscope camera can be inserted inside a lettuce plant to measure the light environment in the plant. Methods called deep learning (Guo W, Chapter 23 of this book), shape-from-silhouette (SfS) (Golbach et al. 2016), lidar (Omasa et al. 2006) and structure-from-motion (SfM) (Li et al. 2014) are useful tools to reconstruct 3D models of plant canopy from multiple camera images of plant canopy.

26.5.7 Increase in the Number of Trays at Stages 2 and 3

In the case of leaf lettuce production in the PFAL, transplanting for spacing is necessary around Day 14 (Stage 2) and around Day 24 (Stage 3) after sowing, using 25-cell trays and 6-cell trays, respectively (Fig. 26.10) (the number of cells per tray varies with different factors). In Fig. 26.10, the total number of trays required is 63 ($= 1 + 12 + 50$), and the total cultivation area occupied by the trays is 630 ($= 1 \times 10 + 12 \times 10 + 50 \times 10$) tray spaces. Then, the area for 500 trays at Stage 3 accounts for about 80% ($= 100 \times 500/630$) of the area for 630 trays at Stages 1, 2, and 3. Therefore, it is important to minimize the number of days at Stage 3 and maximize the number of days at Stage 1 to maximize the annual productivity per cultivation area for a case in which the fresh weight of the produce at harvest is not decreased. Optimal days of cultivation at Stages 1, 2, and 3 can be found relatively easily using the data obtained by the CSM.

Table 26.6 Fresh weights (g) of marketable, root, and trimmed parts of romaine lettuce grown in the PFAL, 34 days after sowing

No.	Marketable	Roots	Trimmed	Total	% marketable
1	82.7	9.8	3.3	95.8	86.3
2	75.2	8.4	2.9	86.5	86.9
3	76.8	8.8	2.7	88.3	87.0
4	95.0	9.0	5.0	109.0	87.2
5	72.5	7.8	2.5	82.8	87.6
Ave.	**80.4**	**8.8**	**3.3**	**92.5**	**87.0**

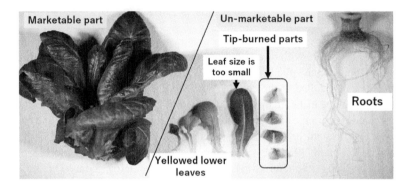

Fig. 26.11 Marketable, trimmed, and root parts of romaine (or cos) lettuce (*Lactuca sativa* L. var. longifolia) 34 days after sowing in the PFAL. See Fig. 26.11 for the photo of the plant No. 5 in the above table

26.5.8 Annual Productivity per Tray

In the above example, by assuming a fresh weight of 100 g per plant of marketable produce and 35 cropping (10 days for Stage 3 per cropping) per year, the annual productivity is estimated to be 21 kg ($= (100\ g \times 6\ plants \times 35\ cropping)/1000$)) per tray or 217 kg/m^2 ($= 21\ kg/(0.3\ m\ wide \times 0.6\ m\ long)$), although in actuality there should be a significant loss during cultivation due to many different reasons.

26.5.9 Percent Marketable Part of Plants

Percent marketable parts of five romaine lettuce plants grown in the PFAL for 34 days from seeding are shown in Table 26.6. Figure 26.11 is a photograph of the plant No. 5 in Table 26.6. The productivity can be improved by maximizing the percent marketable part or minimizing the fresh weight of yellowed lower leaves and tipburn leaves. Trimming reduces the percent marketable part and increases the labor cost. Trimming is the most difficult part of automation/robotization.

26.5.10 Numbers of Photo Images of Trays and Plant Individuals Taken at Stages 1–3

In Sect. 26.5, Item 4 "Phenotyping the germinated seeds in trays," it is shown that 686,000 photo images are obtained for 35 trays at Stage 1 (14 days) in 1 year. By spacing the seedlings at Stage 2 (10 days) from 35 trays to 420 trays (= 35 × 12), 8,232,000 (= 686,000 × 12 × 10/14) photo images are obtained. At Stage 3 (10 days), 24,500,000 (= 686,000 × 50 × 10/14) photo images are obtained. Then, 27,381,000 (= 686,000 + 8,232,000 + 24,500,000) digital images are obtained in 1 year in the PFAL producing 10,000 plants daily. The numerical data for each photo image is tagged in Tables 26.2, 26.3, and 26.4. Thus, the total volume of data is large enough for general data mining and deep learning in AI.

By sowing 10,000 seeds daily for 350 days, nearly 3,500,000 (=10,000 × 350) records of individual history of seed growth can be collected together with all the related data in 1 year. This amount of dataset is large enough for plant cohort research.

26.5.11 Data Warehouse and Internet of Things (IoT) for PFALs

In the plant cohort research, all the data shown in Fig. 26.12 is stored in a data warehouse, mostly in an open-source data warehouse, which might be called the Internet of Things (IoT) for PFALs. The "things" include the data on phenomes, environments, genomes, and human/machine interventions.

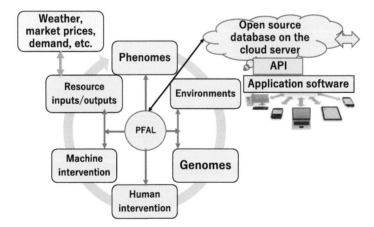

Fig. 26.12 Internet of Things (phenomes, environments, genomes, and human/machine interventions) for PFALs. All the data is systematically stored in an open database and shared with people. API denotes the application software interface

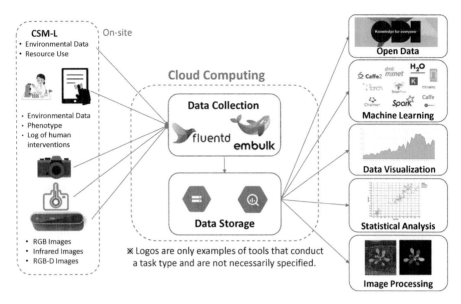

Fig. 26.13 Configuration of data warehouse
Image processing: https://www.plant-phenotyping.org/CVPPP2015-challenge
Statistical analysis: https://www.datacamp.com/community/tutorials/machine-learning-in-r

The requirements for a data warehouse in the context of plant cohort research differ slightly from traditional data warehouses and IoT platforms. However, there is much knowledge to be gleaned from the areas of business intelligence and machine learning.

Many of the same problems faced in setting up a data warehouse in web companies, like data storage, graphical visualization, and access control, are seen in the context of plant cohort research (Fig. 26.13). We can reuse the solutions that have been developed by machine learning engineers in web companies for the problems which are overlapping.

Since most web companies conduct their analysis on server logs, some work needs to be conducted in the data input and logging layer to complement the existing solutions. Unlike traditional data warehouses which store server logs, PFALs have different kinds of data sources which are streaming data to the database. We need to accommodate the many and sometimes esoteric sources of data which can be related to the environmental conditions, human labor, machine operation, resource use, or phenotype data.

In the context of data regarding plants, there exist a few open datasets regarding plants. However, most datasets have a few dimensions of data types which were selected for a particular experiment. The goal of a plant cohort study is to collect many dimensions of data such that many organizations can conduct analysis which may provide different kinds of insight using the same dataset. In order to make this possible, we must take care to bind each observation that we make to an individual plant or a group of plants at the observation's maximum resolution. For example, if there is only a single CO_2 sensor per CSM, we must apply the CO_2 measurement to all plants within the CSM, and note that the measurement resolution is at the CSM level.

Once this database is constructed, different kinds of applications become possible. Some examples include statistical analysis, graphical visualizations in dashboards, machine learning (deep learning, gradient boosting, etc.), publishing the dataset openly, and hosting machine learning competitions like Kaggle. The open database can function as the base platform for creating application software for PFAL operators.

26.6 Application Software for the CSM

To make full use of the CSM, phenotyping unit, and data warehouse, application software such as the ones listed in Table 26.7 needs to be developed. Application software for showing the hourly resource inputs (supply rates), their cost, and

Table 26.7 Examples of application software to be developed for the CSM

Level	Application software
1	Three-dimensional structure of plant canopy
1	Vertical distribution of leaf area, number of leaves, and PPFD in the plant canopy
1	Ratio of photosynthetic photons received by plants to those emitted by lamps
1	Vertical distribution of air current and diffusion coefficient in the plant canopy
1	Vertical distribution of net photosynthetic and transpiration rates
1	Alarm system for malfunctions, accidents, and damage of plants and equipment
1	Tracking the position of plant individuals in the CSM
1	Tracking the growth of plant individuals in the CSM
1	Average growth rates of leaf area, height, fresh weight, and their variances
1	Histograms of fresh weight of produce, trimmed leaves and roots, and their causes
1	Hourly resource inputs (supply rates), their cost, and resource use efficiencies
1	Motion analysis of workers, equipment, and tools and motion improvements
2	Location(s) of tipburn spots on leaves and its causes and solution
2	Plant growth curves under various conditions
2	Parameterization of mechanistic, statistical, and AI models using actual and predicted resource inputs and output rates
2	Dates of first and second transplanting to maximize the cost performance of the PFAL
2	Hourly resource output (production) rates of produce, waste, heat energy, etc.
2	Prediction of physiological disorders, spread of insects and microorganisms
3	Vertical distribution of specific secondary metabolite production
3	Cost and benefit relationships under various growth curves
3	Plant production schedule to maximize the cost performance of the PFAL
3	Prediction of plant canopy structure and secondary metabolite production

Notes: *CSM* cultivation system module (Chap. 5), *PPFD* photosynthetic photon flux density

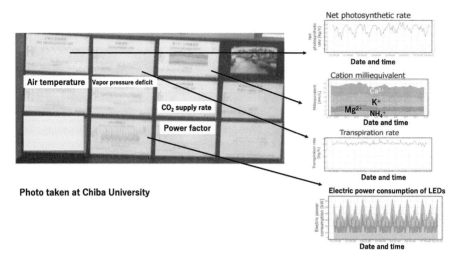

Fig. 26.14 Dashboard (12 display screens) of the PFAL management system (SAIBAIX) developed by PlantX Corporation and installed at a commercial-use PFAL. The power factor in the lower display screen indicates the electric energy use efficiency of the air conditioners

resource use efficiencies (Level 1) has already been developed and applied in a commercial-use PFAL (Fig. 26.14) (see Chap. 9 for details).

26.7 Conclusion

Plant cohort research is a new area in PFAL research, and few experimental results have been obtained so far. However, such research has the potential to reveal the relationships between plant phenomes, genomic traits, resource inputs/outputs, and human/machine interventions, and to breed new cultivars and to improve plant productivity in terms of electric energy, cultivation area, and labor hours. It is hoped that this chapter will contribute to the development of such next-generation smart PFALs.

Acknowledgments This chapter is based on results obtained from a project commissioned by the New Energy and Industrial Technology Development Organization (NEDO). The authors express their deep appreciation to Alexander Feldman, Toshiji Ichinosawa, Nozomi Hiramatsu, Toyoichi Meguro, Fumihiro Tanaka, Toshitaka Yamaguchi, Shin Watanabe, and Yu Zhang of the Japan Plant Factory Association for their technical and administrative support.

References

Floridi L (2014) The fourth revolution: how the infosphere is reshaping human reality. Oxford University Press, Oxford, 248 pages

Golback F, Kootstra G, Damjanovic S, Otten G, van de Zedde R (2016) Validation of plant part measurements usign a 3D reconstruction method suitable for high-throughput seedling phenotyping. Mach Vis Appl 27:663–680

Harper C, Siller M (2015) OpenAG: a globally distributed network of food computing. IEEE Pervasive Comput 14(4):24–27

Li L, Zhang Q, Huang D (2014) A review of imaging techniques for plant Phenotyping. Sensors 14(11):20078–20111

Omasa K, Hosoi F, Konishi A (2006) 3D lidar imaging for detecting and understanding plant responses and canopy structure. J Exp Bot 58(4):881–898

Shen Y, Zhang S, Zhou J, Chen J (2017) Cohort research in "omics" and preventive medicine. Adv Exp Med Biol 1005:193–220. https://doi.org/10.1007/978-981-10-5717-5_9

Chapter 27
Concluding Remarks

Toyoki Kozai

Abstract The author's view about frequently encountered negative comments on plant factories with artificial lighting (PFAL) is briefly described regarding the initial and operation costs, economic viability, environmental sustainability, usefulness to small-scale growers, usefulness in local areas far from large cities, and applicability for seedling production, seed propagation, and breeding. PFAL is considered to contribute to achieving some of the 17 Sustainable Development Goals (SDGs) set by the United Nations.

Keywords Initial cost · Production cost · Sustainability · Sustainable Development Goals (SDGs)

27.1 Introduction

Frequently encountered negative comments by citizens and business people on plant factories with artificial lighting (PFAL) are (1) its business would not be economically viable due to the high initial and operation costs; (2) it would not be commercially viable unless it is operated on a large scale or it produces extremely high-value crops such as medicinal plants; (3) it is not environmentally sustainable because it uses a large amount of electric energy instead of free solar light energy; and (4) it is unnatural to grow plants in the PFAL, so that PFAL-grown vegetables would neither be tasty nor nutritious. In this chapter, the author's view about those comments are briefly described. For a more detailed description, refer to the preceding chapters of this book, Kozai et al. (2015) and Kozai et al. (2016).

T. Kozai (✉)
Japan Plant Factory Association (NPO), Kashiwa, Chiba, Japan
e-mail: kozai@faculty.chiba-u.jp

© Springer Nature Singapore Pte Ltd. 2018
T. Kozai (ed.), *Smart Plant Factory*, https://doi.org/10.1007/978-981-13-1065-2_27

27.2 Are Initial Set-Up Costs and Operating Costs of the PFAL Too High?

27.2.1 Recent and Future Cost Reductions

In terms of the relevant technology, it is possible that both initial costs and operating costs of many PFALs (but not most PFALs) will come down by half by 2022–2025 in comparison with the averaged costs in 2017, as is discussed in Chap. 5 of this book. In fact, several PFALs in Japan reduced their operation costs by 30–40% during 2014–2017. It should be noted that the costs per kg of produce have been reduced and will be reduced steadily.

27.2.2 Per Land Area Initial Set-Up Cost and per Output Initial Set-Up Cost

As for the initial costs, we need to consider both the per land area initial set-up cost and the per output (production or yield) initial set-up cost. Very simply put, the per land area initial set-up cost for a PFAL is over 100 times that for an open field farming operation and around 10 times that of a greenhouse fitted out with environmental control units. Initial set-up costs for open field farming include the costs of land itself; field preparation and soil property improvement, and machines and facilities for irrigation, drainage, tilling, sowing or planting, harvesting, transporting, sorting or grading. These costs are often partially or largely supported in many countries by local, national, and/or international governments. In many cases, the initial investment for open field farming with a history of few decades has already been recovered.

As described above, production capacity or output per land area of the PFAL is over 100 times that of an open field farming operation, and around 10 times that of a greenhouse fitted with an environmental control system. So, if the initial set-up cost is measured against production capacity, there is no real difference between the three.

27.2.3 More Efficient Operation and Management Are Required

Current PFALs often achieve an output of only 60–70%, which is one of reasons that many PFALs operate at a loss, because even if a PFAL is working at 60–70% capacity, its production costs have not been reduced by the same proportion. These low capacity rates are predominantly due to design and construction flaws in the PFAL in question, followed by a lack of production management experience on the part of factory managers, and finally a lack of overall competence in product planning, business operations, and sales.

27.2.4 Introducing New Technologies

It should be noted that (1) more and more PFALs are a making profit recently and technologies and business models employed in these PFALs will be spread within several years and (2) PFAL technologies are advancing every year and new PFAL business models are created every year and will be created in the forthcoming coming years, with use of light emitting diodes (LEDs), artificial intelligence (AI), Internet of things (IoT), virtual reality (VR), augmented reality (AR), big data mining, phenotyping units, robots, and omics. Online estimation, visualization, and management of productivities in terms of electricity consumption, working hours, and cultivation area (or initial investment) will further enhance the increase in productivities and reduction in production costs. More and more engineers and researchers are in fact entering this field, and within several years many PFALs will come to use those technologies.

27.2.5 Is the Sustainability of the PFAL Improving in Terms of Energy and Natural Resources?

The number of PFALs operating that use more natural resources per production quantity will increase in the near future. It is in theory quite possible to design, construct, and operate PFALs that are self-sustaining in terms of energy and natural resources. Industry, government, and academia need to work together to that end.

27.3 Are PFALs Useful for Small-Scale Growers?

27.3.1 PFALs Are for Functional Plants, Not for Staple Crops

The world's main staple or food crops such as rice, wheat, and maize with dry matter percentage of 80–85% will not be produced economically in PFALs and greenhouses, because its price per kg dry weight is roughly 1/100th of that for PFAL-grown lettuce plants with dry matter percent of about 5%. Light energy required for plant production is proportional to its dry weight, not to its fresh weight. Staple crops require a higher photosynthetic photon flux density (PPFD), a longer cultivation period from seeding to harvest, and a lower planting density, compared to leafy vegetables. Besides, weight ratio of grains to the whole plants (or harvest index) is low.

27.3.2 Markets of PFAL Produce Are Different from Those of Produce by Open Fields and Greenhouses

The species of plants that PFALs will produce will mostly be either species that are already being produced in greenhouses today or medicinal herbs and wildflowers that are currently gathered in the wild. However, cultivation methods, weight per plant, quality, and appearance of PFAL-grown plants must be different from those of open field- and greenhouse-grown plants. Otherwise, the PFAL-grown plants compete with open field- and greenhouse-grown plants in the commercial market. A new market for PFAL-grown plants is expected to be created.

27.3.3 PFAL-Grown Vegetables Would Become More and More Tasty and Nutritious

Taste, texture, and/or concentrations of nutritional and/or functional elements in vegetables vary with environmental factors. Thus, those in field-grown vegetables vary with soil, weather and season. Those in PFAL-grown vegetables are, on the other hand, relatively stable because they are grown under controlled environments including light environment. However, it does not mean that PFAL-grown vegetables always taste better and more nutritious than field-grown plants. The same applies for medicinal and other functional plants.

More and more researches have recently been conducted for many plant species on effects of environmental factors on the taste, texture, and concentrations of functional or nutritional components. Once the optimal environments for the same are found, high-quality plants can be produced at high yield throughout the year.

27.3.4 Affordable Application Software and Training Program Need to Be Developed

The lack of suitable, easy-to-use training software and of free or inexpensive application software and open databases downloadable online is a factor behind the shortage of skilled managers and workers for PFALs, which is the biggest problem that the sector faces at present. An absence of strategies to overcome this problem is holding back the greater take-up of PFALs. We need public supports at this aspect.

27.3.5 PFAL Technology Can Be Applied for Seedling or Transplant Production, Vegetative Propagation and Micropropagation

There is an increase in demands for PFALs producing seedlings and transplants of fruit trees, industrial trees, tuberous crops, disease-free transplants, and field and greenhouse crops. This is because high-quality seedlings and vegetatively propagated transplants at low costs are indispensable for economic cultivation of high yield and quality of produce in the open fields and greenhouses.

27.3.6 PFALs Can Be Integrated with Other Biological Systems

With the exception of the shiitake and matsutake varieties, today the majority of mushrooms are produced in a kind of factories. Besides, many of the principal commercial fish varieties have also come to be farmed in recent years, and the value of farmed fish (aquaculture) production is going to be exceeded the value of wild-caught fish production. There will be an increase in full-lifecycle aquaculture, where every step of the process from cultivation and egg collection to maturity is undertaken inside a facility. But as those developments take place, people living in cities feel nostalgic about traditional farming and fishing practices, and due respect should be paid to their wishes to see those practices continued.

On the other hand, those traditional practices and businesses do not provide an income commensurate to the grueling and unstable nature of the work that they entail. This means that we have to look straight in the eye of the reality today, namely, that many farmers feel compelled to give up farming, forestry, or fishing as they currently practice it, seeing it as something they would hesitate to pass on to their own children. On the other hand, traditional ways of culturing plants, mushrooms, and fish need to be conserved as a cultural heritage in local regions.

27.3.7 PFALs Can Be used for Efficient Breeding and Seed Propagation

Preferable characteristics of plants for commercial production in the PFAL include (1) growing fast under low photosynthetic photon flux (PPFD), high CO_2 concentration, and high planting density and 2) negligible physiological disorders such as tipburn at a high growth rate.

On the other hand, plants to be grown in the PFAL do not need resistances to diseases caused by pathogens and to environmental stresses such as draught, low/high temperatures, heavy rain, and strong wind. These resistances of plants are essential for growing plants in the open fields. Thus, targets for breeding for PFAL-grown plants are considerably different from those for open field and greenhouse plants. This is a big challenge for us all, because humans have never tried to breed plants neglecting any resistances to pest insects and environmental stresses during 10,000 years of history of agriculture.

Besides, PFALs can be used for efficient breeding of field and greenhouse crops and seed propagation of those crops. Integration and analysis of the data on plant traits obtained by phenotyping, omics, environments, and human/machine interventions is a key in breeding using PFALs.

27.4 PFALs Are Worth in any Regions

27.4.1 Sub-Saharan Africa, Asia, South America, and Other Regions

For many years, and in a very diverse range of political, economic, and social circumstances, almost all developed nations have provided aid to these regions with the aim of solving issues there relating to food, resources, and the environment, and this approach has resulted in a certain degree of success.

Now, however, people have become aware of the widespread take-up of smartphones in these regions, which has been taking place without any of this aid. Through the use of smartphones, people in these regions can get education, if they wish, either formally or through self-study, and more and more businesses and communities are being built. Many of these people never had access to traditional fixed line forms of telephony.

Even in regions without commercial sources of electricity, smartphones can be recharged and operated using just small (1 m^2 or much less) solar cell panels. What is driving the spread of smartphones is the low cost of the devices themselves and easy-to-use free application software that can be used without the need for any intermediary or even a manual. This means that the smartphones can be a handy terminal unit of a PFAL via the Internet.

What should be noted here is that the smartphones are not destroying a community's unique culture. In a sense, the spread of smartphones is actually strengthening local cultures. While the hardware and basic software for smartphones are essentially the same the world over, the application programs and personal data on any smartphone are peculiar to individuals, their friends, and their country or community. At the same time, if they want, they can use the Internet to connect with people anywhere around the world.

The PFALs that have started to attract people's attention are PFALs the size of bar fridges and bookshelf cases installed in homes, schools, restaurants, offices, community welfare centers, and aged care homes. It is these PFALs that hint of a potential different to traditional backyard and balcony vegetable gardens or to horticultural therapy.

If these mini PFALs could be offered through nongovernmental organization (NGOs) and nonprofit organization (NPOs) to members of communities in sub-Saharan Africa, Asia, South America, and many other countries, these group members could access free application programs from anywhere in the world and develop methods for using them that suit their community.

While blueprints for mini PFALs are available to the public for free, people can also draft their own. Application programs that people have developed can be studied or downloaded via the Internet anywhere in the world. Managers, educators, researchers, and local proponents for large-scale PFALs would then emerge from among these active members, giving rise to hopes that creative solutions could be discovered for issues relating to food, resources, and the environment in local communities, tailored to local conditions as they are put into practice.

27.4.2 Value to Villages in Mountainous and/or Remote Regions

In villages in mountainous or remote regions far from any major city in any country, while it is possible for people to produce fresh vegetables in yards or facilities for their own home consumption, it is often difficult for them to produce enough to sell.

If villagers had a mini PFAL inside or adjacent to their home, however, an individual or family could farm the plants that go to make up dietary supplements and traditional herbal medicines, which immediately after harvesting they could dry and put in storage, before shipping off a delivery to customers every few months or so. These individuals and families could get advice online via the Internet for any problems they had with how to produce, dry, or store the produce. The electricity for operating the PFAL can be generated by natural energy such as solar energy, wind power, hydropower, biomass energy, and geothermal energy.

This production model could be introduced into mountainous and/or remoted regions that experience heavy snowfalls, into small settlements in arctic regions, and even into extremely hot and/or arid regions. They have in fact been used by research teams spending the winter in Antarctica. Lastly in the wake of a disaster anywhere in the world, a prefab container-type PFAL could be set up and provide fresh vegetables in the disaster area within 1 or 2 weeks, using ultra water conservation methods. Furthermore, they can produce clean drinkable water by collecting condensed water at cooling panels of air conditioners installed in the PFAL.

27.5 PFALs Can Contribute to Achieve Sustainable Development Goals (SDGs) Set by the United Nations

A primary objective of PFAL is to contribute to solving the trilemma on food, resource, and environment on small, medium, and large scales in various aspects. It is a challenge of PFALs to contribute to achieve most of the Sustainable Development Goals (SDGs). At least, the PFALs can play a role to some of the SDGs.

The Sustainable Development Goals (SDGs) are a collection of 17 global goals set by the United Nations (2018). The broad goals are interrelated though each has its own targets to achieve. The total number of targets is 169. The SDGs cover a broad range of social and economic development issues for all people at all levels, as is summarized below.

(1) End poverty. (2) End hunger, achieve food security and improved nutrition, and promote sustainable agriculture. (3) Ensure healthy lives and promote well-being. (4) Ensure inclusive and equitable quality education, and promote lifelong learning opportunities. (5) Achieve gender equality and empower all women and girls. (6) Ensure availability and sustainable management of water and sanitation. (7) Ensure access to affordable, reliable, sustainable, and modern energy. (8) Promote sustained, inclusive, and sustainable economic growth, full and productive employment, and decent work. (9) Build resilient infrastructure, promote inclusive and sustainable industrialization, and foster innovation. (10) Reduce inequality within and among countries. (11) Make cities and human settlements inclusive, safe, resilient, and sustainable. (12) Ensure sustainable consumption and production patterns. (13) Take urgent action to combat climate change and its impacts. (14) Conserve and sustainably use the oceans, seas, and marine resources for sustainable development. (15) Protect, restore, and promote sustainable use of terrestrial ecosystems, sustainably manage forests, combat desertification, halt and reverse land degradation, and halt biodiversity loss. (16) Promote peaceful and inclusive societies for sustainable development, provide access to justice, and build effective, accountable, and inclusive institutions. (17) Strengthen the means of implementation, and revitalize the global partnership for sustainable development.

27.6 Conclusion

This chapter has set out the author's views with respect to several issues relating to PFALs. Research into PFALs is only very recent, and researchers in this field were overwhelmingly few in number until recently. PFALs are only at the initial stages of research and development (R&D) and commercial uptake. Much is expected of the broad perspective and acute vision, as well as the ample creativity, of the people in this field.

As with any technology, and that includes PFALs, the areas where PFALs come to be used and their worth to the community will be impacted by the philosophies, sense of mission, goals, and methods of those individuals who will develop and spread the management of PFALs. It is a field where much is riding on the involvement of individuals with the necessary resolve.

References

Kozai T, Niu G, Takagaki M (eds) (2015) Plant factory: an indoor vertical farming system for efficient quality food production. Academic, 405 pages

Kozai T, Fujiwara K, Runkle E (eds) (2016) LED lighting for urban agriculture. Springer, Singapore, 454 pages

United Nations (2018) Sustainable development goals. https://sustainabledevelopment.un.org/sdgs

Index

A
Absorptance, 170–172, 189–190
Absorption coefficient, 189–190
Absorption spectra, 170, 230
Action spectrum
  absorptance, 171
  plant canopy in PFALs
    green light reflection, 175
    leaf area index, 173
      PAR energy flux density, 174
      PPFD, 174–176
    vs. quantum yield, 173
Advanticsys IAQM-THCO2, 274
Aeroponics system, 35
Agricultural output, 243
Agrobacterium tumefaciens, 313
AI-based agricultural data science, 410, 411
Air circulation, 40–41, 47, 159, 162
Air conditioning (AC) system, 40–41, 154
Air current speed and direction
  on photosynthesis or/and transpiration, 156–157
  on tipburn prevention, 157–158
Air distribution system
  air current speed and direction
    on photosynthesis or/and transpiration, 156–157
    on tipburn prevention, 157–158
  assessment
    air exchange effectiveness, 164
    coefficient of variation, 165
    heat removal, ventilation efficiency for, 165
    local age of air, 164
  climate uniformity, 164
  computer modeling and simulation-based approach, 154
  leaf boundary layer, 154–156
  leaf boundary layer resistance, 156
  limitation, 154
  operational costs and resource-use efficiency, 153
  overall control
  PFALs
    air movement, 158–159
    localized control, 161–163
    overall control, 159–161
  system designs, 154
Air exchange effectiveness, 164
Airflowing, 155
Airflow pattern, 159, 160
Air handling units, 158
Air infiltration/exfiltration, 158
Air movement
  to affect heat and mass transfer, 156
  crop growth and physiology, 154
  in cultivation room and CSM
    cultivation racks, 71
    law of similarity in fluid dynamics, 72, 73
    over plant canopy, 72
    plant growth, laboratory-, medium-and large-scale cultivation systems, 72, 73
  forced, 157
  in PFALs, 158–159
Algae, 22, 47, 397
Aminodeoxychorismate synthase (ADCS), 318, 319
Anthocyanin, 39, 320, 329, 330, 338

Antioxidants, 307, 320, 327, 329, 336, 356, 360
Arabidopsis rotundifolia 3 (rot3) mutant, 303
Aromatic plants, 353, 355
  See also Medicinal plants
Artificial intelligence (AI) technology, 10, 11, 20, 21, 24–25, 145–146, 244
Automatic CF measurement system, 372–373
Automation technology, 241, 242

B

Baby leaf growers, 85
Backcrossing breeding methods, 307
Basic Local Alignment Search Tool (BLAST), 309, 310
Big data, 20, 21, 68, 132, 244, 406, 407, 409–411, 414, 416
Big data mining, 147, 148, 416
Big dataset, 26, 417
Bioeconomy, 345
Bioinformatics, 20, 21
Biological rhythms, 367–368
  control of, 378–382
  high-throughput growth prediction systems using, 370–372
    automatic CF measurement system, 372–373
    machine learning, 373–374
  molecular timetable method, 374–375
    body time estimation, transcriptome, 376–379
    circadian rhythm analysis using, 375–376
    transcriptome, environmental variations effects, 375
  seedling diagnosis using, 368–370
Biomass production, 130
Blue light, 215–216
Boundary layer resistance (r_a), 155
Bowtie2, 375
Boxcar method, 271
Brassica species, 308–309
Breeding
  acceleration of, 291
  applications in, 290–291
  crop, plant factories, 305–306
    conventional and modern plant-breeding strategies, 305, 306
    lettuce, breeding practices of, 306–308
  pedigree, 307
Building-integrated agriculture (BIA), 247, 248
Building-integrated CEA setup, 259, 260
Buoyancy, 158

Business planning
  business profitability, 105–106
  correlation of efficiency and productivity to profitability, 113, 116
  man hour productivity and profitability, 113–115
  operational productivity, 102–105
  people management, 105, 106
  PPF productivity and profitability, 106–113
  sensitivity analysis, 117
Business planning sheet
  analysis, 101–102
  capital funding for plant factories, 117
  conversion factor and energy conversion efficiency, 91
  cultivation area elements, 87–89
  daily light integral planning, 91
  efficiency and productivity indexes, 117
  electricity, 104–105
  initial investment, 87, 90–91
  lighting elements, 87, 89
  management, 101–102
  profit/loss statement and cash flow, 87, 95, 99, 100
    budget 5-year operations, 87, 96–98, 100
    standard operations, 87, 92–94, 100
  sensitivity and risk analysis for funding, 117
  transportation cost, 100
  trial and error process, 102
  unit prices, 100
  vertical firming systems, 100
Business profitability, 105–106
  efficiency and productivity, 113, 116
  and man hour productivity, 113–115
  PPF productivity, 106–113
Business to business to consumer (B2B2C) business model, 106

C

Calcium transporter genes, 315
Canopy coverage estimation, 387–388
Carbon dioxide (CO_2), 255–256, 304, 361
  and fertilizer efficiencies, PFALs, 17
  plant growth model, 267–270
CCPPS, *see* Completely closed plant production systems
Ceiling-based ventilation, 160–161
Cell death, 304
Chlorophyll fluorescence (CF)
  automatic measurement system, 372–373
  biological rhythm, seedling diagnosis using, 368–370

Index   445

circadian rhythm features, 373
fresh weight and index of, 369, 372
high-throughput growth prediction system, 373, 374
Chloroplasts, 305
Chrysanthemum
  flowering, 234–235
  growth conditions, 231–232
  morphogenesis, 233–234
  photoperiodic light treatments, 232
  plant materials, 231–232
Circadian clock, 374
Circadian rhythms, 367–368
  biological rhythms, molecular timetable method, 375–376
  spatial control of, 381–382
City life, 245
Climate uniformity, 164
Closed plant roduction system, 6
Clustered regularly interspaced short palindromic repeats (CRISPR), 312
CMOS, *see* Complementary metal-oxide-semiconductor
Coefficient of variation (CV), 165
Colony-forming units of microorganisms (CFU), 19
Color rendering index (CRI), 179, 188, 203–205
Color space-based segmentation, 387
Complementary metal-oxide-semiconductor (CMOS), 368, 372–373
Completely closed plant production systems (CCPPS), 340–341, 345
Computer modeling approach, 154
Computer vision-based techniques, 385, 386
  canopy coverage estimation, characterizing plant growth by, 387–388
  image segmentation, 386–387
  paddy rice, characterizing flowering dynamics, 388–390
Conduction heat and mass exchange model, 262
Controlled environment agriculture (CEA), 247, 259, 260
Control models
  data storage and model, 53
  design elements of smart plant factory, 55
  fundamental design elements of smart plant factory, 55
  hierarchical control model, 54
  model-based control, 53
  smart plant factory,controlled target, 52–53
  updation, PDCA cycle, 54–55
Control system theory, 51–52

Convection heat and mass exchange model, 263–264
Correlation color temperature (CCT), 179, 203–205
CRISPR/Cas9 technology, 302, 312, 321
  use of, 313–315
Crop breeding
  genetic engineering and, 305–306
    conventional and modern plant-breeding strategies, 305, 306
    lettuce, breeding practices of, 306–308
  outfield and PFAL conditions, 303
Crop yield, 268
Crude drugs, 355
Cryptochrome, 215
CSM, *see* Cultivation system module
CSM-L, *see* Cultivation system module L
Cultivation system module (CSM), 24, 408, 413, 414, 429–430
  air exchange and flow types, 73, 74
  air flow scheme, 73, 74
  characteristics, types A–D, 73, 75
  classification, air flow pattern and RUE estimation, 73, 74
  cost performance
    annual sales increase, 60–61
    electricity and labor costs reduction, 59–60
    and production cost, 63–64
  CSM-L
    plant production process measurement and control, 68–71
    for scalable PFALs, 73–79
  cultivation room
    air movement, 71–73
    cultivation, 65
    facilities and equipment, 65
    firmware components, 65, 66
    hardware components, 65, 66
    software components, 65, 66
    spatial components, 65
  definition, 64
  function and configuration, 67–68
  with horizontal cultivation panels/trays, 73
  payback period, 64
  production cost
    and cost performance, 63–64
    and initial investment, 57
  productivity increase, 57
  properties, 67
  resource element
    consumption per kg of produce, 63
    productivity, 61

Cultivation system module L (CSM-L)
  plant production process measurement and control
    environmental factors, 68, 69
    labor, cultivation area and electricity productivity, 68
    plant growth models, parameter values, 68
    plant phenotypic traits, 68, 70
    production rates, 68, 69
    resource elements supply rates, 68, 69
    resource use efficiencies, 68, 71
    signal inputs/outputs, equipment/actuators and sensors, 68
  for scalable PFALs (*see* Scalable PFALs, CSM-L)
Cuticular resistance (r_s), 155

D

Daily light integral (DLI), 91, 188, 342
Darwin sea, 242
Data warehouse, 427–429
Day-neutral plant (geranium), 229–230
Decision-based process design (DPD) analysis, 146
Decision tree based vegetation segmentation model (DTSM), 387
Deep flow technique (DFT) system, 33
Deep learning model with mechanistic and multivariate statistical models, 26–27
Deep learning unit, phenome, genome and environment datasets, 26
Dihydroneopterin triphosphate (HMDHP), 318
Disease, microbe and insects, 48
Drip irrigation system, 36
Dual (virtual/actual) PFAL, 27, 28
Dutch glass greenhouse technology, 15
  vs. PFALs, 16

E

Easy Plant Coverage Calculation tool (EasyPCC), 388
Ebb and flow system, 35
Eelectrical power consumption, 205–206
Electrical conductance (EC), 140, 192
Electric vehicles (EVs), 120, 121
Electron impact (EI), 362
Electrospray ionization (ESI), 363
Emerson effect, 171, 173
Endogenous circadian clock, 378

Energy and mass (substance) balance, PFAL, 21
Energy per photon (E), 172
Environmental control technology, 31
Equivalent or mol equivalent (Eq), 191
Evaporation of plants indoors, 127–128

F

Farquhar-von Caemmerer-Berry biochemical leaf photosynthesis model, 268
Far-red (FR) light, 217
  additive effects, 199
  characteristics, 216
  in indoor farming, 128–130
  LEDs emit, 173
  photosynthesis and production, 198, 212
Feed-in tariff (FIT) system, 122
Fertilizer, 190, 191
Flame ionization detector (FID), 362
Flavr Savr, 306
Floor layout, plant factory, 44
Floricultural plants
  data collection and analysis, 226–227
  day-neutral plant (geranium), 229–230
  growth conditions, 226
  long-day plant (petunia), 227–229
  photoperiodic treatments, 226–227
  plant materials, 225–226
  short-day plant (chrysanthemum), 230–231
Flowering dynamics, 388–390
Food security, 12, 31, 134, 440
Food urge agriculture, 125
Forestry and Fisheries Research Council (AFFRC), 328–329

G

Gas chromatography (GC), 362–363
Gene duplication, 310
Genetically modified organisms (GMOs), 306, 321
Genetic engineering
  and crop breeding, plant factories, 305–306
    conventional and modern plant-breeding strategies, 305, 306
    lettuce, breeding practices of, 306–308
  integrated pipeline for plant factories
    genome editing, 312
    indoor farming, CRISPR/Cas9 technology, 313–315
    transgene and disadvantage, 311
  for vitamin biofortification, 318–320

Index

447

Genetic modification, target genes
 Arabidopsis, 308
 bioinformatics pipeline, 309–311
 leafy vegetables, evolutionary relationships of, 308–309
Genome editing, 312
Genomics, 20, 21
Gibberellic acid (GA), 303
Global Good Agricultural Practice (GAP), 19
Global price of the solar panels, 119
Green light effect
 absorption rates, 212
 in cultivation area, 175
 disease resistance, 21
 growth and development, 21
 growth promotion, mechanism of, 216
 human health, 21
 lighting of PFALs, 198
 monochromatic LEDs, 200
 photosynthesis, 21
 for photosynthesis, 173
 quantum yield, 172
 secondary metabolite production, 21
Growth management of plants, 142
Guanosine triphosphate cyclohydrolase I (GTPCHI), 318
Guide RNA (gRNA), 312, 313

H

Hardware/software unit, spatial variations minimization, 25
Hazard analysis and critical control point (HACCP), 19
Heat and mass exchange model
 conduction, 262
 convection, 263–264
 latent heat transfer, 264–266
 overall model, 266–267
 radiation, 261–262
Heat balance, 267
Heat removal ventilation efficiency, 165
Heat sterilization method, 42–43
Hemagglutinin (HA), 344
Herbal medicine
 active compounds in, 355–356
 phytochemicals, 356–358
 reference standard of, 363–364
Hierarchical control model, 54
High-performance liquid chromatography (HPLC), 363
High-pressure sodium (HSD) lamps, 254

High-throughput growth prediction systems,
 biological rhythms, 370–372
 automatic CF measurement system, 372–373
 machine learning, 373–374
High-wire tomato production
 Axiany and Axiradius, 293
 dry matter partitioning over time, 295, 296
 edema, 292
 environmental conditions and irrigation control, 297–298
 experimental layout, 298
 leaf area index, 294
 light intensities, 292, 293
 Plant Balance Model yields, 291
 yields, 296
Horticultural crops, functional compounds
 in value-added plant production, 327
 classification, 326, 327
 environmental conditions, 329–330
 in plants, 327–328
 target concentrations of, 328–329
Human resource development for PFAL, 22
Hydroponic systems
 application purposes, 32
 growing plants, water-based, nutrient-rich solution, 32
 irrigation systems, 32–37
 nutrient solutions
  advantages, 396
  air, 46
  electrical conductivity, 38, 396
  element balance, 46
  ion balance, 401–402
  nutrient composition, 37, 38
  nutrient element uptake, 397, 398
  parameter estimation, 398
  pH adjustment, 46–47
  pH management, 38, 396
  plant growth, 397, 398
  properties, 397
  sterilization, 42–43
  temperature, 38–39, 46
  variables, 398, 399
  water balance, 400–401
  water transpiration, 397, 398
  water uptake, 397, 398
 PFAL, 24
Hydroponic urban farms, 246
 resource needs for, 250–252
  airflow, 254
  carbon dioxide, 255–256
  co-benefit and trade-offs, 257

Hydroponic urban farms (*cont.*)
　　heat, 252–253
　　light, 254–255
　　optimal conditions, 256–257
　　water, 254

I
Image analysis techniques
　　detecting and counting fruits, 389, 391
　　leafy vegetables, growth dynamics characterization, 389
Image segmentation, 386–387
Indoor farming
　　climate and associated technology, 128
　　climate in, 126
　　dry matter production, breakdown of, 132, 133
　　far-red light in, 128–130
　　flower induction, 130
　　growing crops without daylight, 126
　　growing tomato, 130
　　large-scale production facility, 133, 134
　　next generation of agriculture, 126
　　plant balance, 131–133
　　plant evaporation, 127–128
　　plant production, 126–127
　　technology, 290, 291
Indoor production system, 157
In field and greenhouse evaporation of plants, 127
Intelligent control, 200
Interberry alpha, 341
Internet of Things (IoT), 10, 11, 20, 21, 427
Ion balance, 401–402
Ion concentration control unit, hydroponic system, 25
Ion-specific nutrient solution control
　　ion balance, 401–402
　　nutrient element uptake, 397, 398
　　parameter estimation, 398
　　plant growth, 397, 398
　　variables, 398, 399
　　water balance, 400–401
　　water transpiration, 397, 398
　　water uptake, 397, 398
Ion use efficiency (IUE), 402
IoT technology infrastructure, 243
Irrigation systems
　　DFT system, 33
　　drip irrigation system, 36
　　Ebb and flow system, 35
　　modified hybrid system, 34
　　NFT system, 32, 33
　　spray system, 35
　　Wicking system, 36, 37

J
Japanese honeysuckle, flower buds of, 335, 336
Japanese quality agricultural crops, 243
Japan Plant Factory Association (JPFA), 85–86, 405, 406

K
Klebsormidium, 310
Koshihikari, 389

L
Labor-intensive-type industry, 241
Lactuca sativa, 368
Land area use efficiency, 17–18
Large-scale production
　　facility, 133, 134
　　indoor farming (*see* Indoor farming)
Latent heat flux, 254, 255
Latent heat transfer, 261, 262, 264–266
Law of similarity in fluid dynamics, 72, 73
Leaf area index (LAI), 133, 173, 174, 292, 294, 387
Leaf boundary layer (LBL), 154–156
Leaf boundary layer resistance (LBLR), 156
Leaf boundary layer thickness (LBLT), 156
Leafy vegetables, evolutionary relationships of, 308–309
LEDs, *see* Light-emitting diodes
Lettuce, 308
Light
　　absorptance, 189
　　absorption coefficient, 190
　　color rendering index, 188
　　conventional and proposed technical terms, 186, 187
　　daily light integral, 188
　　light intensity, 45–46, 185, 186
　　light period, 39
　　light spectrum, 39–40
　　lumen, 188
　　metrics of
　　　　photometry, 184, 185
　　　　photonmetry, 184, 185
　　　　photosynthetic photon number efficiency, 185–186

Index

photosynthetic radiation energy
  efficiency, 185–186
radiometry, 184, 185
SI units, 184, 185
photon yield, 189
photosynthetically active radiation,
  186–187
photosynthetic radiation and photon, 189
photosynthetic radiation flux, 187–188
PPFD, 39
  vs. PPF, 186, 188
PRFD, 188
quantum yield, 189
wavelength and photon number, 188–189
Light-emitting diodes (LEDs), 10, 11, 76–78,
  254, 276
  advantages, 198
  broad-spectrum white LEDs
    blue basal color with complementary
      phosphor-conversion yellow,
      200, 201
    color rendering index, 203–205
    correlation color temperature,
      203–205
    phosphor technology, 200, 202
    power consumption, 205–206
    R/B and R/FR ratios, 202–204
    spectral differences on vegetable
      production, 200–201, 203, 204
    three light primary color, 200, 201
  efficient design and operation, 169
  fundamental properties, 176–177
  future aspects, 208–209
  indoor use, 127
  mainstream agricultural LED
    applications, 198
  maximum photosynthetic photon number
    efficacy, 177–178
  mechanism, 83
  monochromatic facilitates plant growth
    far-red effects, 199
    green light effects, 200
    lettuce, growth features, 198–199
  morphology and other reactions
    light quality effect, 215–217
    light receptors, 214–215
    PPFD, 217–218
  night interruption light treatments,
    226, 227
  photosynthesis
    angle of leaves, 212–213
    light absorption rates, 212
    PPFD, 213–214
    stomatal opening, 213

plant cultivation
  adequate PFD dose, 206–207
  damage control, 207–208
  growth phase and troubles, 208
  pulsed light (intermittent light), 218
  red drop and Emersion effect, 173
  SPFD setting
    on duration of cultivation, 220
    on time of day, 218–220
  white LEDs
    do not emit white light, 178, 179
    for plant lighting, 178–179
Light energy use efficiency, PFALs, 17
Lighting system, 39–40
Light intensity, 45–46
Light manipulation, 329
Light quality effect
  blue light, 215–216
  far-red light, 217
  green light, 216
  red light, 216–217
Light receptors, 214–215
Light spectrum, 39–40
Liquid chromatography/mass spectrometry
  (LC-MS), 363
Liquid crystal display (LCD) projector, 381
Lithium-ion manufacturing capacity, 120
Local age of air, 164
Localized control (LC)
  airflow or cooling fan, 162
  perforated air tube, 162–163
Long-day plant (petunia), 227–229
Longitudinal research, *see* Plant cohort research
Lumen (lm), 188
Luminous efficacy (lm W 255^{-1}), 179

M

Machine learning, 147, 373–374
Marketing management, 31
Mass and energy balance models, 407
Mechanical force, 158
Mechanistic models, 407
Mechanization agriculture, 242–243
Medicinal plant production, value-added
  environmental conditions, 334–338
    light, 335–336
    temperatures, 336–338
    UV light, 336
  industry, world market and condition,
    330–332
  merits and demerits of, 332–333
  superiority of, 333–334
  variety, 334

Medicinal plants, 355
    chemical analysis of, 360–361
        extraction, 361–362
        gas chromatography, 362–363
        HPLC, 363
        organic molecules, 361
        separation, 362
    cultivation of, 358–359
    for high-quality health care and cosmetics, 21
    issue of, 354, 355
Mesophyll resistance (r_m), 155
Metabolic engineering method, 320
Microbiological ecosystems, 22
Ministry of Education, Culture, Sports, Science and Technology (MEXT), 327
Model-based control, 53
Modified hybrid system, 34
Molecular timetable method, biological rhythms, 374–375
    body time estimation, transcriptome, 376–379
    circadian rhythm analysis using, 375–376
    transcriptome, environmental variations effects, 375
Mollier diagram, 127

N

Natural products, 355
Net photosynthesis, transpiration and dark respiration of plants, 21
Neural networks, 373
New Energy and Industrial Technology Development Organization (NEDO), 346–347
Next generation smart PFALs, see "Smart" plant factories with artificial lighting
Nicotiana benthamiana, 344
Night interruption light (NIL) quality
    morphogenesis and flowering chrysanthemum, 231–235
    in floricultural plants, 225–231
    transcription of photoreceptor genes, 225
NIL, see Night interruption light (NIL) quality
Nonlinear mathematical models, 147
Nonpolar organic solvents, 361
Non-sustainable harvest of plants, 356–358
Nutrient elements, 190, 191
Nutrient film technique (NFT), 32, 33, 75–76, 133, 297

Nutrient ions, 190, 191
Nutrient solution
    air, 46
    circulation unit, PFAL, 24
    electrical conductivity, 38, 192, 396
    element balance, 46
    Eq kg^{-1}, 192
    equivalent concentration, 191
    equivalent/mol equivalent, 191
    fertilizer, 190, 191
    in hydroponics system
        advantages, 396
        ion balance, 401–402
        nutrient element uptake, 397, 398
        parameter estimation, 398
        plant growth, 397, 398
        properties, 397
        variables, 398, 399
        water balance, 400–401
        water transpiration, 397, 398
        water uptake, 397, 398
    ions, 190, 191
    nutrient composition, 37, 38
    nutrient elements, 190, 191
    parts per million, 192–193
    pH adjustment, 46–47
    pH management, 38, 396
    respiration rate, 402
    sterilization, 42–43
    temperature, 38–39, 46
    valence, 191

O

Off-grid operation, 122
Open-source business planning and management system, 22
Operational efficiency, business planning, 113, 116
Operational productivity, business planning
    crop transplantation, 105
    economic depreciation and LED fixtures cost, 104
    electricity consumption, 105
    major productivity indexes, 102, 103
    productivity indexes, 102, 103
Optical (spectral) sensing methods, 79, 80
Optimized multitier system design, 154
Organic fertilizer, 190, 191
Organic hydroponic systems, 21
Organic molecules, 361

Overall control (OC)
  airflow pattern, 159
  ceiling supply and extract, 160–161
  definition, 159
  side wall supply and extract, 159–160
Oxygen (O_2) bubbling method, 43
Ozone (O_3) sterilization method, 42

P

Paddy rice, 388–390
p-aminobenzoate (pABA), 318
Parts per million (ppm), 192–193
Pedigree breeding, 307
Penman-Monteith (P-M) formula, 255, 258, 264
People management, business planning, 105, 106
Perilla frutescens, 334, 336, 360
Periodic movement of plants, PFAL, 25
PFALs, *see* Plant factories with artificial lighting
Phase response curves (PRCs), 378–379
Phenolic flavonoids, 320
Phenomics, 10, 11, 20, 21
Phenotyping-and AI-based smart PFALs
  AI-based agricultural data science, 410, 411
  big data analysis of phenotype, 406, 407
  breeding, 408–409
  commercialization of research achievement, 409
  cultivation system module, 408
  environmental control, 406–409
  genotype, 406, 407
  Kashiwanoha, agri-smart city, 410
  location, 406
  mechanistic models, 407
  plant management, 406, 407
  project research, 408
Phenotyping technology, 145–146
Phenotyping unit, 10, 11
  environmental control and breeding for PFAL, 24–25
Photometry, 184, 185
Photonmetry, 184, 185
Photon yield, 189
Photoperiodic light treatments
  chrysanthemum, 232
  floricultural plants, 226–227
Photosynthesis, 10, 155, 267, 268
  action spectrum, 170–171, 173
    absorptance, 171
    green light reflection, 175
    leaf area index, 173
    PAR energy flux density, 174

PPFD, 174–176
  vs. quantum yield, 173
angle of leaves, 212–213
β-carotene, 170
Emerson effect, 171, 173
energy per photon, 172
isolated chlorophyll a and b, 170
light absorption rates, 212
PPFD, 213–214
quantum yield
  vs. action spectrum, 173
  definition, 171–172
  for green light, 172
red drop, 171, 173
stomatal opening, 213
Photosynthetically active radiation (PAR), 127, 186–187, 254, 261
Photosynthetic efficacy, spectrum, 104
Photosynthetic photon flux (PPF), 84, 175–176, 184, 186, 188
  productivity and profitability, business planning, 106–113
Photosynthetic photon flux density (PPFD), 174–176, 184, 417
  concave function, 213
  vs. cost of LED fixtures, 107, 112
  and light period, 39
  net photosynthetic rate, 213–214
  vs. operational productivity, 107–111
  vs. PPF, 186
  and relative SPFD, 217–218
Photosynthetic photon number efficiency, 185–186
Photosynthetic radiation energy efficiency, 185–186
Photosynthetic radiation flux (PRF), 187–188
Photosynthetic radiation flux density (PRFD), 188
Phototropin, 215
Photovoltaic power generation, 119–120
Phylogenetic tree, 310
Physics-based numerical simulation, 257
Physiological disorder, 419
Phytochemicals, 353, 356–358
Phytochemistry, 359–360
Phytochrome, 215
Plan-do-check-act (PDCA) cycle, 54–55, 61
Plant Balance Model, 131–134, 292
Plant canopy
  action spectrum of, PFALs
    green light reflection, 175
    leaf area index, 173
    PAR energy flux density, 174
    PPFD, 174–176
  coverage ratio, 387

Plant circadian system, 379
Plant cohort research
  application software, 429–430
  cultivation system module, 413, 414
  democratization, 414, 415
  phenotyping unit, 413
  plant growth stages, 416–417
  plant traits, 417, 418
  resource input and output variables, 417, 418
  socialization, 414, 415
  in social sciences, 415
  transparentization, 414, 415
  variation, in plant traits
    annual productivity per tray, 426
    data warehouse, 428
    environmental factors, 419
    fresh weight (g) per plant at stages 2 and 3, 422–423
    genome, 419
    germination rate and cumulative germination percentage of seeds, 420, 421
    human intervention, 419
    to increase number of trays at stages 2 and 3, 425
    initial plant growth rate, after seed germination, 420
    Internet of Things, 427
    machine intervention, 419
    percent marketable parts, 426
    phenotyping unit, 421, 422, 424–425
    trays and plant individuals data at stages 1–3, 418–419, 427
    unavoidable marginal error, 419
Plant cultivation
  germination stage, 44
  growth stage, 44
  knowledge, 31
  nursery stage, 44
Plant factories with artificial lighting (PFALs), 301
  actual working/virtual models, 23
  advantages, 4
  affordable application software, 436
  air ventilation/distribution system
    air movement, 158–159
    localized control, 161–163
    overall control, 159–161
  artificially controlled environment, desired plant traits for, 302–303
    cell death, 304
    crop-breeding strategies for, 302
    cultivation time and tipburn occurrence, 304
    low light intensity, preference for, 305
    plant height and leaf area, 303–304
  characteristics, 6
  commercial production, 9, 57
  components (or units), 7–8
  components standardization, 23
  cost performance, 23
  CSMs, 414, 415
  cultivating high-quality crops, 315–317
  cultivating high-value crops, 317
    pharmaceuticals, genetically modified plants for, 317
    plants with high antioxidant contents, 320
    vitamin biofortification, genetic engineering for, 318–320
  cultivation knowledge, 31
  design and environment control, 20, 58
  vs. Dutch glass greenhouse technology, 16
  economic suitability of plants, 9
  efficient breeding, 437–438
  efficient operation and management, 434–435
  energy and natural resources, 435
  estimated number, 8
  floor plan and equipment layout, 23
  food-resource-environment trilemma, 3, 4
  food security, 31
  fresh vegetables, 5
  functional plants, 435–436
  genetic engineering and crop breeding, 305–306
    conventional and modern plant-breeding strategies, 305, 306
    lettuce, breeding practices of, 306–308
  genetic engineering, integrated pipeline for genome editing, 312
    indoor farming, CRISPR/Cas9 technology, 313–315
    transgene and disadvantage, 311
  genetic modification, target genes for
    Arabidopsis, 308
    bioinformatics pipeline, 309–311
    leafy vegetables, evolutionary relationships of, 308–309
  herbal medicine
    active compounds in, 355–356
    reference standard of, 363–364
  high adaptability for the location, 19

high annual productivity, 17–18
high controllability of sanitary
    conditions, 19
high-quality plants, 18
high reproducibility and predictability of
    yield and quality, 18–19
high resource use efficiency, 16–17
high traceability, 19
high weight percentage of marketable
    parts, 18
integration with biological system, 437
issue control
    air and nutrient solution temperature, 46
    algae, 47
    disease, microbe and insects, 48
    element balance in nutrient solution, 46
    light intensity, 45–46
    pH adjustment, 46–47
    seed quality and storage, 48
    tip burn, 47
for large-scale production and breeding, 24
LED use, 16
light and safe work, 19
local culture and technology, 12–13
longitudinal research (see Plant cohort
    research)
long shelf life, 19
management and environment control,
    software components, 66
marketing, 23
medicinal plants
    chemical analysis of, 360–363
    cultivation of, 358–359
    issue of, 354, 355
micropropagation, 437
mission and goals, 4
non-sustainable harvest of plants,
    356–358
objectives, 9–10
open field-and greenhouse-grown plants,
    436
personnel development and training
    programs, 57–58
PFAL-grown vegetables, 436
vs. PFSL, 6
phytochemistry with plant factories,
    359–360
plant canopy, action spectrum of
    green light reflection, 175
    leaf area index, 173
    PAR energy flux density, 174
    PPFD, 174–176
plant cultivars breeding, 23
plant environment, high controllability, 18

problem solving
    advanced technologies, 20, 21
    initial investment and operation costs
        reduction, 20
    medicinal plants for high-quality health
        care and cosmetics, 21
    organic hydroponic systems, 21
    robotics and flexible automation, 20
    sustainable production, 20
    worldwide active organizations, 21
production capacity, 8, 57
production cost, 8, 18
production management system
    plant growth, 138–139
    process, 139
    sales management, 139–140
profitability
    cultivation environment, 137, 138
    ion uptake rate per unit area, 141
    mineral-rich vegetables or medicinal
        plants, 140
    photosynthetic rate per unit area, 141
    plant growth rate, 137
    state and rate variables measurement,
        140–142
quality concentrations, 353
recent and future cost reductions, 434
in regions, 438–439
research and development, 16, 22–23
seedling/transplant production, 437
seed propagation, 437–438
smart PFALs
    expected ultimate functions, 11–12
    management system, 10, 11
    phenomics, 10, 11
    software/hardware units, 24–27
    to solve food, resource, and environment
        issues, 4
software with database
    electricity costs minimization, lighting
        and air conditioning, 22–23
    environmental control, plant species,
        22–23
source-plant species, conservation of, 365
stomatal opening, 213
Sustainable Development Goals, 440
technologies and business models, 435
training program, 436
value-added plant production (see Value-
    added plant production)
value chain, 57–58
vegetative propagation, 437
waste produced in urban areas, 5–6
wholesale price, 8

Plant factory business, 83
Plant genome recombinant technologies, 347
Plant growth
    computer vision-based techniques, 385, 386
        canopy coverage estimation, characterizing plant growth by, 387–388
        image segmentation, 386–387
        paddy rice, characterizing flowering dynamics, 388–390
    and development models, 407
    image analysis techniques
        detecting and counting fruits, 389, 391
        leafy vegetables, growth dynamics characterization, 389
Plant growth-environment model, 26–27
Plant growth model
    CO_2 exchange, 267–270
    plant growth and crop development, 271
Plant growth stages, 416–417
Plant-made pharmaceuticals (PMPs), 339
Plant phenotyping
    camera types, 68, 70
    traits, 68, 70
PPFD, *see* Photosynthetic photon flux density
Production process management, 146, 147
Protospacer adjacent motif (PAM) sequence, 312, 314
Pulsed light (intermittent light), 218
Pulse width modulation (PWM), 218

Q
Quantitative trait loci (QTL), 305–307
Quantum yield (Q_p)
    vs. action spectrum, 173
    definition, 171–172, 189
    for green light, 172

R
Radio frequency identifier (RFID) system, 372, 373
Radiometry, 184, 185
Reactive oxygen species (ROS), 327
Recombinant inbred lines (RILs), 307
Red/blue (R/B) ratio, 202–204
Red drop, 171, 173
Red/far-red (R/FR) ratios, 202–204
Red light, 216–217
Region of interest (ROI), 386
Relative standard deviation (RSD), *see* Coefficient of variation

Renewable energy, 119–123
Resource use efficiency (RUE), 406
    CO_2 and fertilizer efficiencies, 17
    and cost performance, 21
    light energy use efficiency, 17
    online estimation, 142–144
    online measurement and rate variables control, 21, 22
    water use efficiency, 16–17
Ribonucleic acid (RNA), 374
Ribonucleic acid interference (RNAi), 306
Robotics and flexible automation, 20
Robot industry, 241
Robot technology, 241
Rosmarinic acid, 360
Roundup Ready crops (RR crops), 311
RuBisCo, 259
RUE, *see* Resource use efficiency

S
SAIBAIX, 142–144, 151
    commercial applications, 148–149
    for educational purposes, 149, 150
    effectiveness of, 149
Salinity control, 41–42
Sand filtration method, 43
Scalable PFALs, CSM-L
    air exchange and flow types, 73, 74
    air flow scheme, 73, 74
    automation and robotization, 79
    batch production and push/pull production, 79
    characteristics, types A–D, 73, 75
    classification, 73–75
    classification, air flow pattern and RUE estimation, 73, 74
    with horizontal cultivation panels/trays, 73
    LED lighting system, 76–78
    nutrient solution flow types, hydroponic unit, 75–76
    optical (spectral) sensing methods, 79, 80
Secondary metabolites, 328, 329, 348
Seed quality and storage, 48
Sequence similarity network (SSN), 311
Set points
    of air current speed, 40
    of air temperature, 40
    of CO_2 concentration, 40
    of VPD, 40
Short-day plant (chrysanthemum), 230–231
Shortening cycles, 291
Silicon photovoltaic cells, 120

Silver/titanium oxide utilization method, 43
Simulation-based approach, 154
Smart cell industry society, 345
Smart Cell Project, 326, 346, 347
Smart energy for smart plant factory, 121–122
Smart LED lighting system, 5, 25
"Smart" plant factories with artificial lighting (smart PFALs), 4
   CSM (*see* Cultivation system module)
   design elements, 55
   expected ultimate functions, 11–12
   experimental calculation, 122–123
   management system, 10, 11
   phenomics, 10, 11
   phenotyping-and AI-based
      AI-based agricultural data science, 410, 411
      big data analysis of phenotype, 406, 407
      breeding, 408–409
      commercialization of research achievement, 409
      cultivation system module, 408
      environmental control, 406–409
      genotype, 406, 407
      Kashiwanoha, agri-smart city, 410
      location, 406
      mechanistic models, 407
      plant management, 406, 407
      project research, 408
   to solve food, resource, and environment issues, 4
Smart plant factory control model, 52–53
Software/hardware unit, DNA expressions/markers, 26
Software units
   environmental factors, 25
   spatial variations of environment, 25
Solar power system, 120–121
Spectral photon flux density (SPFD), 211, 214
   on duration of cultivation, 220
   and PPFD, 217–218
   on time of day, 218–220
Speed breeding, 27
SPFD, *see* Spectral photon flux density
Spray hydroponic system, 35
Stable transgenic plants, 339
   environmental control for, 342–344
Stand-alone CEA setup, 259, 260
Stomatal conductance, 258
Stomatal resistance (r_s), 155, 265
Stress-induced flowering, 129, 291
Supercritical fluid extraction (SFE), 361–362
Support vector machine (SVM), 388
Sustainable Development Goals (SDGs), 440

T

Target leaf index (LAI), 84–85
Temperature and humidity control, 83
Temperature control mechanism, 126
Tipburn, 47, 304, 315
   prevention, 157–158
Tomato model, 271
Traditional mechanization, 243
Transcriptome
   body time estimation of, 376–379
   environmental variations effects, 375
Transcriptome analysis, 374
Transgene, 311
Transgenic crop breeding method, 321
Transgenic plants, value-added plant production
   methods of, 339, 340
   PMPs production, CCPPS for, 340–341
   stable transgenic plants, 342–344
   transient gene expression method, 344–345
Transgenics, 311
Transient gene expression method, 344–345
Transpiration, 264, 316
Tree topology, 310
Turbulent flow, 155

U

Ultraviolet (UV)
   radiation, 320
   sterilization method, 42
Ultraviolet-visible (UV-VIS) detectors, 363
Urban farm(ing), 246
   derelict and vacant space, 247–250
   growth of, 248
   potential benefits of, 246
Urban heat island (UHI) effects, 246
Urban-integrated agriculture
   historical context, 257–259
   simulation models for, 259–261
      environmental management of tunnels, 271–272
      heat and mass exchange model, 261–267
      model description, 272
      numerical simulation, 257
      to optimise environmental conditions, 276–281
      plant growth model, 267–271
      verification and outcomes, 272–276
Urban-integrated CEA setup, 259, 260

V

Vaccine-producing plants, 318
Valley of death, 242

Value-added plant production, 326
  bioactive compounds production, 326, 327
  bioeconomy and smart cell industry society, 345, 346
  classification of, 326, 327
  horticultural crops, functional compounds in, 327
    environmental conditions, 329–330
    in plants, 327–328
    target concentrations of, 328–329
  medicinal plant production
    environmental conditions, 334–338
    industry, world market and condition, 330–332
    merits and demerits of, 332–333
    superiority of, 333–334
    variety, 334
  NEDO, 346–347
  secondary metabolites, 328
  transgenic plants, pharmaceutical and functional protein production
    methods of, 339, 340
    PMPs production, CCPPS for, 340–341
    stable transgenic plants, environmental control for, 342–344
    transient gene expression method, environmental control for, 344–345
Vapor pressure deficit (VPD), 83
Vegetative growth process, 415
Vertical hydroponic systems, 246
Verticillium, 307
Vitamin B9, 318, 319
Vitamin biofortification, genetic engineering for, 318–320

W

Water balance, 400–401
Water diffusion, 127
Water transpiration (WT), 397, 398
Water uptake (WU), 397, 398
Water use efficiency (WUE), 16–17
White LEDs
  do not emit white light, 178, 179
  efficient use, 21
  for plant lighting, 178–179
Wicking system, 36–37
Wind pressure, 158
Work-process optimization, DPD, 146

X

Xylem, 292

Z

Zero Carbon Food, 271